Green Energy and Technology

For further volumes:
http://www.springer.com/series/8059

John T. S. Irvine · Paul Connor
Editors

Solid Oxide Fuels Cells: Facts and Figures

Past, Present and Future Perspectives for SOFC Technologies

Springer

Editors
John T. S. Irvine
School of Chemistry
University of St. Andrews
Fife
UK

Paul Connor
School of Chemistry
University of St. Andrews
Fife
UK

Additional material to this book can be downloaded from http://extras.springer.com/

ISSN 1865-3529 ISSN 1865-3537 (electronic)
ISBN 978-1-4471-4455-7 ISBN 978-1-4471-4456-4 (eBook)
DOI 10.1007/978-1-4471-4456-4
Springer London Heidelberg New York Dordrecht

Library of Congress Control Number: 2012949079

Printed on acid-free paper

Springer is part of Springer Science+Business Media (www.springer.com)

Preface

SOFC Facts and Figures: Past Present and Future Perspectives for SOFC Technologies

To mark the contribution of Ulf Bossel in initiating and driving the first nine European SOFC Forums, a special series of talks on SOFC Facts and Figures were presented in 2010 European SOFC Forum in Luzern. These talks reflected the critical impact of Ulf's seminal papers that provided an appropriate grounding on SOFC processes and requirements, most notably in the IEA report "Final Report on SOFC Data, Facts and Figures", Swiss Federal Office of Energy, Berne, 1992, Appendix 1. This book presents a collation of papers based upon these talks, addressing past, present, and future perspectives for SOFC technologies. The objective of these chapters is more focused on highlighting critical issues for the future of SOFC technology, rather than presenting comprehensive reviews of the development of this technology.

"Solid Oxide Fuel Cells: Past, Present and Future" presents an initial historical background to Solid Oxide Fuel cells, then describes the materials and component parts required to make a working stack. It then compares different cell designs, particularly comparing planar with tubular concepts. Examples of SOFC units for different applications are then presented.

Ni-based anodes are comprehensively discussed in "Ni-Based Solid Oxide Cell Electrodes", detailing their electrical conductivity, electrochemical performance, thermal and chemical compatibility, and interactions with real fuels. Finally, the electrode mechanism is discussed using real and model data.

"Strategies for Perspective Cathode Materials for IT–SOFC" focuses upon perovskite materials that show promise as intermediate temperature SOFC cathode materials. This chapter discusses perovskites properties and then has a closer look at LSCF ($La_{1-x}Sr_xCo_{0.2}Fe_{0.8}O_{3-\delta}$) and how its structural and transport properties vary with A or B site doping.

"Critical Issues of Metal-Supported Fuel Cell" relates to the integration of metals as supports for solid oxide fuel cells. It includes the advantages of using a metal support, metal alloy choice, and use of extra protective layers. Methods of manufacture of the metal supports are covered followed by some real cell data.

"Challenges Imposed by Thermochemical Expansion of Solid State Electrochemical Materials" covers the topic of chemical expansion, which is due to the creation of point defects in a material due to changes in oxygen activity/ stoichiometry. The chapter covers the measurement of chemical expansion, the analysis of various phases, and finally the stress induced in materials from both thermal and chemical expansions.

"Current State of Models for the Prediction of Mechanical Failures in Solid Oxide Fuel Cells" discusses the modeling of the mechanical failure of the cells. It starts with an overview of the causes of failure in the various components of a stack, and the properties of the materials that lead to stresses in the stack. They then go on to discuss the various methodologies of modeling stacks, with some applications, and results.

Material choice is crucial to produce a successful fuel cell. "Alternative Materials for SOFCs, Opportunities and Limitations" touches on materials and their suitability for both electrolyte and anode applications. The chapter finishes on the use of redox stable anodes to improve overall fuel utilization, and the in situ formation of nano-structure to improve electrode performance.

"Materials for Next Generation SOFCs" relates two important strategies for improving fuel cell performance. The first section covers a nano-engineering solution, to produce thin layers to get surface (interfacial) ionic conductivity. The second section covers the use of new electrode materials based on Perovskite-related phases, Ruddleston–Popper phases, and double and triple perovskites.

In "Future Fuel Cells", we are taken through the various drivers for the design of future fuel cell systems. Some of the topics covered are operating temperature, cell design, fuel choice, system design, durability, and reliability. The topics of metal supported solid oxide fuel cells and the use of nano-structured materials in cells are also covered. The chapter finally discusses with how fuel cells will fit into the global market and can be integrated into the existing energy flows.

John T. S. Irvine
Paul Connor

Contents

Solid Oxide Fuel Cells: Past, Present and Future

Subhash C. Singhal

Abstract Although fuel cells have been known for over 150 years, research on solid oxide fuel cells (SOFCs) based on an oxide ion conducting electrolyte only accelerated in the last 30 years. This chapter, after a brief history of SOFCs, reviews the materials for different cell components (electrolyte and the two electrodes) and cell stacks (seals and interconnects); novel materials and their structures have been investigated and developed to improve electrochemical performance. Different SOFC designs and their relative advantages and disadvantages are then discussed. Finally, various applications of SOFC power systems and the status of their demonstration and commercialization are reviewed.

1 Introduction

A solid oxide fuel cell (SOFC) is an energy conversion device that converts chemical energy of a fuel (such as hydrogen or methane) into electricity through a series of electrochemical reactions; no combustion process is involved. As a result, fuel cell efficiencies are not limited by theoretical Carnot efficiencies since the electrical work is directly converted from a substantial fraction of the enthalpy associated with the electrochemical oxidation of the fuel into water and/or carbon dioxide. Hence, fuel cells can deliver higher electrical conversion efficiencies when compared with traditional technologies such as coal-fired power plants and electrical generators based on internal combustion engines. In addition to high

S. C. Singhal (✉)
Pacific Northwest National Laboratory, 902 Battelle Boulevard, Richland, WA 99352, USA
e-mail: singhal@pnl.gov

J. T. S. Irvine and P. Connor (eds.), *Solid Oxide Fuels Cells: Facts and Figures*, Green Energy and Technology, DOI: 10.1007/978-1-4471-4456-4_1,

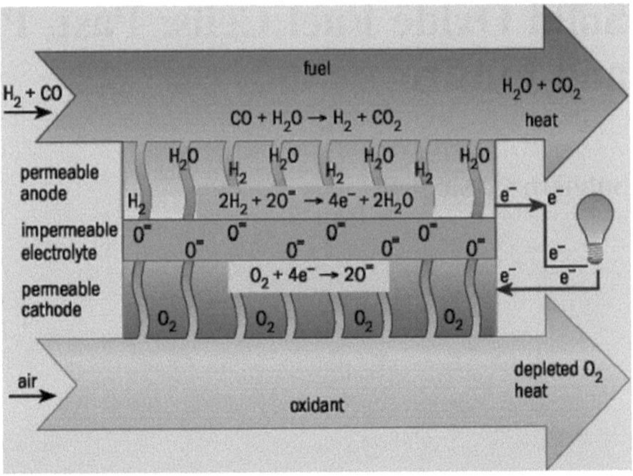

Fig. 1 Operating principle of a solid oxide fuel cell

efficiency, SOFC, when compared with other fuel cells, has two particular advantages, owing to its high temperature operation; first, SOFC allows the use of a variety of fuels ranging from hydrogen to CO to hydrocarbons, and second, an SOFC produces significant amount of exhaust heat, which can be used in combined heat and power systems (CHP). In addition, quiet, vibration-free operation of fuel cells also eliminates noise usually associated with power generation systems. These fuel cells also produce no or very low levels of SO_x and NO_x emissions.

Shown in Fig. 1 is a typical SOFC, which essentially consists of two porous electrodes, separated by a dense, oxide ion conducting electrolyte. Oxygen gas molecules on the cathode side react with incoming electrons coming from the external circuit to form oxygen ions, which migrate through the oxide ion conducting electrolyte to the anode. At the anode, oxide ions react with H_2 or other fuels to form H_2O (and/or CO_2), liberating electrons, which flow from the anode through the external circuit to the cathode to produce electricity. Provided that both fuels and oxygen are supplied constantly, the continuous electrochemical reactions can steadily generate electricity.

Quantitatively speaking, the electromotive force (EMF) of a cell is determined by the chemical potential of oxygen (i.e., oxygen activity) which is expressed by Nernst Equation as:

$$\mathrm{EMF} = \Gamma \frac{RT}{4\mathrm{F}} \ln\left(\frac{(\mathrm{pO}_2)_a}{(\mathrm{pO}_2)_b}\right) \quad (1)$$

where Γ is the ionic transference number (ionic conductivity/total conductivity), T is operation temperature, F is Faraday constant, $(pO_2)_a$ is the oxygen activity on the oxidant side and $(pO_2)_b$ is the oxygen activity on the fuel side. In the case without external circuit, the EMF corresponds to open circuit voltage (OCV). For instance, an OCV $\sim$1.1 V can be calculated for an SOFC operating at 800 °C when air and room-temperature-humidified H_2 are used.

Under cell operating conditions, i.e., when a current passes through it, a fingerprint characteristic of fuel cells is the relationship between voltage and current density. The cell voltage (V) is given by:

$$V = E^{\circ} - iR - \eta_A - \eta_F \quad (2)$$

where i is the current passing through the cell, R is the electrical resistance of the cell, and η_A and η_F are the polarization voltage losses associated with the air electrode (cathode) and the fuel electrode (anode), respectively. These polarizations are related to three physical processes: gas diffusion, gas–solid interaction, and ionic migration.

1.1 Brief History

Solid oxide fuel cells have come a long way to become practical power generation devices since the initial discovery of a ceramic material consisting of 85 % ZrO_2 and 15 % Y_2O_3, the so-called "Nernst Mass", by Walther Nernst in late 1890s that laid the foundation for the electrolyte material for these cells [1, 2]. The first conceptual SOFC is believed to have been demonstrated in 1937 by Bauer and Preis [3]. However, the more focused studies on SOFCs only began after the pioneering 1943 work by Carl Wagner who attributed the electrical conductivity in mixed oxides such as doped ZrO_2 to the presence of oxygen vacancies [4]. In 1962, scientists at Westinghouse Electric Corporation published a paper titled simply as "a solid electrolyte fuel cell" [5]. This initial effort became the foundation of Westinghouse's cathode-supported tubular SOFCs. Based on this cell design, Westinghouse successfully produced and tested several 5–250 kW sized SOFC power systems from the 1980s to the 1990s. ZrO_2-based materials remain the most commonly used electrolyte for SOFCs even today.

From the mid-1990s to the present, several other cell designs and materials have been explored; in particular, anode-supported planar SOFCs have become quite popular because of performance and cost considerations. Today, SOFCs in many different designs and containing different cell materials are being explored and produced for power generation in small (few watts) to large (several hundred kWs) sizes in residential, commercial, and central power station applications. Although still fairly small in number, most demonstrations of SOFC power systems to date have been made of the 1–5 kW sized residential combined heat and power (CHP) units. The biggest drawback to the large-scale commercialization of SOFC power systems remains their relatively high cost relative to other power generation technologies. Several technical issues have been identified and are being addressed in the development of SOFC products for practical applications. These issues mainly relate to the two drivers to enable commercialization: competitive cost and reliable performance with desired operating characteristics.

The progress in SOFC technology and its commercialization status has been detailed in various scientific publications, books [6, 7] and in proceedings of two international conferences devoted solely to solid oxide fuel cells; first, the biennial international symposia on solid oxide fuel cells starting in 1989 [8–18], and second, the SOFC Forums of the European fuel cell forum [19–27]. These two series of proceedings give an excellent snapshot in time and progress achieved in SOFC technology over the last twenty years.

This chapter briefly reviews the cell and stack materials, cell designs, and present status of SOFC demonstrations, and concludes with expectations for the near future; more detailed description of these subjects is available in the subsequent chapters of this book.

2 Cell and Stack Materials

Materials research for SOFCs is driven by the desire to improve SOFC stability and to reduce various polarization losses. Initially, SOFCs were developed for operation primarily in the 900–1000 °C temperature range to overcome the high ohmic loss from a thick electrolyte. However, reduction of the SOFC operating temperature by 200 °C or more allows the use of a broader set of materials, is less demanding on the seals and the balance-of-plant components, simplifies thermal management, aids in faster start-up and cool down and should result in less degradation of the cell and stack components. Because of these advantages, activity in the development of SOFCs capable of operating in the intermediate temperature (IT) range of 650–800 °C has increased dramatically in the last decade, which necessitates advanced processing and novel materials in order to achieve high electrode kinetics and a low ohmic loss in the IT region. The following sections briefly review the cell (electrolyte and electrodes) and stack (sealant and interconnect) materials as detailed in Ref. [28].

2.1 Electrolyte Materials

In an SOFC, the electrolyte is exposed to both oxidizing (air side) and reducing species (fuel side) at high temperatures; hence, long-term successful SOFC operation requires that the electrolytes have the following: (1) Sufficient ionic conductivity—the electrolyte materials must have an ionic transference number close to unity, i.e., the electronic conductivity in the electrolyte must be sufficiently low in order to provide a high energy conversion efficiency. Also, the oxide ion conductivity must be high to minimize the ohmic loss; (2) Dense structure—in order to produce maximum electrochemical performance, the electrolyte must be gas tight; and (3) Stability—the electrolyte is exposed to both the air and the fuel at

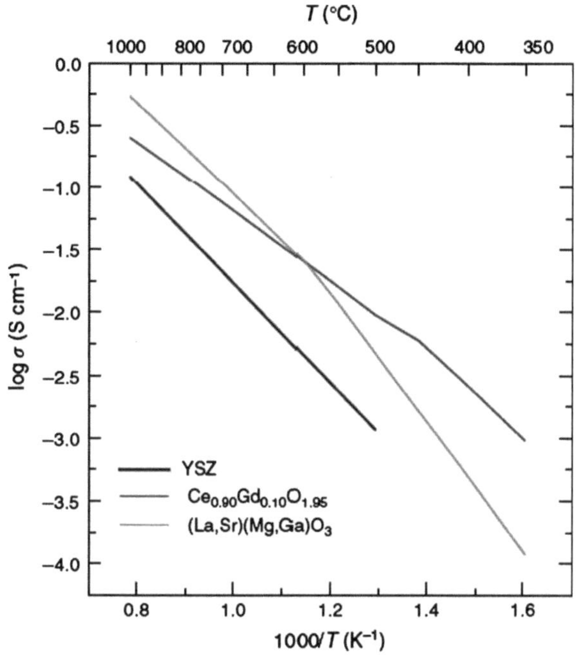

Fig. 2 Conductivity as a function of temperature for yttrium-stabilized ZrO_2 (YSZ), gadolinium-doped CeO_2, and (La,Sr)(Mg,Ga)O_3 (LSMG)

elevated temperature and must be chemically stable in these. Also, the thermal expansion coefficients must match at the electrolyte/electrode interfaces.

Typical electrolyte materials for SOFCs are oxides with low valence element substitutions, sometimes named acceptor dopants, which create oxygen vacancies through charge compensation. Yttrium-stabilized ZrO_2 (YSZ) is the most common material that satisfies these requirements; acceptor doped CeO_2 and perovskite structure oxides have also been investigated as electrolyte materials.

A plot of ionic conductivity as a function of temperature for common SOFC electrolyte materials is shown in Fig. 2. Although YSZ shows the lowest ionic conductivity in Fig. 2, it remains as the only material that has been demonstrated to provide long-term stability under cell operation conditions up to about 80,000 h at 1000 °C. In order for YSZ to be used in the IT regime, a typical approach to minimize ohmic loss is to decrease the thickness of the YSZ electrolyte. Another approach to reduce ohmic loss is to improve the ionic conductivity of a ZrO_2-based electrolyte. Sc-doped zirconia shows higher conductivity than yttrium-doped ZrO_2; however, the cost and known aging of Sc-doped ZrO_2 present challenges in using this material for commercial SOFCs.

Doped CeO_2 materials are candidates for the electrolyte for cell operation at temperatures below about 600 °C as discussed by Steele [29] and utilized by Ceres Power Inc. (UK), due to their higher oxide ion conductivity ($Ce_{0.9}Gd_{0.1}O_{1.95}$: 0.025 $\Omega^{-1}cm^{-1}$ at 600 °C) compared to YSZ (<0.005 $\Omega^{-1}{\cdot}cm^{-1}$). Gd or Sm doped CeO_2 provides the highest ionic conductivity in CeO_2-based materials due to similar ionic radii between Gd^{3+}/Sm^{3+} and Ce^{4+}. The principal challenge with doped CeO_2 is the onset of electronic conduction in reducing conditions at temperatures above about 650 °C due to the reduction of Ce^{4+} to Ce^{3+} to compensate the formation of oxygen vacancies. Further reduction of Ce^{4+} results in lattice expansion and often creates microcracking in the electrolyte. Cell temperatures below about 600 °C seem necessary for successful use of doped CeO_2 in SOFCs for long-term operation.

The perovskite structure $(La,Sr)(Mg,Ga)O_3$ (LSMG) and other similar materials have also been developed as oxide ion conductors. The use of LSMG is attractive because it has reasonable oxide ion conductivity and is compatible with a variety of cathodes, in particular the highly active ones; hence excellent electrochemical performance has been reported with LSMG electrolytes. The challenges for LSMG are the uncertain costs of Ga sources and its chemical and mechanical stability.

2.2 Cathode Materials

The cathode material is crucial because electrochemical reduction of O_2 on the cathode requires a series of elementary reactions and involves the transfer of multiple electrons. The cathode must have: high electronic conductivity, chemical and dimensional stability in environments encountered during cell fabrication and cell operation, a thermal expansion match with other cell components, compatibility and minimum reactivity with the electrolyte and the interconnection with which it comes into contact, and a high catalytic activity for oxygen molecule dissociation and oxygen reduction. Finally, the cathode must have a stable, porous microstructure so that gaseous oxygen can readily diffuse through the cathode to the cathode/electrolyte interface. These stringent electrochemical and mechanical requirements greatly restrict the number of suitable candidate materials.

In addition to noble metals, conducting perovskite family oxides are the preferred cathode materials. Lanthanum manganite ($LaMnO_3$), which, when substituted with low valence elements, such as Ca or Sr, has a superior p-type electronic conduction due to the formation of large amount of Mn^{4+}. Moreover, doped $LaMnO_3$ possesses adequate electrocatalytic activity for oxygen reduction, a reasonable thermal expansion match to YSZ, and stability in the SOFC cathode operating environment. Under the cathode operation conditions, doped $LaMnO_3$ has oxygen excess, or precisely speaking, has cation vacancies, and the oxide ion conductivity is relatively low in Sr doped $LaMnO_3$ (LSM), of the order of 10^{-7} S/cm at 800 °C.

For SOFCs operating in the IT regime, alternative cathode materials have been developed, since LSM is not the ideal candidate, owing (at least in part) to its low ionic conductivity and slow surface oxygen exchange kinetics. The desirable cathode for IT SOFCs is a mixed ionic and electronic conductor (MIEC). In order to design the MIEC with a single chemical composition, numerous perovskite compositions, typically, containing La on the A site, and transition metals such as

Co, Fe, and/or Ni on the B site, for instance (La,Sr)(Co,Fe)O_3—the so-called LSCF, have received attention. In general, compared with LSM, these materials offer higher oxide ion diffusion rates and exhibit faster oxygen reduction kinetics. Promising results have been reported using these materials; though in many cases the improved cathodic performance is found to decrease during cell lifetime as a result of chemical or microstructural instability. The decreased electrochemical performance of the cathode often is related to the interdiffusion of LSCF/LSF toward YSZ. A thin interlayer, mostly of a CeO_2-based material, is generally used to reduce interdiffusion between the cathode and YSZ.

Shao and Haile [30] have reported on a new cathode material, $Ba_{0.5}Sr_{0.5}$-$Co_{0.8}Fe_{0.2}O_3$ (BSCF), which at ~600 °C possesses fast bulk diffusion coefficients. With BSCF as the cathode, they obtained an excellent cell electrochemical performance of ~1 W/cm^2 at 600 °C. They attributed the improved cathode performance to both fast surface exchange and bulk diffusion coefficients. Although thermal and chemical stability of BSCF remains a challenge, the material itself is of interest for use in intermediate temperature SOFCs.

In addition to chemical composition, electrode microstructure has a profound effect on the cathode polarization.

2.3 Anode Materials

The anode must be an excellent catalyst for the oxidation of the fuel, and must be: stable in the reducing environment, electronically conducting, and have sufficient porosity to allow gas species in and out of the electrolyte/anode interface where the fuel oxidation reaction takes place. Other requirements include matching of the thermal expansion coefficient with that of the electrolyte and the interconnect material; chemical stability with the electrolyte and the interconnect; and applicability for use with various fuels and impurities (such as sulfur). In addition, cost effectiveness is always a factor in the commercialization of SOFCs. During the early stages of SOFC development, the anode materials were precious metals (such as Pt and Au) and transition metals (such as Fe and Ni). These candidates, however, cannot meet the stringent aforementioned requirements for the anode in SOFCs. For instance, Ni is an excellent catalyst; however, it possesses a high thermal expansion coefficient (13.4×10^{-6}/°C), and exhibits a coarsening of microstructure due to metal aggregation through grain growth. A composite of Ni-YSZ, was found capable of preventing sintering of the nickel particles, decreasing the effective thermal expansion coefficient of Ni, and providing better adhesion of the anode with the electrolyte. Currently, Ni-YSZ cermet is the most commonly used anode material.

In a Ni/YSZ anode, nickel has the dual role of being a catalyst for hydrogen oxidation as well as being an electrical current collector. In addition to being an excellent catalyst for the oxidation of hydrogen, nickel is also highly active for the steam reforming of methane. This catalytic property is exploited in the so-called internal reforming SOFCs that can operate on fuels composed of mixtures of methane and water. However, Ni can catalyze the formation of carbon from hydrocarbons under reducing conditions. Unless adequate amount of steam is present along with the hydrocarbon to remove this carbon from the nickel surface, the anode activity may suffer. As a result, even when using methane as the fuel, relatively high steam-to-carbon ratio is needed to suppress this deleterious carbon deposition reaction. Unfortunately, due to the high catalytic activity of nickel for hydrocarbon cracking, this approach does not work for higher hydrocarbons. Therefore, a pre-reformer is necessary when using higher hydrocarbon-containing fuels.

Incorporating oxide-based catalysts is an approach that has been tried to overcome the limitations of the nickel-based anodes to directly utilize hydrocarbons, and has met with some success. A composite of Cu/CeO_2 has been investigated as the SOFC anode for the direct utilization of hydrocarbon fuels. Unlike Ni, Cu is not an active catalyst for carbon formation. The addition of CeO_2 provides active catalytic sites. However, sintering of copper at cell operating temperatures has limited its use in practical SOFCs. In terms of sulfur tolerance, ceramic oxides based on ceria or strontium titanate, have yielded some promising results, but the benefits obtained are counterbalanced by other limitations (such as high electrical resistivity and the difficulty of integrating such materials with existing cell and stack fabrication processes and materials). Research is progressing on further development of such oxide-based anodes.

2.4 Sealing Materials

In planar SOFC stacks, seals are critical and must prevent (or at least minimize sufficiently for acceptable performance) leakage of both fuel and oxidant gases from the stack to the outside environment, as well as mixing of the fuel and oxidant gases within the stack. Successful development of sealing materials and the concepts for planar SOFCs are probably the most important issues for the long-term performance stability and lifetime of planar SOFC stacks and hence their commercialization at competitive costs. The primary challenge in developing seals resides in a variety of strict requirements that effective seals need to meet. First, an effective seal must be thermochemically stable under the operational conditions of the stack, they must be electrically insulating, and must have a thermal expansion match to the fuel cell components. Additionally, the seal should be compatible with other components, and should be formed at a low enough temperature to avoid damaging the cell components and also should not migrate or flow from the designated sealing region during sealing or cell operation. It is also necessary to develop a sealing system that can withstand thermal cycling between the cell operational temperature and room temperature. In order to satisfy these requirements, a number of different sealing approaches have been developed, which have met with some success, including rigid, bonded seals (e.g., glass-ceramics and brazes), compliant seals (e.g., viscous glasses), and compressive seals (e.g., mica-based composites); multiple sealants may be used in any given stack design between different components.

Rigid seals typically rely on a glass that softens and "glues together" the adjacent stack components during stack fabrication (at a temperature above the operating temperature), but then become rigid and immobile when cooled to the operating temperature. A wide variety of glasses and glass-ceramics have been developed, including alkali silicate, alkaline earth silicate, borosilicate, aluminoborosilicate, and others. Glass-based seals represent a relatively straightforward means of sealing an SOFC stack (at least initially), but the brittle nature of glasses (below the glass transition temperature) and glass–ceramics makes these seals vulnerable to cracking (resulting from stresses related to thermal expansion mismatches with the adjacent components). As a consequence, the coefficient of thermal expansion (CTE) must be similar to that of the other components. While glass compositions can be tailored to optimize their physical properties, the selection of glasses offering appropriate thermal expansion behavior is relatively narrow, and the selection is further limited by the need for the glass to have appropriate wetting behavior and viscosity at the sealing temperature. The best results to date have been obtained using compositions based on silica. Alkali silicate glasses tend to be very reactive toward other SOFC components, though alkaline-earth aluminosilicate glasses have yielded promising results. One of the primary advantages of the latter (particularly those containing BaO) is the ability to tailor their CTE to match that of other SOFC stack components.

In addition to glass seals, brazing has been developed as a possible sealing approach, whereby a filler metal, the liquidus of which is well below that of the materials to be joined, is heated to a point at which it is molten, is allowed to flow and fill the gap between the two joining pieces under capillary action, and then cooled to solidify the joint. Although most commercial brazes do not wet ceramic surfaces, the addition of a reactive metal, such as Ti or Zr, results in the reduction of the ceramic phase at the joining interface to give an intermediate layer that is in chemical equilibrium with both the ceramic and braze-metal phases. Once this reduced phase forms, wetting of the ceramic surface by the filler metal is improved greatly. While active metal brazing solves the wetting issue, preferential oxidation of the active element in the braze at SOFC operating temperatures can lead to rapid deterioration of the joint. The desire to accomplish sealing in an inexpensive oxidizing environment greatly complicates the search for an appropriate joining material. Use of a reducing environment at joining temperatures of the order of 800 °C or greater adds cost and can damage cathode materials, resulting in a severe loss in cell performance. This processing restriction reduces the list of candidate joining materials to noble metal-based brazes.

Compressive seals, which typically utilize materials such as sheet-structure silicates to act as a gasket between components, can help to improve the stack's tolerance to thermal expansion mismatch between various stack components. A compliant high-temperature material is placed between the two sealing surfaces and compressed, using a load frame external to the stack, to accomplish sealing.

Because the seal conforms to both sealing surfaces and is under constant compression during use, it forms a dynamic structure that reduces the need for CTE matching. It is also possible for the sealing surfaces to slide past one another without disrupting the hermeticity of the seal. The primary challenge associated with this technology is the need for a compliant high-temperature sealing material that functions adequately as a reliable compressive seal. A number of materials have been considered for compressive seals, including mica, nickel, and copper. Each has been found to have limitations when used in a simple gasket configuration, ranging from oxidation in the case of the metals to relatively high leak rates in the case of the mica. The route to enhanced performance probably lies with new hybrid and composite seal approaches, in which mica is combined with other materials.

2.5 Interconnect Materials

An interconnect provides the electrical connection between cells and ensures air and fuel separation within the cell stack. Typical requirements of an interconnect material include: high electronic conductivity with low ionic conductivity; chemical stability in both fuel and air; a thermal expansion match to other cell components; high mechanical strength; high thermal conductivity and chemical stability with regard to other cell components. In addition, when considering a stack design, one must take into account ease of fabrication. To satisfy these requirements, doped lanthanum chromite has been used as the interconnect for cells intended for operation at about 1000 °C. Lanthanum chromite is a p-type conductor; its conductivity is due to small polaron hopping from room temperature to 1400 °C at oxygen pressures as low as 10^{-18} atm. The conductivity is enhanced as lower valence ions (e.g., Ca, Mg, Sr, etc.) are substituted on either the La^{3+} or the Cr^{3+} sites. In cells intended for operation at lower temperatures (<800 °C), it is possible to use oxidation-resistant metallic materials for the interconnect. Compared to lanthanum chromite ceramic interconnects, metallic alloys offer advantages such as improved manufacturability, significantly lower raw material and fabrication costs, and higher electrical and thermal conductivity. But to be useful for the interconnect application, the metallic alloys must satisfy additional requirements, particularly resistance to surface oxidation and corrosion in a dual atmosphere (exposure to oxidizing atmosphere on one side and reducing atmosphere on the other side). An oxide protective layer, formed in situ during oxidation or ex situ by a coating technique, can provide resistance to surface oxidation and corrosion. As a result, the oxide must have the following characteristics: slow growth kinetics, sufficient electrical conductivity, strong adhesion to the underlying alloy, even in severe thermal cycling conditions, and good chemical compatibility with the electrodes, with no significant diffusion or evaporation of components that can influence electrode performance or oxide integrity.

Ferritic stainless steels are the most promising materials as the interconnect, owing to the fact that some alloys in this family offer a protective and conductive Cr-based oxide scale, appropriate thermal expansion behavior, ease of manufacturing, and low cost. Several new ferritic stainless steels, such as Crofer22 APU, have been developed specifically for the SOFC interconnect application. Although these alloys demonstrate improved performance over traditional compositions, several critical issues remain. Among these are the formation of chromium containing species at electrolyte/cathode interface and subsequent poisoning of cathodes; corrosion and spalling under interconnect exposure conditions, and compatibility with the adjacent components such as seals and electrical contact layers. As an alternative approach to developing new alloys, surface modification can result in enhanced performance. For example, electrically conductive perovskites and spinels can be applied onto metallic interconnects to minimize scale growth, electrical resistance, and Cr volatility.

Volatility of Cr species from the metallic interconnect plays a key role in the degradation of the electrochemical performance of SOFCs. Metallic interconnects with Cr_2O_3 scale give out Cr-containing volatile species that deposit on the electrolyte and electrode surfaces and interfaces, and lead to rapid deterioration of the cathode's oxygen reduction reaction rate and thus to a significant decline in cell performance. The mechanism of Cr poisoning is not clear, although reactions of the cathode with $CrO_2(OH)_2$ are considered the likely cause. Possible chromium poisoning mitigation strategies include the use of suitable surface layers on alloy interconnects to reduce Cr evaporation; La-chromite, Mn-chromite, and non-Cr-containing spinels such as $(Mn,Co)_3O_4$ can reduce Cr flux during the operation of cell or stack. Such layers could potentially be applied either in a separate fabrication step (e.g., by thermal spray) or grown in situ through suitable modification of the alloy bulk or surface.

3 SOFC Designs

Over the years, several SOFC designs have been investigated and developed by configuring the three active components of the cell, the electrolyte and the two electrodes, in different geometric configurations. The two most common designs are planar and tubular, and their many variants. These are reviewed below.

3.1 Planar SOFC Design

In the planar design, a series of cell components are configured as thin, flat plates, then electrically connected to build up desirable electrochemical performance. A schematic of a generic planar SOFC design is shown in Fig. 3. The planar cells can be either electrolyte-supported, electrode-supported, or metal-supported.

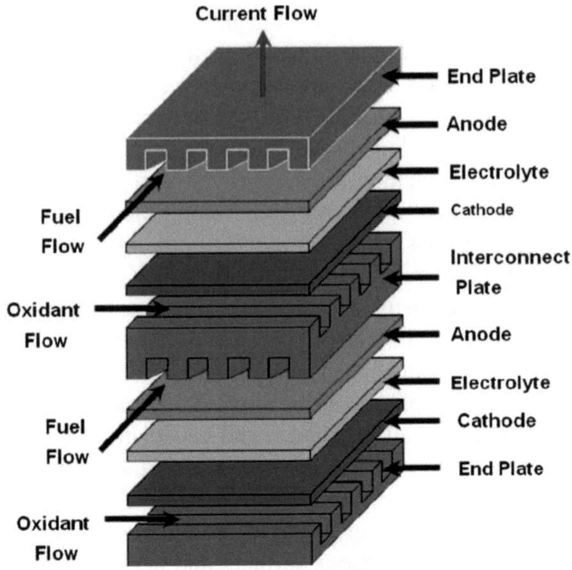

Fig. 3 Planar solid oxide fuel cell design

Each of these designs can also have a number of interesting variants; for example, the planar SOFC may be in the form of a circular disk fed with fuel from the central axis, or it may be in the form of a square or rectangular plate fed from the edges. Planar designs offer several potential advantages, including simpler and less expensive manufacturing processes and higher power densities than tubular cells described in the next section. However, planar designs need high-temperature gas-tight seals between the components in the SOFC stack; such seals are not necessary with tubular cells. Seal development remains as the most challenging area in successfully commercializing planar SOFCs. An important feature of the planar design resides in the configuration of gas flow and gas manifold, which play a crucial role in minimizing the use of seals, improving fuel utilization, managing uniform distribution of temperature and current to reduce thermal stresses, and improving the stability of the stack. Generally, the interconnect is ribbed on both sides to allow cross-flow, co-flow, or counter-flow configurations.

The initial planar SOFC configurations employed a thick electrolyte as the support, which required an operating temperature often higher than 900 °C. Advances in ceramic processing have allowed fabrication of thin electrolytes, 10 μm or thinner, by low-cost conventional ceramic processing techniques such as tape casting, tape calendaring, slurry sintering, screen printing, or by plasma spraying. As a consequence, anode-supported planar cell stacks have been extensively fabricated and tested by a number of developers for long-term operation. Electrochemical performance of an anode-supported single planar cell at 600, 700, and 800 °C is shown in Fig. 4; the cell was optimized with an anode thickness of 0.5 mm and porosity ~57 %. The anode interlayer was ~20 μm. The electrolyte was ~8 μm and cathode interlayer ~20 μm. The flow rates of

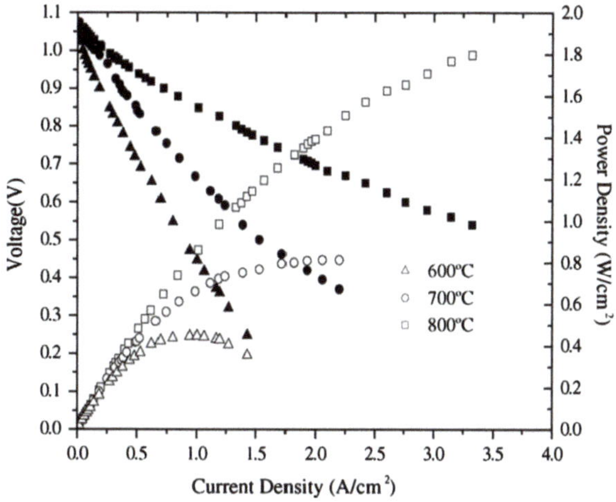

Fig. 4 Electrochemical performance for an optimized anode-supported single cell measured at 600, 700, and 800 °C [31]

humidified hydrogen and air were 300 and 550 ml/min, respectively. Zhao and Virkar [31] have shown that the electrochemical performance of planar SOFCs is highly dependent on cell materials, electrode microstructures, and cell geometric parameters, with power densities as high as 1.8 W/cm^2 at 800 °C being possible as shown in Fig. 4.

3.2 *Tubular SOFC Design*

Tubular SOFCs may be of a large diameter (>15 mm), or of much smaller diameter (<5 mm), the so-called microtubular cells. Also, the tubes may be flat and joined together to give higher power density and easily printable surfaces for depositing the electrode layers. In a typical tubular SOFC, as illustrated by the Siemens Westinghouse design shown in Fig. 5, the cell tube is porous doped lanthanum manganite fabricated by extrusion/sintering and is closed at one end. The cell components, dense YSZ electrolyte, porous Ni-YSZ anode, and doped lanthanum chromite interconnect, are deposited in the form of thin layers by atmospheric plasma spraying.

For cell operation, oxidant (air or oxygen) is introduced through an alumina injector tube positioned inside the cell. The oxidant is discharged near the closed end of the cell and flows through the annular space formed by the cell and the coaxial injector tube. Fuel flows on the outside of the cell from the closed end and is electrochemically oxidized while flowing to the open end of the cell generating electricity. At the open end of the cell, the oxygen-depleted air exits the cell and is combusted with the partially depleted fuel. Typically, 50–90 % of the fuel is utilized in the electrochemical cell reaction. Part of the depleted fuel is recirculated in the fuel stream and the rest combusted to preheat incoming air and/or fuel. The electrochemical performance of a tubular cell (2.2 cm diameter, 150 cm active

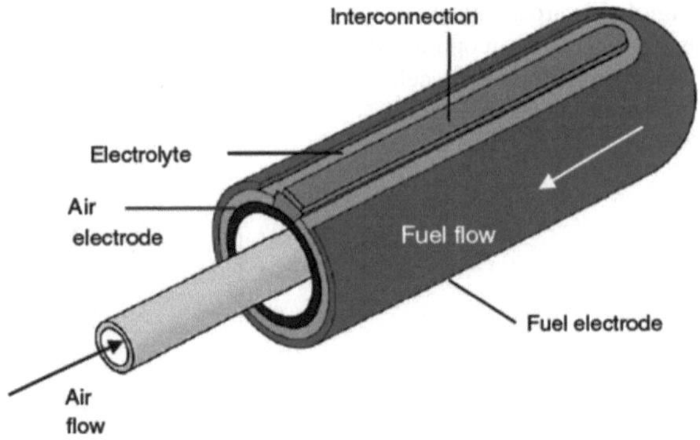

Fig. 5 Tubular solid oxide fuel cell design

length) at 800, 900, and 1000 °C with 89 % H_2 + 11 % H_2O fuel (85 % fuel utilization) and air as oxidant is shown in Fig. 6. Such tubular cells have a power density at 1000 °C of about 0.2 W/cm^2.

The single biggest advantage of tubular cells over planar cells is that they do not require any high-temperature seals to isolate oxidant from the fuel, and this makes performance of tubular cell stacks very stable over long periods of times (several years). However, their areal power density is much lower (about 0.2 W/cm^2) compared to planar cells (up to 2 W/cm^2 for single cells and at least 0.5 W/cm^2 for stacks), and manufacturing costs higher. The volumetric power density is also lower for tubular cells than for planar cells. To increase the power density and reduce the physical size and the cost of tubular SOFC stacks, alternate tubular geometry cells have been investigated. Such alternate geometry cells combine all the advantages of the tubular SOFCs, such as not requiring high-temperature seals, while providing higher areal and volumetric power densities. Figure 7 shows several of such designs investigated by Siemens; the performance of these new design cells is higher than that of cylindrical tubular cells, but still lower than that of anode-supported planar cells.

A single SOFC, whether planar or tubular, generally produces a voltage of less than about 1 V. A series connection of single cells, a stack, is therefore formed to obtain higher power. Unlike a single cell, which deals with mostly material aspects of the components, cell stacks must consider many other issues to generate maximum stable power. The design of SOFC stacks must take into account the electrochemical performance to generate more power from the stack; structural and mechanical integrity to allow the stack to be operated at high temperature for sufficiently long duration and possibly under thermal cycles; gas manifolding for supplying reactant gas to a large cell area, removing reaction products and providing gas hermeticity; and ease of fabrication.

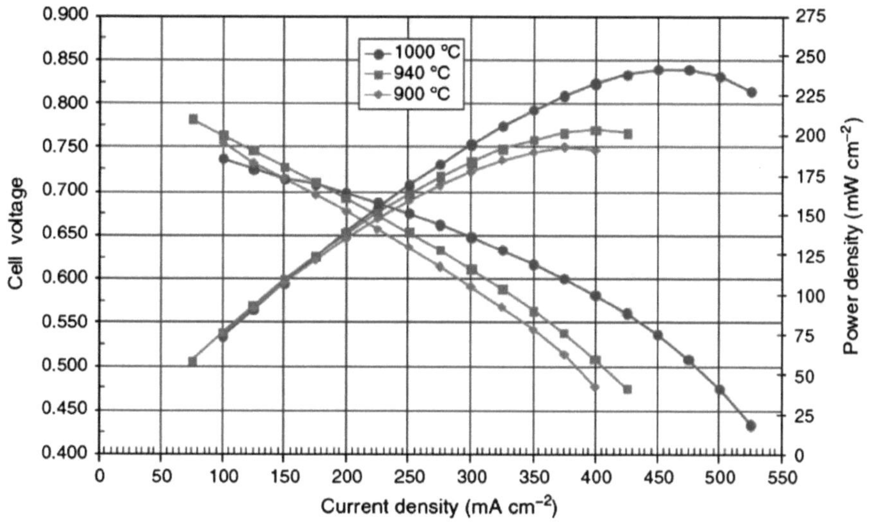

Fig. 6 Voltage-current density and power-current density plots of a tubular SOFC

Fig. 7 Alternate geometry tubular cells investigated by Siemens

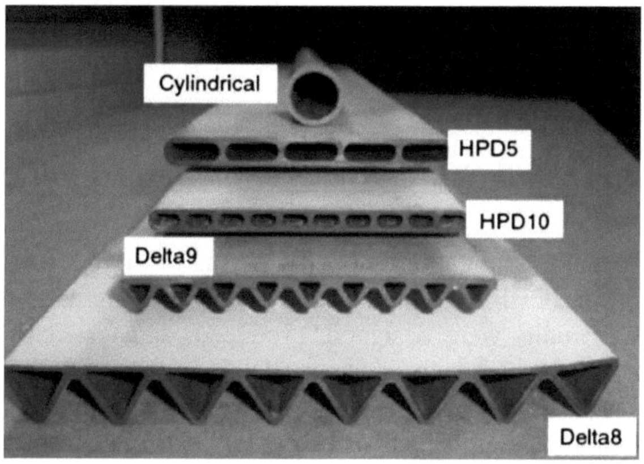

4 SOFC Applications and Power Systems

SOFCs have high efficiency and fuel flexibility because of the high operation temperature (550–1000 °C). On one hand, operating SOFCs at elevated temperatures produces high quality exhaust, which can be used to provide heat sources for stationary applications, or to heat the reformer for endothermic steam reforming reactions, or even to fire a secondary gas turbine. Hence, relative to other fuel cells, SOFCs have a high electrical efficiency, even in small size systems. On the other hand, over the operating temperature regime (550–1000 °C), CO can be used as the fuel, rather than poisoning the anode. Therefore, a variety of fuels can be reformed within the cell stack (internal reforming) or through a separate fuel reformer (external reforming). This flexibility allows use of fuels, such as biogas, liquid hydrocarbon fuels, and landfill gas, etc., and enables SOFCs to be well-suited especially for stand-alone and remote applications. These fuels can be reformed to a mixture of hydrogen and carbon monoxide. The direct internal reforming represents reformation of the fuel directly on the fuel cell anode, thus electrochemical reaction and fuel reformation simultaneously take place at the anode. This is a simple and very efficient design with least loss of energy; however, carbon deposition and temperature inhomogeneity due to endothermic cooling may present problems in direct internal reforming. Indirect reforming uses a separate fuel reforming catalyst which is integrated within the SOFC stack upstream of the anode side and typically, utilizes heat and water from the SOFC stack. Indirect reforming may not be as efficient as direct reforming, but it provides more stable cell performance. Any external reformer is physically separated from the fuel cell stack; hence it can be operated at different pressure and temperature if necessary. This is of particular importance because it eliminates the problem of carbon deposition from fuel decomposition that deactivates the cell anode.

Because of their superior electrical efficiency and fuel flexibility, SOFC-based power systems, compared to other fuel cell systems, enable numerous applications at various power levels, from a few-watt to megawatt size systems. The following sections describe various applications of SOFC-based power systems that have been demonstrated or are under development.

4.1 Small SOFC Systems for Residential CHP Applications

A major application for SOFCs is at 1–5 kW level to supply combined heat and power (CHP) to residential buildings utilizing natural gas as the fuel. Early SOFC CHP units were designed, produced, and tested by Hexis (Switzerland). These units are based on a planar SOFC system, with an inner round aperture in the center serving as a channel for the supply of fuel. Fuel flows toward outside in a radial manner. The air is preheated and flows from outside through four channels onto the metallic interconnect inside the cell stack, then is diverted there and flows toward outside in a radial manner. The excess fuel that has not electrochemically converted is afterburned at the edges of the cell stack. Based on this design, a system, named "Galileo 1000", has been developed to supply an electrical power of 1 kW and a thermal capacity of about 2.5 kW; over 20 such units are presently under demonstration. Ceres Power (UK), in partnership with British Gas, has developed a metal-supported planar SOFC system, utilizing ceria-based electrolyte, for generating electricity and central heating requirement (including hot water) for a typical home. Metallic supported SOFC stack operates at relatively low temperatures ($\sim$550 °C) and tends to be lightweight. As a result, Ceres Power's unit is compact and wall-mountable, which may provide access to both the boiler replacement and new residential applications. Ceramic Fuel Cells Ltd.

Fig. 8 A 1 kW SOFC CHP unit (*left*: SOFC system; *right*: hot water tank) utilizing Kyocera's anode-supported flat tubular cells

(Australia) utilizes planar SOFCs for residential CHP units; over 50 of its 1 kW units have been or are being demonstrated in various parts of the world.

Tubular SOFC design is also attractive for residential units because of its reliable and stable performance over long periods. Kyocera (Japan) has developed anode-supported flat tubular cells, and these cells are being utilized by several organizations in Japan (Osaka Gas, Tokyo Gas, etc.) to fabricate and test SOFC-based residential CHP systems. The first trial operation of such a 1 kW unit in a residential house was conducted in 2005–2006 for a period of 2000 h; the average electric efficiency and hot water heat recovery efficiency were 44.1 % (LHV) and 34 % (LHV), respectively. Over 75 such units, illustrated in Fig. 8, are now installed and operating in homes throughout Japan, and are expected to be commercialized in the next two years.

Toto Ltd. (Japan) is producing and testing 2 kW size CHP units using cathode-supported tubular cells. Siemens (USA) working with Fuel Cell Technologies (Canada) also produced and tested about a dozen 3–5 kW size CHP units using cathode-supported tubular cells; these units performed well with very stable performance for up to one year; unfortunately, both of these organizations are now out of the SOFC business.

Several other organizations are also developing residential CHP units using either planar or tubular SOFCs. Development of SOFC systems for residential CHP applications is accelerating, particularly in Japan and Europe, through formation of partnerships with appliance makers. Residential CHP units will probably be the first commercial application of SOFCs.

Fig. 9 Siemens Westinghouse's 100 kW SOFC cogeneration system

4.2 Large SOFC Systems for Distributed Power Generation

Starting in 1986, Westinghouse (whose fossil energy business was acquired by Siemens in 1998) successfully fabricated about 15 integrated SOFC systems with a power range from 0.4 to 220 kW and tested them on customer sites. The longest test was conducted on a 100 kW atmospheric power generation system, shown in Fig. 9, using tubular SOFCs. This 100 kW cogeneration system was the first demonstration of the tubular cells in a full scale SOFC module. The stack consisted of 1152 cells (2.2 cm diameter and 150 cm active length) in 48 cell bundles of 24 cells each. This system operated for over 36,750 h in USA, Netherlands, Germany, and Italy on desulfurized natural gas without any detectable performance degradation at an efficiency of $\sim$46 %. This was the first successful demonstration of the solid oxide fuel cells for large-scale power generation.

Siemens also produced a 220 kW pressurized SOFC/gas turbine hybrid system, which was installed and tested at the National Fuel Cell Research Center on the campus of the University of California-Irvine (USA). This system was the world's first demonstration of an SOFC coupled with a microturbine generator and the first demonstration of a pressurized SOFC generator. The system accumulated nearly 3400 h of runtime, while operating at a calculated net AC electrical efficiency of 53 %. Analysis indicates that with such pressurized SOFC/gas turbine hybrids, an electrical efficiency of $\sim$70 % is achievable in large MW sized systems.

As mentioned previously, Siemens is no longer in the SOFC business. However, other organizations are continuing work on the utilization of SOFCs for large-scale power generation. Chief among these are Versa Power Systems (in collaboration with FuelCell Energy, USA), United Technologies (in collaboration with Delphi Corporation, USA), and Rolls Royce Fuel Cell Systems (UK/USA). These organizations are supported by the US Department of Energy's Solid State Energy Conversion Alliance (SECA) program that was initiated in the year 2000 to reduce the cost of SOFC power generation systems. Another company, Bloom Energy (USA), started in the year 2001 with venture capital, recently announced the sale, delivery, and installation of several 100 kW sized SOFC power systems, known as Bloom Boxes shown in Fig. 10, to commercial customers such as Adobe

Fig. 10 Five 100 kW sized SOFC systems (Bloom Boxes) installed at eBay Headquarters

Systems, Bank of America, Cox Enterprises, Coca Cola Company, eBay, FedEx, Google, Safeway, Staples, Walmart, etc.

Extensive research is presently being conducted worldwide to decrease the cost of such power systems, particularly in the areas of cell and stack materials, DC to AC power conditioning systems, and other balance-of-plant components. With decreased cost, large-scale SOFC power systems are expected to provide a low- or no-pollution technology to produce electricity.

4.3 Portable SOFC Power Systems

The portable applications generally require power in the range from milliwatts to a few hundred watts. Proton exchange membrane fuel cells (including direct methanol fuel cells) are fuel cells of choice for such portable applications due to their lightweight and low operating temperature; however, SOFC-based units are also being developed, primarily for certain military, leisure, emergency, and transportation applications because of their superior fuel flexibility, not requiring hydrogen. SOFCs can be successfully operated on fuels such as propane, gasoline, diesel, kerosene, JP-8 military fuel, ethanol, and other biofuels. Challenges arising for SOFCs in portable applications are that stacks must be light, have a short start-up time, and be thermally sustaining. Quick start-up time is particularly crucial for portable applications; however, it is very difficult to achieve because of relatively low thermal shock resistance of ceramic components. One approach is to use microtubular SOFCs, which are capable of tolerating thermal shock resistance. Moreover, microtubular design can give reasonable volumetric power densities because the power density increases inversely proportional to the tube diameter.

Microtubular SOFCs have been successfully integrated into portable power units, primarily by companies such as Adaptive Materials Inc. (USA) and Protonex Technology Corporation (USA). Figure 11 shows Adaptive Materials (AMI) 50 W and 250 W systems that use propane as the fuel to produce continuous power. The 50 W system, running on propane, provides power for ground sensors, unmanned aerial vehicles, and robots. The 250 W system is fueled by propane or LPG and is

Fig. 11 AMI's 50 W (*left*) and 250 W (*right*) portable SOFC systems

used to extend military mission durations and delivers off-grid power for electronics, radios, and computers. The use of globally available fuels in such portable SOFC systems eliminates complicated logistics.

4.4 SOFC-Based Transportation Auxiliary Power Units

Another application of SOFC systems is in the transportation sector. As for portable applications, the proton exchange membrane fuel cell is generally regarded as the fuel cell of choice for transportation applications, particularly for propulsion to replace the internal combustion engine. Proton exchange membrane fuel cells require pure hydrogen, with no carbon monoxide, as the fuel to operate successfully. However, presently no hydrogen infrastructure exists, and onboard reformer systems to produce hydrogen from existing fuel base (gasoline, diesel) are technically challenging, complex, and expensive. Furthermore, it is difficult to eliminate all carbon monoxide from the reformate stream. In contrast, SOFCs can use carbon monoxide along with hydrogen as fuel, and their higher operating temperature and availability of water on the anode side make on-cell or in-stack reformation of hydrocarbon fuels feasible. Also, no noble metal catalysts are used in SOFCs, reducing the cost of the cells. Although not practical for propulsion, the application of SOFCs in the transportation sector will be for onboard auxiliary power units (APUs). Such APUs, operating on existing fuel base, will supply the ever-increasing electrical power demands of luxury automobiles, recreational vehicles, and heavy-duty trucks for various comfort items, such as refrigerators, televisions, stereos, even computers, and microwaves. The challenges for SOFC in APUs, similar to the aforementioned portable applications, reside in achieving SOFC systems with compact size, light weight, short start-up time, mechanical robustness, and capability for thermal cycling.

Delphi Cooperation has developed a 3–5 kW size SOFC APU system using anode-supported planar cells. This unit is intended to operate on gasoline or diesel, which is reformed through partial oxidation within the APU unit. The building

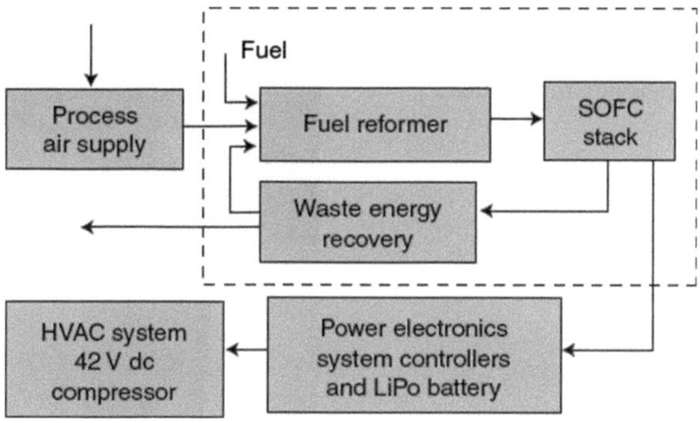

Fig. 12 Basic building blocks of an SOFC auxiliary power unit (APU)

blocks of such an APU consist of an SOFC stack, fuel reformation subsystem, energy recovery unit, thermal management subsystem, process air supply subsystem, control subsystem, and power electronics and energy storage (battery) subsystem, arranged as illustrated in Fig. 12. In 2008, Delphi Corporation (USA) and Peterbilt Motors Co. (USA) successfully demonstrated the operation of a Delphi's SOFC APU in powering a Peterbilt Model 386 truck's "hotel" loads (Fig. 13). The Delphi SOFC APU provided power for the Model 386′s electrical system and air conditioning and maintained the truck's batteries, all while the truck's diesel engine was turned off. Delphi hopes to commercialize such SOFC APUs in the next few years. Several other companies are also developing similar SOFC APUs.

5 Summary

Fuel cells have been known for over 150 years. Solid oxide fuel cells, in particular, utilize an electrolyte material that was first mentioned by Nernst in the late 1890s. Research, development, and demonstration of SOFCs accelerated in the last 30 years and they are now on the verge of commercialization for at least certain applications. SOFCs offer potential for various applications with a wide power range from milli-watts to mega-watts because of fuel flexibility and combined high quality heat and power. Novel SOFC materials and their structures are being investigated and developed to reduce polarization and thus improve electrochemical performance, especially in the intermediate temperature regime. Stable performance over long durations sometimes is overlooked, but remains a crucial factor in SOFC commercialization. With respect to planar SOFCs, successful development of high-temperature sealing materials and concepts is of paramount importance for the long-term performance stability and lifetime of stacks and hence for their eventual commercialization at competitive costs. On the contrary, a number of tubular SOFC systems have been successfully demonstrated up to 220 kW size for long-term stable performance. Extensive research is being conducted to decrease the cost of these power systems, particularly in the areas of cell

Fig. 13 Delphi's SOFC APU mounted underneath a Peterbilt's truck cabin

and stack materials, DC to AC power conditioning systems, and other balance-of-plant components. With decreased cost, SOFC power systems are expected to provide a widespread low- or no-pollution technology to produce electricity. The first SOFC commercialization is expected to be for residential CHP for which many successful demonstrations by several companies have been accomplished. Large-scale distributed power generation also got a recent boost this year when Bloom Energy (USA) sold, delivered and installed many 100 kW sized SOFC systems for commercial customers.

References

1. W. Nernst, Electrical glow-light. U.S. Patent 623,811 (1899)
2. W. Nernst, Uber die elektrolvtische Leitung Fester Korper bei sehr hohen Temperaturen. Z. Electrochem. **6**, 41–43 (1900)
3. E. Bauer, H. Preis, Über Brennstoff-Ketten mit Festleitern. Z. Electrochem. **43**, 727–732 (1937)
4. C. Wagner, Über den mechanismus der elektrischen Stromleitung im Nernststift. Naturwissenschaften **31**, 265–268 (1943)
5. J. Weissbart, R. Ruka, A solid electrolyte fuel cell. J. Electrochem. Soc. **109**, 723–726 (1962)
6. S.C. Singhal, K. Kendall, *High Temperature Solid Oxide Fuel Cells: Fundamentals, Design and Applications* Oxford, Elsevier (2003)
7. K. Huang, J.B. Goodenough, Solid oxide fuel cell Technology. Woodhead Publishing Ltd, Cambridge (2009)
8. S.C. Singhal (ed.), SOFC-I, PV89-11, *The Electrochemical Society Proceedings Series*, Pennington (1989)
9. F. Grosz, P. Zegers, S.C. Singhal, O. Yamamoto (eds.) SOFC-II, Commission of the European Communities, Luxembourg (1991)
10. S.C. Singhal, H. Iwahara (eds.), SOFC-III, PV93-4, The Electrochemical Society Proceedings Series, Pennington (1993)
11. M. Dokiya, O. Yamamoto, H. Tagawa, S.C. Singhal (eds.), SOFC-IV, PV95-1, The Electrochemical Society Proceedings Series, Pennington (1995)

12. U. Stimming, S.C. Singhal, H. Tagawa, W. Lehnert (eds.), SOFC-V, PV97-40, The Electrochemical Society Proceedings Series, Pennington (1997)
13. S.C. Singhal, M. Dokiya (eds.), SOFC-VI, PV99-19, *The Electrochemical Society Proceedings Series*, Pennington (1999)
14. H. Yokokawa, S.C. Singhal (eds.), SOFC-VII, PV2001-16, *The Electrochemical Society Proceedings Series*, Pennington (2001)
15. S.C. Singhal, M. Dokiya (eds.), SOFC-VIII, PV2003-07, *The Electrochemical Society Proceedings Series*, Pennington (2003)
16. S.C. Singhal, J. Mizusaki (eds.), SOFC-IX, PV2005-07, *The Electrochemical Society Proceedings Series*, Pennington (2005)
17. K. Eguchi, S.C Singhal, H. Yokokawa, J. Mizusaki (eds.), SOFC-X, the Electrochemical Society Transactions, vol. 7 (2007)
18. S.C. Singhal, H. Yokokawa (eds.), SOFC-XI, the Electrochemical Society Transactions, vol. 25 (2009)
19. U. Bossel (ed.), *First European Solid Oxide Fuel Cell Forum Proceedings* (European Fuel Cell Forum, Oberrohrdorf, 1994)
20. B. Thorstensen (ed.), *Second European Solid Oxide Fuel Cell Forum Proceedings* (European Fuel Cell Forum, Oberrohrdorf, 1996)
21. P. Stevens (ed.), *Third European Solid Oxide Fuel Cell Forum Proceedings* (European Fuel Cell Forum, Oberrohrdorf, 1998)
22. A.J. McEvoy (ed.), *Fourth European Solid Oxide Fuel Cell Forum Proceedings* (European Fuel Cell Forum, Oberrohrdorf, 2000)
23. J. Huijsmans (ed.), *Fifth European Solid Oxide Fuel Cell Forum Proceedings* (European Fuel Cell Forum, Oberrohrdorf, 2002)
24. M. Mogensen (ed.), *Sixth European Solid Oxide Fuel Cell Forum Proceedings* (European Fuel Cell Forum, Oberrohrdorf, 2004)
25. J. Kilner (ed.), *Seventh European Solid Oxide Fuel Cell Forum Proceedings* (European Fuel Cell Forum, Oberrohrdorf, 2006)
26. R. Steinberger-Wilckens (ed.), *Eighth European Solid Oxide Fuel Cell Forum Proceedings* (European Fuel Cell Forum, Oberrohrdorf, 2008)
27. J.T.S. Irvine (ed.), *Ninth European Solid Oxide Fuel Cell Forum Proceedings* (European Fuel Cell Forum, Oberrohrdorf, 2010)
28. X.-D. Zhou, S.C. Singhal, Fuel cell: solid oxide fuel cells: overview. In: Garche J, Dyer C, Moseley P, Ogumi Z, Rand D, Scrosati B (eds.). Encyclopedia of electrochemical power sources, Amsterdam, Elsevier **3**, 1–16 (2009)
29. B.C.H. Steele, Appraisal of $Ce_{1-y}Gd_yO_{2-y/2}$ electrolytes for IT-SOFC operation at 500°C. Solid State Ionics **129**, 95–110 (2000)
30. Z.P. Shao, S.M. Haile, A high-performance cathode for the next generation of solid-oxide fuel cells. Nature **431**, 170–173 (2004)
31. F. Zhao, A.V. Virkar, Dependence of polarization in anode-supported solid oxide fuel cells on various cell parameters. J. Power Sources **141**, 79–95 (2005)

Ni-Based Solid Oxide Cell Electrodes

Mogens Mogensen and Peter Holtappels

Abstract This paper is a critical review of the literature on nickel-based electrodes for application in solid oxide cells at temperature from 500 to 1000 °C. The applications may be fuel cells or electrolyser cells. The reviewed literature is that of experimental results on both model electrodes and practical composite cermet electrodes. A substantially longer three-phase boundary (TPB) can be obtained per unit area of cell in such a composite of nickel and electrolyte material, provided that two interwoven solid networks of the two solid and one gaseous phases are obtained to provide a three - dimensional TPB throughout the electrode volume. Variables that are used for controlling the properties of Ni-cermet electrodes are: (1) Ni/YSZ volume ratio, and (2) porosity and particle size distribution, which mainly affected by raw materials morphology, application methods and production parameters such as milling and sintering. The various electrode properties are deeply related to these parameters, but also much related to the atomic scale structure of the Ni-electrolyte interface, which in turn is affected by segregation of electrolyte components and impurities as well as poisons in the gas phase. The main emphasis will be on the following subjects: (a) electronic conductivity of cermets, (b) dimensional and thermodynamic stability including redox cycling, (c) thermal expansion coefficient matching, (d) chemical compatibility with stack components and gaseous reactants and (e) electrode reaction mechanism and polarisation resistance. A brief discussion of the main concepts in the modelling literature is given in context of the latter subject.

M. Mogensen (✉)
Risø National Laboratory for Sustainable Energy,
Fuel Cells and Solid State Chemistry Division,
The Technical University of Denmark, 4000 Roskilde, Denmark
e-mail: momo@dtu.dk

P. Holtappels
Department of Energy Conversion and Storage, Technical University of Denmark,
DTU Risø Campus, Frederiksborgvej 399, DK-4000 Roskilde, Denmark

J. T. S. Irvine and P. Connor (eds.), *Solid Oxide Fuels Cells: Facts and Figures*,
Green Energy and Technology, DOI: 10.1007/978-1-4471-4456-4_2,

1 Introduction

The Ni-cermet with either stabilised zirconia, most often yttria stabilised zirconia (YSZ), or doped ceria has so far been the most successful anode in SOFCs in spite of the many problems associated with this electrode. The main reason for this is the excellent catalytic and electrocatalytic property of Ni for steam reforming of natural gas and for electrochemical oxidation of H_2 and CO. These properties have been so good that they have overshadowed several drawbacks such as the sensitivity to sulphur poisoning and mechanical instability in case of redoxing. Furthermore, the electrode is fully reversible, i.e. works equally well in fuel cell and in electrolysis mode.

This paper presents a brief review of the literature of nickel-based cermet electrodes for application in solid oxide cells at temperature from 500 to 1000 °C. The applications may be fuel cells or electrolyser cells. Variables that are used for controlling the properties of Ni-cermet-electrodes are: (1) Ni/electrolyte volume ratio, (2) additives, e.g. alloying of the Ni or infiltration of the composite with nanoparticles of other elements or compounds, (3) the chemical composition of the electrolyte component and (4) porosity and particle size distribution, which is mainly affected by raw materials morphology, application methods and production parameters such as milling and sintering possibly followed by infiltration of nanosized electrocatalytic active particles. The various electrode properties are deeply related to these parameters, but also much related to the atomic scale structure of the Ni-electrolyte interface, which in turn is affected by segregation of electrolyte components and impurities as well as poisons in the gas phase.

The following properties are reviewed: microstructure, electronic conductivity in cermets, dimensional and thermodynamic stability including redox cycling, thermal expansion coefficient matching, chemical compatibility with stack components and gaseous reactants, sensitivity towards impurities in the fuel, and electrode reaction mechanism and polarisation resistance including reports on model electrodes. The literature covered is a rather limited subjective selection of experimental results on both practical composite cermet electrodes and model electrodes with the purpose of revealing some important facts and figures. Searching the Ni-cermet electrode literature using Scifinder reveals that significantly more than 2500 articles are published in the recognised scientific literature. Thus, this paper is naturally restricted to give an overview on the most important structural and functional aspects and performance indicators for Ni-cermet-based electrodes.

2 Microstructure of Cermets

The electrochemical reaction of gaseous reactants in solid state cells involves three phases, the gas phase, the ceramic electrolyte and the metallic electrode. Thus, the concept of three-phase boundary (TPB) has to be introduced before

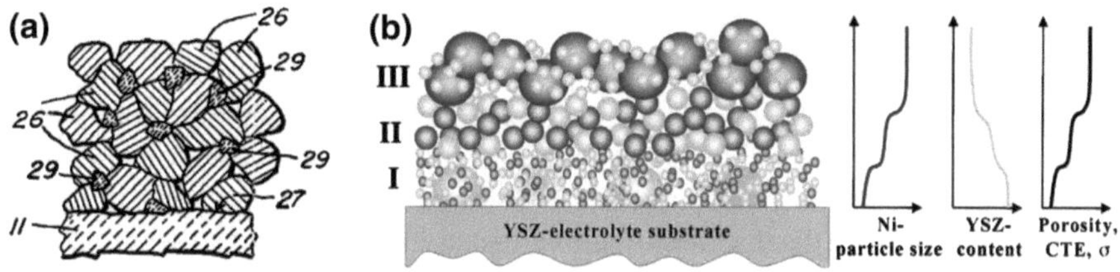

Fig. 1 **a** The drawing of H.S. Spacil [3] that shows the interconnected layer of Ni particles (26) at the interface (27) of the zirconia electrolyte (11) with zirconia particles (29). **b** Modern multilayer Ni-YSZ-cermet anode, CTE is coefficient of thermal expansion, s is electron conductivity [6]. Reproduced here with kind permission from © Elsevier 2002

starting the review. The TPB is the line at which the three reactants can meet each other. They are: (1) the fuel gas, e.g. H_2, (2) the ion in the electrolyte, here O^{2-} and (3) the electron in the Ni. As a line mathematically speaking has no area, the concept of TPB zone has been introduced to recognise that a certain minimum of area on each surface must be involved. As hydrogen may diffuse through the Ni electrode and some ceramic electrolytes, the two-phase boundary may be of importance, but even in such cases the reaction preferentially takes place near the TPB. Thus, the longer and more ideal the TPB, the better the performance (lower polarisation resistance) expected for the Ni electrode.

Considering a closed packed coverage of an electrolyte surface by Ni-spheres, the TPB length is inversely related to the Ni particle radius. Thus, small Ni grain sizes are desired to achieve long TPBs. Ni is a very good bond breaker of both the H–H and the C–H bond (as in CH_4) [1]. In spite of this, the polarisation resistance, R_P, is unacceptably high for a pure Ni electrode. At temperatures of 800–1000 °C pure nickel metal has a high surface mobility [2], causing it to sinter and agglomerate into a coarse structure with a limited TPB against the electrolyte surface. To overcome this drawback, the Ni–YSZ-cermet was invented by H.S. Spacil [3]. The effect of the zirconia particles in the cermet was in the original patent only to prevent the coarsening of the Ni as shown in Fig. 1a. In a modern Ni–YSZ, a substantially longer TPB can be obtained per unit area of cell by interweaving solid networks of the three phases (two solids and a gas) to provide TPBs throughout the electrode volume as sketched in Fig. 1b. A substantially longer TPB can be obtained in this way (per unit area of cell) in a composite of nickel and electrolyte material compared to a simple porous Ni electrode with a two-dimensional contact to the electrolyte. The Ni particles tend to sinter and grow in both cermet structures until the YSZ particles or network prevent their further growth. This means that the structure of the YSZ controls the total structure of the composite including the structure of the porosity. There seems to be a general agreement about this in the literature. It is also agreed that the longer the TPB length per unit cell area, the lower the polarisation resistance [4, 5].

3 Electron Conductivity of Cermets

Variables that are used for controlling the properties of Ni-cermet-electrodes with a random distribution of Ni and zirconia particles are first of all the Ni/electrolyte volume ratio, and next the porosity and particle size distribution. Naturally, these parameters may be varied through the thickness of the cermet as indicated in Fig. 1b, because the requirements of the cermet are different for the anode at and near the electrolyte surface and for outer structure, having the function of electrochemically active area and current collector combined with support structure, respectively. Good in-plane conductivity and easy gas permeability are mandatory throughout the cermet structure.

The Ni–YSZ support is the current collection structure in a so-called anode supported SOFC, and as such it must have a good electronic conductivity in order to keep the internal resistance of the SOFC stack low. For the bipolar flat plate design, a typical lateral current path length of about 1 mm can be assumed. If the resistive contribution should be negligible, say less than 10 mΩ cm^2 for a 200 μm thick 30 % porous structure, this implies a specific electronic conductivity of about 100 S/cm. Other cell designs containing substantially longer lateral current paths in the electrodes may need a much higher specific conductivity, as the conductivity cannot just be overcome by a thicker electrode due to gas diffusion limitations [7, 8]. To illustrate this point the diffusion resistance of a porous layer of 1 mm thickness is estimated (assuming 30 % porosity, pores larger than 1 μm, i.e. no Knudsen diffusion, and tortuousity factor is 3) to 50 mΩ cm^2 [9, 10]. This is 50 % of the acceptable electrode polarisation.

The electronic conductivity of a simple Ni–YSZ cermet structure with randomly mixed particles of similar size seems to follow the rules of percolation with fair approximation [11, 12]. This means that by gradually varying the Ni content from below 30 vol % of solids to above 50 vol %, there is an abrupt increase in the electronic conductivity. The inflexion point of the S-shaped curve is usually in the range of 30–40 vol % Ni as shown in Fig. 2. If a significant fraction of the YSZ particles are much bigger than the Ni particles, then the inflexion point may be moved down to 15 vol % Ni [13]. However, in order to get a long TPB length in the active anode layer in the vicinity of the electrolyte, a significant amount (20–30 %) of the YSZ must have very fine particles, below 1 μm diameter. Fine YSZ particles must be evenly distributed around the Ni particles in order to keep both a stable TPB length and electronic conductivity over long periods of time. Otherwise the Ni particles will grow and become too coarse. The porosity of the simple cermet with 40 vol % Ni (of the solids) must be lower than about 30–35 % after the reduction of the NiO to Ni [14, 15] in order to keep the connectivity between the Ni particles.

4 Thickness of the Electrochemically Active Layer

As the specific electronic conductivity of Ni is orders of magnitude higher than the specific ionic conductivity of YSZ (Fig. 2), the electrochemical active electrode volume is expected to reside near to the electrolyte surface. The distance from

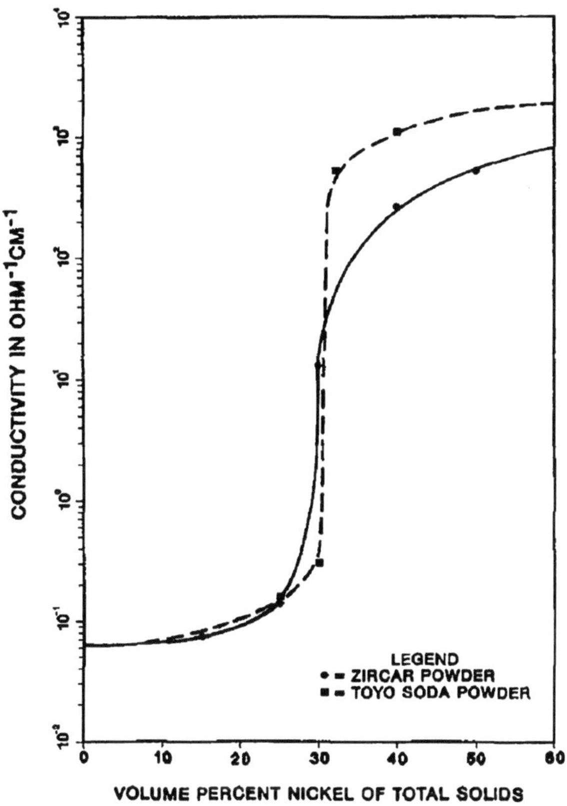

Fig. 2 Conductivity of two Ni-YSZ-cermets as a function of Ni vol % of total solids at 1000 °C [16]. Reproduced here with kind permission from © The Electrochemical Society 1987

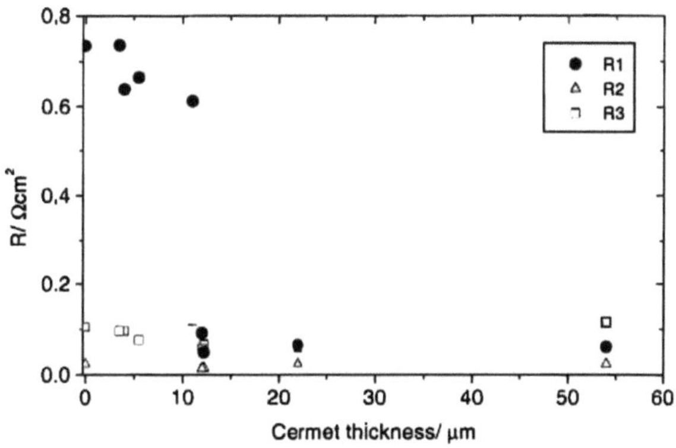

Fig. 3 Dependence of anode polarisation resistance on cermet thickness for the fine cermet anodes at 1000 °C in 97 % H_2 + 3 % H_2O at OCV. *R*1 is electrode impedance, *R*2 and *R*3 are gas diffusion and gas conversion outside the anode structure, respectively [19]. Reproduced here with kind permission from © The Electrochemical Society 2000

the YSZ electrolyte surface into the cermet, where the electrochemical reactions take place is called the active thickness. It has been found to be in the order of 10 µm at 1000 °C for an Ni–YSZ-cermet [17, 18] as shown in Fig. 3. Naturally, the precise active thickness must at least depend on structure, gas composition, temperature and conductivity of the electrolyte, because the TPB line-specific resistance (LSR) will depend on these parameters. The active thickness is reflecting the ratio between the LSR and the ionic resistance of the porous YSZ in the cermet layer next to the electrolyte.

5 Electrochemical Performance, Durability and Stability

The lowest reported values of initial electrode polarisation resistance corrected for mass transport, R_P, for the H_2/H_2O/Ni–YSZ electrode are about 20–40 mΩ cm^2 at 1000 °C [20]. Typical ranges seen from leading SOFC developers are in the range of 20 mΩ cm^2 $< R_P <$ 60 mΩ cm^2 at 1000 °C and 40 mΩ cm^2 $< R_P <$ 80 mΩ cm^2 at 850 °C, 60 mΩ cm^2 $< R_P <$ 100 mΩ cm^2 at 750 °C [21, 22].

The durability of the Ni–YSZ anode performance is to a large extent linked to the microstructure in the same way as attainable performance. For a cermet with coarse and fine YSZ sintered together with NiO, a negligible degradation over 2500 h at 1000 °C with 200 mA/cm^2 has been demonstrated [23]. For cells with an Ni–YSZ cermet fabricated by fixation of the Ni metal using electrochemical vapour deposition of YSZ under reducing conditions no measurable degradation was found over 4800 h at 1000 °C with 450 mA/cm^2 and a fuel utilisation of 85 % [24]. Cermets with fine Ni particles are usually less stable. Such Ni–YSZ electrodes tend to degrade to some extent during the first few hundreds of hours of operation [25, 26]. This has been ascribed to Ni particle growth, which naturally will be limited by the YSZ particles. Sehested et al. studied Ni particle growth (in catalyst with microstructures and composition different from typical Ni/YSZ electrodes) and found the relation $d_{Ni} = d_{Ni,0} + k(1-\exp(-t/\tau))$ at temperatures of 550–750 °C and pH_2O/pH_2 of 0.2, 1 and 50, where d_{Ni} is the Ni particle diameter with index 0 indicating the starting value, k is a constant and τ is the time constant for the Ni particle growth [27, 28]. Similar trends for mean Ni diameter development over time have been reported for SOFCs [29–32]. The findings of Hauch et al. that the increase over time, t, of the three phase boundary resistance, $R_{Ni,TPB}$, which reflects the electrode performance degradation, followed the same type of relation, i.e. $R_{Ni,TPB}(t) = R_{Ni,0} + \Delta R\cdot(1 - \exp(-t/\tau))$, [33] might be taken as a further support for the hypothesis of the Ni particle growth being responsible for the electrochemical degradation of the cermet. The Ni particle growth rate and the increase rate of $R_{Ni,TPB}$ seem to increase with increasing steam partial pressure and temperature.

However, recent results indicate that the Ni–YSZ electrode degradation rate is dependent on several other parameters than the Ni particle growth rate. Microstructural degradation can also occur by removal of Ni within the cermet. The stability of Ni in the anode is affected by high water vapor pressures at operation temperatures. $Ni(OH)_2$ has been calculated to have a significant vapor pressure at high pH_2O, and depletion of Ni from cermet anodes at the gas inlet has been demonstrated in durability tests [34]. The gradients in pH_2O at active TPBs are supposed to be responsible for this Ni dissolution and a following deposition elsewhere, as the pH_2O decreases and the $Ni(OH)_2$ decomposes.

The restriction of the electrochemically active area to a zone around the TPB has triggered more detailed investigation regarding the nanoscopic structure and chemical properties of this area. Impurities and the finer details of the cermet structure depending on fabrication procedure and pretreatments may influence this particular area and thus play important roles [35–37].

6 Thermal Expansion Coefficient Matching

The difference in thermal expansion coefficient (TEC), also called coefficient of thermal expansion (CTE), between the electrolyte and the Ni–YSZ cermet electrode layers may result in buildup of stresses. The CTE of TZ8Y is between 10.6 and $11.0 \times 10^{-6}\,K^{-1}$ [38]. A CTE of $14 \times 10^{-6}\,K^{-1}$ can be found for NiO, and a CTE of $12.3 \times 10^{-6}\,K^{-1}$ can be measured for an NiO/YSZ composite with 53 vol % NiO [39], i.e. about what would be found by linear interpolation between the values of the two components. For an Ni–YSZ cermet with 40 vol % Ni the CTE value was found to be close to that of YSZ. This seems low considering the CTE of pure Ni is about $17 \times 10^{-6}\,K^{-1}$. This means that if a well sintered YSZ network is established, this influences heavily the average CTE of the cermet, but it may cause a buildup of internal stresses. The reason for the low average CTE is not sure, but it probably reflects the relative degree of interconnection of the phases. (The explanation given in [39] that the elastic modulus 137–157 GPa for YSZ is very high compared to the range of only 14.8–21.1 GPa for Ni is not correct, because the Ni values are, in fact, not in this range, but about 10 times higher [40, 41], i.e. similar to the YSZ values).

Both composites of electrolyte + Ni and electrolyte + NiO are of importance, because the cermet is usually produced by sintering of NiO + YSZ. Thus, this composite will usually experience the largest change in temperature. The non-reduced NiO–YSZ composite has a CTE, which is close to the weighted average of the CTEs of the two components because the elastic moduli of the two oxides are approximately equal [39]. Considering cells to be free of stress at the sintering temperature of 1300–1350 °C in air, the CTE difference between the composite and the YSZ causes tensile stresses in the composite at lower temperatures. This may result in serious crack formation in the composite, and these cracks may proceed into the electrolyte sheet causing cell failure during cool down in case of electrolyte supported cells [42]. In contrast, this mismatch is beneficial in case of anode supported cells, because then the thin electrolyte will be in compression at temperatures lower than the sintering temperature.

A composite of 8YSZ with a CTE of $10.8 \cdot 10^{-6}\,K^{-1}$ and NiO with CTE $14.0 \cdot 10^{-6}$ K^{-1} was studied by Müller et al. [6]. Figure 4 (left) shows that the CTE is mainly dependent on the ratio between the two solid components. Furthermore, they found (Fig. 4 right) that the CTE of NiO-8YSZ composites not only depends on the NiO content but also on the particle size ratio of the two components. The CTE exhibited

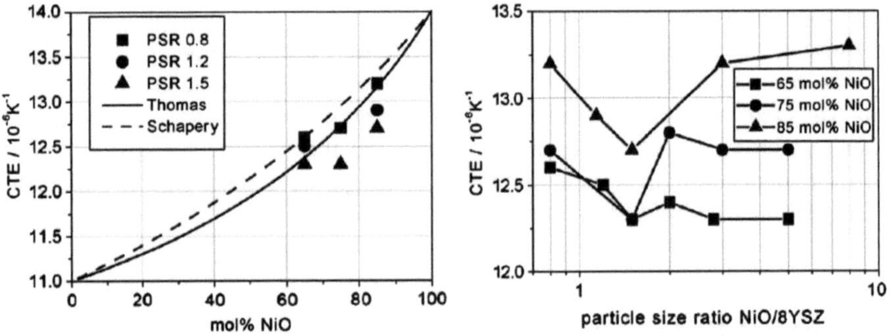

Fig. 4 Dependence of CTE on NiO content and particle size of NiO and 8YSZ; *left* comparison of measured data with theoretical predictions; *right* CTE exhibits a minimum around a particle size ratio of 1.5. Samples were sintered at 1300 °C [6]. Reproduced here with kind permission from © Elsevier 2002

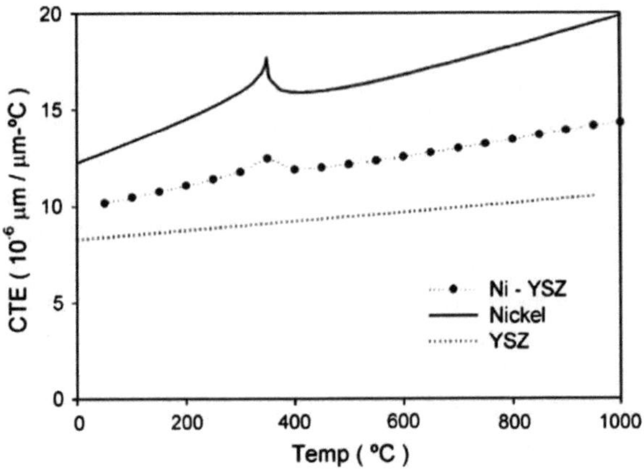

Fig. 5 Differential CTE of Ni–YSZ with respect to temperature for 0–1000 °C. Modelled data benchmarked with experimental data. Vol %: Pore, Ni, YSZ: 40, 25, 35 [43]. Reproduced here with kind permission from © Elsevier 2008

a minimum of $12.3 \cdot 10^{-6}$ K^{-1} for a particle size ratio of NiO/8YSZ = 3:2. A cermet with this composition seems to have had an optimal package configuration for the investigated NiO amounts.

Figure 5 shows the differential CTE of an Ni–YSZ cermet as a function of temperature as found by modelling of a given structure. The modelled data was benchmarked with measurements. The peak is found at the Curie temperature of Ni.

7 Chemical Compatibility with Stack Components and Gaseous Reactants

In general the Ni–YSZ cermet is very stable towards other stack components. No detrimental reactions are detected for silica-based glass seals in contact with the cermet, whereas Ni-depletion by nickel-phosphate formation is observed by reaction with phosphate-based glasses [44].

Fig. 6 Photographs of an Ni–YSZ cermet **a** after heating in H_2 at 800 °C, **b** after heating in 100 % CH_4 at 800 °C for 1.5 h, and **c** after heating in 40 % toluene at 700 °C for 1.5 h. From Gorte and Vohs [45]. Reproduced here with kind permission from © Academic Press 2003

The SOFC anode should be able to tolerate natural available fuels without expensive cleaning. In this respect Ni is far from ideal. If it is exposed to pure methane or other hydrocarbons at SOFC operating temperatures, the hydrocarbons will crack and form hydrogen and carbon because Ni is a good cracking catalyst [1]. If the carbon activity is high enough the carbon tends to precipitate on the interface between the Ni and the YSZ. This may cause precipitation of carbon in the form of carbon nanotubes that destroys the Ni-Cermet mechanically. Figure 6 shows three examples of Ni–YSZ cermets exposed to hydrogen, methane or toluene at typical SOFC operation temperatures. The carbon containing fuels destroys the cermets mechanically within 1.5 h. Thus, the Ni-cermet is mechanically destroyed by dry hydrocarbons.

A way around this problem is to steam reform the hydrocarbons either externally or internally. In the latter case, the reforming takes place on the Ni in the cermet support and electrode, and this may cause another problem. At high temperatures the reforming reaction may be very fast, and the reforming reaction is strongly endothermic [46]. This causes steep temperature gradients over a larger cell area [47], which may result in mechanical failure of the cell [48], but in case of anode supported cell with well sintered Ni–YSZ supports, this seems to be no problem at temperatures below 850 °C. Also, CO_2 may be used as a reforming agent. The source of steam and CO_2 may come from recycling a fraction of the exhaust gas from the SOFC anode. Yet another possibility is partial oxidation (POX) using air.

The fuel composition is thus depending on the kind of reforming process but in general contains a mixture of hydrogen, CO, steam and CO_2. In order to better understand the electrochemical performance of Ni-cermet SOFC anodes in real fuel gas mixtures and co-electrolysis of CO_2 and H_2O into syngas ($CO + H_2$), the CO/CO_2 electrode reaction has been investigated separately. Results range from similar electrode reaction rates for H_2/H_2O and CO/CO_2 [49] to orders of magnitude lower CO/CO_2 rates [50]. The reason for this is still under discussion and further investigations, but a plausible suggestion is that the huge differences in the reported kinetics are related to differences in purity of materials and fuel gases, see below.

8 Sensitivity Towards Fuel Impurities

Sulphur and sulphur compounds known to reside in natural fuels are known to adsorb to the Ni surface and to cause a loss of electrode performance. This effect is in mild cases observed to be reversible [51]. Phosphorus seems to be an even worse poison for the Ni-cermet electrodes than sulphur, as very low (ppb) levels may destroy the cermet [52].

9 Sulphur Poisoning

Sulphur seems to be the most studied fuel impurity that is detrimental to Ni-based electrodes. The reason for this is probably that sulphur is present in most fuels derived from fossil sources either by nature or added as odorant.

Figure 7 shows that already at an H_2S content of 50 ppm, the Ni will be fully covered with sulphur at typical SOFC operation temperatures in the range of 700–850 °C. A full S coverage will stop the steam reforming of CH_4 totally [50], whereas the rate of electrochemical oxidation of H_2 in the Ni-YSZ cermet is decreased with some 20 % only as Fig. 8 reveals.

Figure 9 shows that the composition of the zirconia in the cermet electrode is of major importance, i.e. a scandia stabilised zirconia (Ni–SSZ) is significantly less affected by H_2S than the usual Ni–YSZ anode. This points to the TPB (the contact between Ni and zirconia) being of paramount importance for the electrochemical H_2 oxidation as usually assumed today even though the exact mechanism is not yet determined in detail.

This important difference between SSZ and YSZ has been confirmed by Hagen and Rasmussen over 500 h at Risoe DTU [57]. Thus, the sensitivity to S changes with the composition of the electrolyte in the cermet. In case of the Ni–SSZ cermet it looks as if "NiS" (Ni with S coverage ca. 1) is a fair anode for H_2 oxidation [53].

It should be noted that Ni free electrodes with very high sulphur tolerance has been reported. He et al. studied H_2S tolerance of solid oxide fuel cells (SOFCs) with composite anodes containing Cu, CeO_2 and YSZ. H_2S levels up to 450 ppm had no effect on anode performance when this anode was operated an SOFC on H_2 800 °C. At higher H_2S concentrations, anode deactivation was observed. This was attributed to the reaction of the CeO_2 with H_2S to form Ce_2O_2S [58].

10 Effect of Gas Cleaning

The durability of the Ni–YSZ anode also seems affected by sulphur (and probably also other inorganic impurities [22]). It seems further that the effect of S is dependent on the operation conditions and the composition of the fuel gas, and in

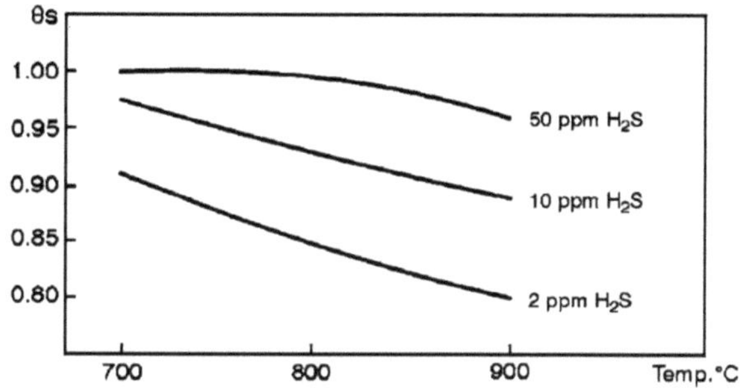

Fig. 7 Chemisorption of H_2S on Ni in CPO (catalytic partial oxidation) gas. Equilibrium coverage with ppm H_2S in feed with dry gas composition (vol. %): H_2 (11.5), CO (10.5), CO_2 (12.5), CH_4 (1.5), N_2 (63.2), [53]. Reproduced here with kind permission from © Springer 2006

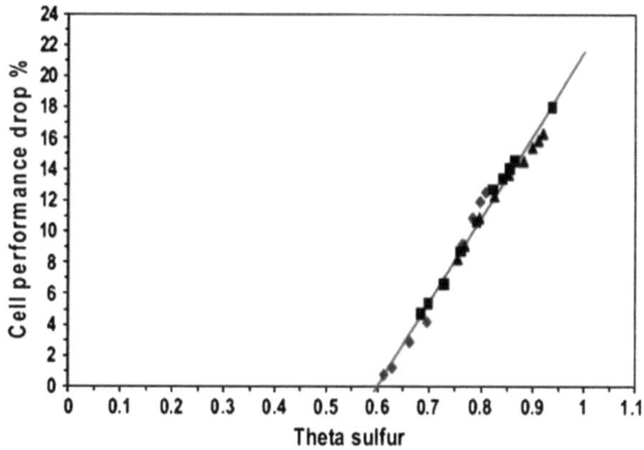

Fig. 8 Cell performance loss as a function of S coverage. Potentiostatic at 0.7 V at 900 °C, 800 °C and 700 °C. From Hansen [54]. Reproduced here with kind permission from © The Electrochemical Society 2008. Measurements from Zha et al. [55]

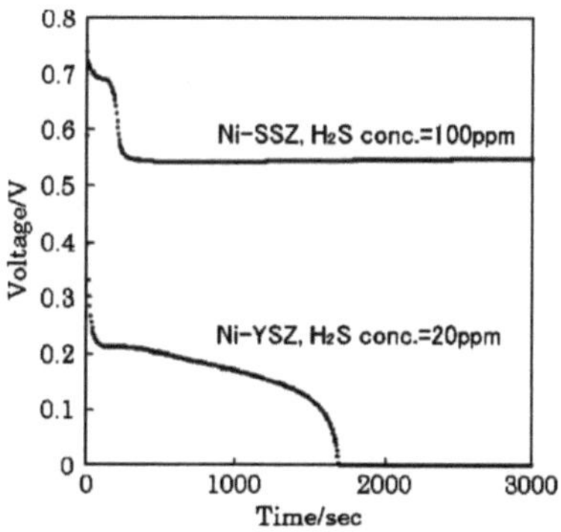

Fig. 9 Cell voltage at 200 mA/cm^2, 800 °C, H_2 + 3 % H_2O, electrolyte: SSZ, anode: Ni–SSZ or Ni–YSZ [56]. Reproduced here with kind permission from © The Electrochemical Society 2006

particular electrolysis mode (H_2O and CO_2 reduction) seems more sensitive to impurities than H_2 oxidation in fuel cell mode. The Ni–YSZ/CO/CO_2 electrode in electrolysis mode seems to be even more sensitive to impurities than the Ni–YSZ/H_2/H_2O electrode. Figure 10 illustrates that, fortunately, it is even in this case possible to clean the gas enough to avoid any measurable degradation. Actually this can be done by relatively simple and inexpensive means [59].

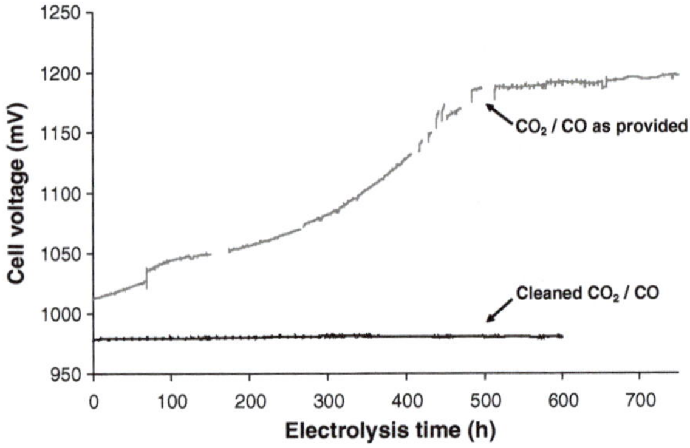

Fig. 10 Cell voltage measured during CO_2 electrolysis (850 °C, −0.25 A cm^{-2}, 70 % CO_2–30 % CO) in the Ni/YSZ-based SOC when applying the inlet gases to the Ni/YSZ electrode as received and when applying cleaned gases [60]. Reproduced here with kind permission from © The Electrochemical Society 2010

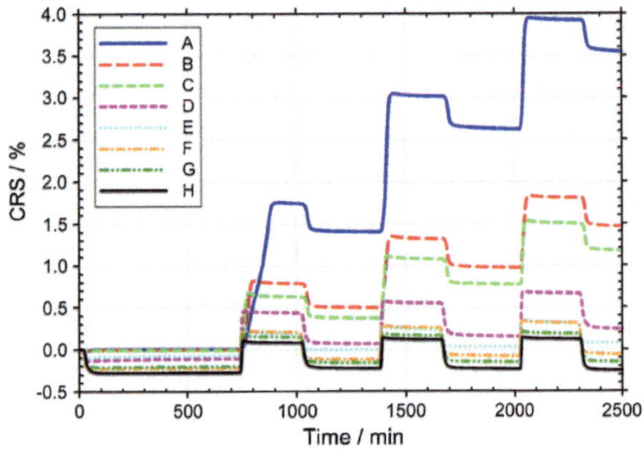

Fig. 11 Cumulative redox strain (CRS) as a function of time from the initial reduction during three isothermal redox cycles of different Ni–YSZ cermets at 850 °C [62]. Reproduced here with kind permission from © The Electrochemical Society 2010

11 Redox Stability

Re-oxidation of the reduced Ni–YSZ cermet electrode is generally found to be detrimental. The volume expansion of the Ni metal by oxidation to NiO typically leads to damage on the surrounding YSZ framework because the molar volume of Ni and NiO is 6.6 cm^3 mol^{-1} and 11.1 cm^3 mol^{-1}, respectively. Thus, bulk expansion of the anode upon oxidation is considered to be responsible for the lack of redox stability of SOFC [61 and references therein] as evidenced by the bulk expansion of nickel–YSZ-anode material. Figure 11 shows isothermal dilatometry as a function of sample geometry, ceramic component, temperature and temperature cycling [62]. The strength of the ceramic network and the degree of Ni redistribution appeared to be key parameters of the redox behaviour. The redox behaviour of nickel–YSZ anode materials was studied as a function of sample geometry, ceramic component, temperature and temperature cycling. Based on the experimental results, models of the redox behaviour in SOFC nickel–YSZ anode materials were proposed. Key parameters for achieving redox stable anodes were indicated to be the ceramic network strength and the degree of restriction of Ni-particle relocation and Ni-particle coarsening, and a figure of merit with respect to redox performance is the cumulative redox strain, i.e. the sum of the strain of several redox cycles. The redistribution of Ni by sintering is very dependent on temperature as well as the creep strength of both cermet components [63–65].

The redox cycling of anode supported cells was investigated by Atkinson et al. in order to determine the critical degree of oxidation that causes mechanical failure in different failure modes. Interrupted oxidation of free-standing cells at 800 °C resulted in failure at a critical degree of oxidation of ca. 50 %, whereas electrochemical oxidation of cells incorporated in metal housings led to fully disrupted cells (the anode substrate also cracked) at ca. 5 % oxidation. For degrees of oxidation < 5 % in the metal housings, the electrochemical performance could be recovered. It was further concluded that in general, non-uniform oxidation will have more catastrophic effects than uniform [66]. It should be noted that the actual values might only be valid for a given type of cell, i.e. will probably vary with thickness of the layers of the cell as well as the cell fabrication details and on the particular temperature at which redox cycling occurred.

Further, Ni infiltration of preformed porous YSZ structures was investigated as a possible method to improve redox stability of Ni–YSZ anode supports. This method was stated to have clear advantages: (1) No dimensional changes were recorded after one redox cycle of an infiltrated 16 wt % Ni/YSZ composite. (2) A 12 wt % Ni/YSZ cermet exhibited conductivity (360 S cm^{-1} at room temperature), similar to alternatively fabricated anodes. (3) The exhibited conductivity degradation (20 %) after one redox cycle is comparable with the best-performing alternative anode materials. However, statement (3) may be disputed. These disadvantages were observed: (1) The infiltration method employed was time-consuming and not suitable for large-scale production. (2) The Ni structure could be susceptible to long-term isothermal degradation [67].

The first disadvantage above is in line with Klemensø et al. [68], who found that the main processing issue with the Ni impregnated anodes is the time-consuming impregnation–decomposition steps. At least ten cycles are necessary, even when the process is optimised by using the solution with the maximum Ni concentration using hot infiltration with molten Ni nitrate, instead of an aqueous nitrate solution that can be infiltrated at room temperature, and minimum heat treatment for decomposing the nitrate. Furthermore, after ten cycles, the Ni content still appeared to be close to the percolation threshold.

12 Polarisation Resistance and Electrode Reaction Rate Limitations

There seems to be general agreement about the observation that normally more than one process contributes to the limitation of the electrode reaction rate. The following picture is often obtained in EIS results: (1) Two concentration impedance arcs, a small diffusion arc (summit freq. 10–100 Hz) arc and a larger conversion

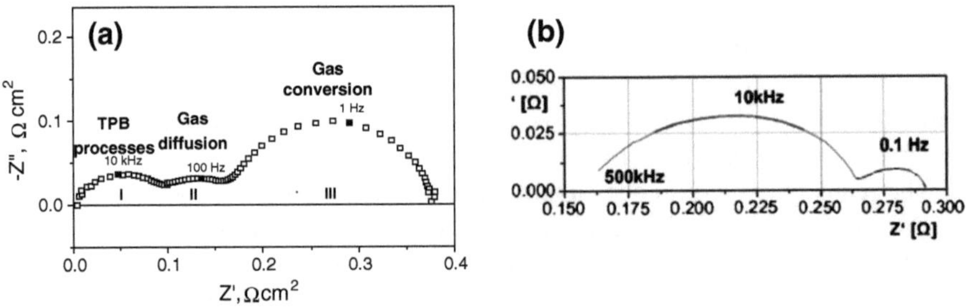

Fig. 12 Examples of impedance spectra of technical Ni/8YSZ cermet electrodes; **a** Impedance spectrum for Ni-YSZ cermets in 3 electrode set-up at 1000 °C in H_2 + 3 % H_2O [9]. Reproduced here with kind permission from © The Electrochemical Society 1999. **b** at 950 °C in H_2 with 5 % H_2O. Electrode area is 1 cm^{-2}. The anode consists of 47 vol % Ni and 53 vol % YSZ of solid content [69]. Reproduced here with kind permission from © The Electrochemical Society 2008

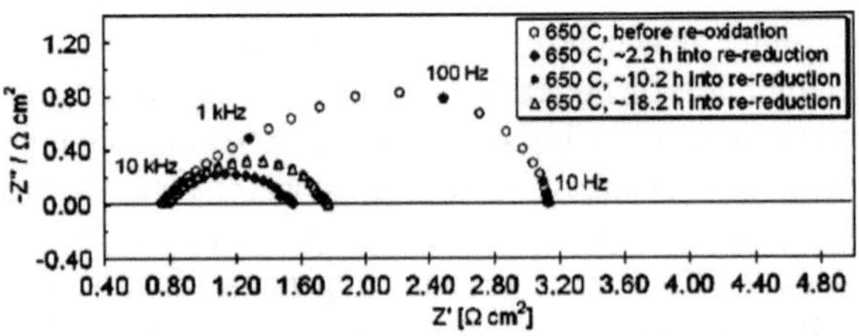

Fig. 13 Impedance spectra obtained for a symmetric cell (Ni-ScYZ/ScYZ/Ni-ScYZ) as a function of re-reduction time at 650 °C after a re-oxidation period of 12 h at the same temperature [62]. Reproduced here with kind permission from © The Electrochemical Society 2009

arc (1–10 Hz) are seen. These arcs are dependent on the cell and test set-up only. No electrochemical reaction is involved. It is not always possible to separate the two arcs. (2) 2 TPB ion transfer process are seen for H_2/H_2O/Ni/YSZ cermet electrodes, see e.g. [69]. An arc reflecting the parallel transport of O^{2-} and e^- in the region near the electrolyte, and an arc due to ion transfer across or around the TPB. Summit frequencies depend on temperature and structure. Also, these two arcs may be difficult to separate and therefore are often taken as one and called R_{TPB}, because usually the main part originates from the ion transfer at the TPB. Figure 12 shows examples from two different laboratories. Figure 13 shows how sensitive the Ni-YSZ electrode impedance is to changes in microstructure, which here were induced by redox cycling at 650 C.

Otherwise, there are big discrepancies in the literature. Activation energies varying from 0.5 to 1.7 eV have been reported. Also, the reported dependencies of R_{TPB} on partial pressures of water and hydrogen vary a lot. The H_2 anodic oxidation rate is most often reported to increase with pH_2O.

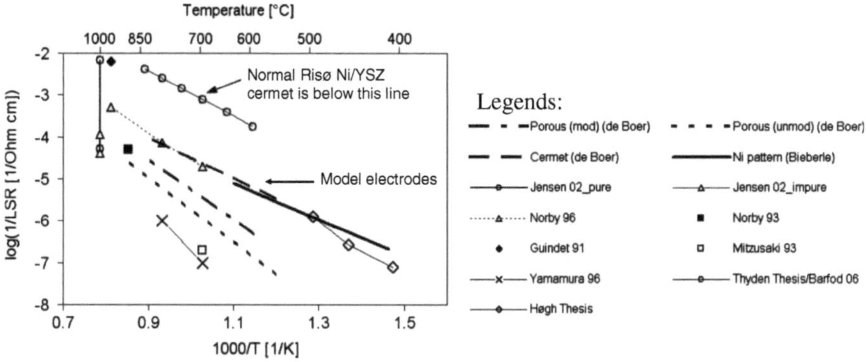

Fig. 14 Arrhenius plot of reaction rate at the Ni-YSZ three-phase boundary in the form of the inverse line-specific resistance (LSR). The model electrode data are from de Boer et al. [70], Bieberle et al. [71], Jensen et al. [72], Guindet et al. [73], Norby [74, 75], Mizusaki et al. [76], Yamamura et al. [77], and Høgh [78]. The upper limit for the Risø Ni/YSZ cermet is obtained by combining the work of Thydén [79], which gives the TPB length per area between the Ni particles and the dense YSZ electrolyte, with the area-specific resistance provided by Barfod et al. [80]

13 Model Electrode Studies

In order to simplify the Ni–YSZ microstructure, several attempts have been made to control the length of the TPL by using simple and well-defined electrode structures, e.g. point electrodes, patterned electrodes and single crystal electrolytes. The hydrogen/water reaction on Ni electrodes and Ni–YSZ cermets is the most widely studied reaction. From these investigations, line-specific resistances are obtained and summarised in Fig. 14. Also, an upper limit for the usual Risø Ni–YSZ cermet is presented in Fig. 14 by combining the measured R_{TPB} with the length of the TPB next to the bulk electrolyte. Thus, we know that the true line-specific resistance, LSR or LSR_{TPB}, must be below the line as there will be some contributions from TPBs away from the bulk electrolyte. The reaction rate, 1/LSR, as well as the temperature dependence (activation energy) differ significantly.

Figure 14 indicates major differences in the model electrodes studied by various groups. Therefore, surface analysis studies were initiated, and a series of investigations at Risø have identified a segregation of a glassy impurity phase both in the Ni/electrolyte interface but especially at the rim of the Ni electrodes, the TPL. The differences observed between the model electrodes (both in terms of polarisation resistance and activation energy) have been attributed to different impurity contents, determined by different precursor materials and different impurity levels in used gases (Fig. 15).

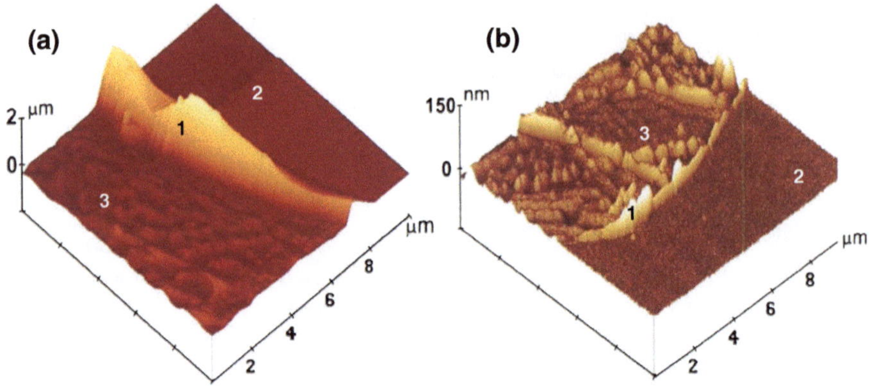

Fig. 15 Atomic force microscope pictures of: rim ridge of impurity phases at TPB of Ni point electrodes on YSZ after ca. 1 week test and removal Ni. **a** "Impure" Ni = 99.8 % Ni. **b** "Pure" Ni = 99.99 % Ni [81]. Reproduced here with kind permission from © The Electrochemical Society 2004

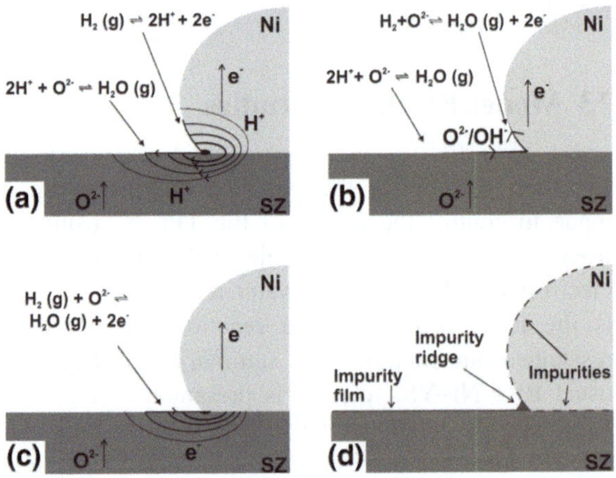

Fig. 16 Example on possible reaction mechanisms (**a**–**c**), and possible barriers (**d**) for the electrode reaction [96]. Reproduced here with kind permission from © The Electrochemical Society 2007

14 Electrode Mechanism

Possible reaction paths are suggested in Fig. 16. The types of processes have been discussed in some detail [82–87]. In addition to the classical electrochemistry viewpoint that the metallic Ni surface also acts as the electrode for the electrode transfer step, the electron transfer may partly take place at the electrolyte surface because the YSZ electrolyte has some electronic conductivity [88, 89], which is probably very dependent on trace impurities. Also, NiO is slightly soluble in YSZ. The solubility varies with temperature from about 1.1 % at 1200 °C to about 1.7 % at 1600 °C [90, 91]. A solubility around 0.5 % at 800 °C is found by extrapolation. In the reducing fuel atmosphere, Ni metal will precipitate in the

YSZ. Probably, neither the electron transfer nor the H_2 bond breaking is rate limiting at the high temperatures. The usual assumption is that either H^+ or O^{2-} transfer near or at the TPB is posing the main rate limitation.

As discussed previously [84, 92, 93], it is reasonable to assume a significant degree of hydrogen coverage on the Ni even at 1000 °C. Also, hydrogen (protons) is slightly soluble in Ni (6×10^{-4} atom %) and highly mobile. The diffusion coefficient is 1.5×10^{-4} cm^2/s at 1000 °C. Finally, protons are to a low extent soluble and mobile in the YSZ (proton concentration up to about 2×10^{-5} atom % and diffusion coefficient 1×10^{-6} cm^2/s). This seems to be enough to allow local current densities of protons through the bulk YSZ of the order of 100 mA/cm^2 at distances of up to 1 μm from the Ni [84]. Thus, it seems that the hydrogen oxidation (and water reduction) may proceed through a number of consecutive steps and through different parallel paths, and in fact evidence of entirely different rate limiting steps at temperatures below 845 °C and above 890 °C has been presented [26].

Experimental evidence exists, which indicate that the reaction zone width on Ni is less than 1 μm [94]. The extension of the electrode process onto the YSZ surface is not well determined, but a value of less than 13 μm at 700 °C has been indicated based on experiments with Ni pattern electrodes [95].

Yet, there are still many uncertainties. The polarisation resistance measured on model electrodes have mainly been carried out using EIS without polarising the electrodes (= OCV conditions in a cell). Thus, the results are related to both the hydrogen oxidation reaction (fuel cell mode) as well as to the hydrogen evolution reaction (electrolysis mode), which can follow different mechanisms and rate determining steps. Considering the complex situation at OCV, the further interpretation of those results regarding possible electrode mechanisms is questionable. Only little usable mechanistic information is available as regards the electrode behaviour under operating conditions. Thus, still more knowledge about the details of the barriers, which limits R_P, is needed in order to optimise the Ni–YSZ performance further. Precise values of the TPB line-specific resistance should be known and preferably also the widths of the reaction zones on the Ni and on the YSZ along the TPB as well as the spatial distribution of the current densities within these zones. If this information is available, the optimal structure and thickness of the cermet may be calculated using statistical models [97]. However, there are discrepancies in the literature.

15 Concluding Remarks

The Ni-based cermet electrode is a becoming quite mature for application in solid-state electrochemical cells, particularly solid oxide fuel cells. We know a lot about how it behaves under various conditions. Thus, we also know its shortcomings much better than those of alternative SOFC anodes/SOEC cathodes. In spite of this, there is still a lot to investigate. The next era of Ni-cermet R&D may be dealing with Ni-alloys and cermet additives in the form of nanoparticles.

The mechanistic studies should concentrate on significantly polarised electrodes, and identifying the width of reaction zones to better design optimal cermet microstructures and the optimal chemical composition (e.g. Sc or Y in the electrolyte) that influences the TPB characteristics in terms of activity and durability.

A further important research activity is still to better understand and predict the long-term stability of the complex cermet structure. Initial relations between microstructure in terms of grain size, connectivity and porosity and structural stability have been developed that allow optimisation as regards redox and thermal cycling as important operation demands. However, more quantitative data is needed to predict several 10000 h of operation for those composite materials.

References

1. J.R. Rostrup-Nielsen, *Catalytic Steam Reforming* (Springer, Berlin, 1984)
2. M. M. Murphy, J. Van herle, A. J. McEvoy, K. R. Thampi, J. Electrochem. Soc. **141**, L94 (1994)
3. H.S. Spacil, Patent US 3,503,809 (1970)
4. J. Mizusaki, H. Tagawa, T. Saito, K. Kamitani, T. Yamamura, K. Hirano, S. Ehara, T. Takagi, T. Hikita, M. Ippommatsu, S. Nakagawa, K. Hashimoto, J. Electrochem. Soc. **141**, 2129 (1994)
5. T. Norby, O. J. Velle, H. Leth-Olsen, R. Tunold, In *Solid Oxide Fuel Cells III*, ed. by S. C. Singhal and H. Iwahara, (The Electrochemical Society, Pennington, 1993) p. 473
6. A.C. Müller, D. Herbstritt, E. Ivers-Tiffée, Solid State Ionics **152–153**, 537 (2002)
7. K.-Z. Fung and A. V. Virkar, In *Solid Oxide Fuel Cells IV*, ed. by M. Dokiya, O. Yamamoto, H. Tagawa and S. C. Singhal, (The Electrochemical Society, Pennington, 1995) p. 1105
8. J.-W. Kim, A.V. Virkar, K.-Z. Fung, K. Metha, S.C. Singhal, J. Electrochem. Soc. **146**, 69 (1999)
9. S. Primdahl, M. Mogensen, J. Electrochem. Soc. **146**, 2827 (1999)
10. M. Mogensen, P. H. Larsen, P.V. Hendriksen, B. Kindl, C. Bagger and S. Linderoth, In *Solid Oxide Fuel Cells (SOFC VI)*, eds. S.C. Singhal and M. Dokiya, Proc. vol. 99–19, The Electrochemical Society, 1999, p. 904
11. M.J. Powell, Phys. Rev. B **20**, 10 (1979)
12. D.W. Dees, T.D. Claar, T.E. Easler, D.C. Fee, F.C. Mrazek, J. Electrochem. Soc. **134**, 2141 (1987)
13. H. Itoh, T. Yamamoto, M. Mori, T. Horita, N. Sakai, H. Yokokawa, M. Dokiya, J. Electrochem. Soc. **144**, 641 (1997)
14. M. Mogensen, S. Primdahl, J. T. Rheinländer, S. Gormsen, S. Linderoyh, M. Brown, In *Solid Oxide Fuel Cells IV*, ed. by M. Dokiya, O. Yamamoto, H. Tagawa, S. C. Singhal, (The Electrochemical Society, Pennington, 1995) p. 657
15. M.S. Brown, N. M. Sammes, M. Mogensen, In *Solid Oxide Fuel Cells V*, ed. by U. Stimming, S. C. Singhal, H. Tagawa, W. Lehnert, Proc. vol. 97–40 (The Electrochemical Society, Pennington, 1997) p. 861
16. D.W. Dees, T.D. Claar, T.E. Easier, D.C. Fee, F.C. Mrazek, J. Electrochem. Soc. **134**, 2144 (1987)
17. S. Sakamoto, H. Taira, H. Takagi, Denki Kagaku **64**, 609 (1996)
18. M. Brown, S. Primdahl, M. Mogensen, N. Sammes, J. Aust. Cer. Soc. **34**, 248 (1998)
19. M. Brown, S. Primdahl, M. Mogensen, J. Electrochem. Soc. **147**, 475 (2000)
20. S. C. Singhal, in *Solid Oxide Fuel Cells V*, ed. by U. Stimming, S. C. Singhal, H. Tagawa, W. Lehnert, Proc. vol. 97–40 (The Electrochemical Society, Pennington, 1997) p. 37

21. R. Barfod, M. Mogensen, T. Klemensø, A. Hagen, Y.-L. Liu, P.V. Hendriksen, J. Electrochem. Soc. **154**, B371 (2007)
22. A. Hauch, M. Mogensen, Solid State Ionics **181**, 745 (2010)
23. H. Itoh, T. Yamamoto, M. Mori, T. Watanabe, T. Abe, Denki Kagaku **64**, 549 (1996)
24. S.C. Singhal, In *High Temperature Electrochemistry: Ceramics and Metals*, eds. by F. W. Poulsen, N. Bonanos, S. Linderoth, M. Mogensen, B. Zachau-Christiansen, (Risø National Laboratory, Roskilde DK, Denmark, 1996) p. 123
25. A. Hagen, R. Barfod, P.V. Hendriksen, Y.-L. Liu, S. Ramousse, J. Electrochem. Soc. **153**, A1165 (2006)
26. A. Hagen, Y.L. Liu, R. Barfod, P.V. Hendriksen, ECS Trans. **7**, 301 (2007)
27. J. Sehested, Catal. Today **111**, 103 (2006)
28. J. Sehested, J.A.P. Gelten, S. Helveg, Applied Catalysis A - General **309**, 237 (2006)
29. S.P. Jiang, J. Mater. Sci. **38**, 3775 (2003)
30. P. Tanasini, M. Cannarozzo, P. Costamagna, A. Faes, J. Van herle, A. Hessler-Wyser, C. Comninellis, Fuel Cells, **9**, 740 (2009)
31. A. Faes, A. Hessler-Wyser, D. Presvytes, C.G.Vayenas, J. Van herle, Fuel Cells, **9**, 841 (2009)
32. L. Holzer, B. Iwanschitz, T. Hocker, B. Mench, M. Prestat, D. Wiedenmann, U. Vogt, P. Holtappels, J. Sfeir, A. Mai, T. Graule, J. Power Sources **196**, 1279 (2010)
33. A. Hauch, M. Mogensen, A. Hagen, Solid State Ionics (2010) http://dx.doi.org/10.1016/j.ssi.2010.01.004
34. A. Gubner, H. Landes, J. Metzger, H. Seeg and R. Stübner, In *SOFC V*, ed. U. Stimming, S. C. Singhal, H. Tagawa and W. Lehnert, (The Electrochemical Society, Pennington 1997) p. 844
35. M. Mogensen, M. Søgaard, P. Blennow, K. Kammer Hansen, In: *Proceeding European solid oxide fuel cell forum*, 2008, on CD-ROM
36. A. Hauch, M. Mogensen, Solid State Ionics **181**, 745 (2010)
37. A. Hauch, P.S. Jørgensen, K. Brodersen, M. Mogensen, J. Power Sources (2011). doi:10.1016/j.jpowsour.2011.01.009
38. R. Männer, E. Ivers-Tiffée, W. Wersing, In *Solid Oxide Fuel Cells (SOFCII)*, eds. F. Grosz, P. Zegers, S.C. Singhal and O. Yamamoto, Commission of the European Communities, Luxemburg, L. EUR-13564-EN, 715 (1991)
39. M. Mori, T. Yamamoto, H. Itoh, H. Inaba, H. Tagawa, J. Electrochem. Soc. **145**, 1374 (1998)
40. R. Farraro, R.B. McLellan, Metall. Transactions A **8**, 1563 (1977)
41. K.R. Pirota, E.L. Silva, D. Zanchet, D. Navas, M. Vázquez, M. Hernández-Vélez, M. Knobel, Phys. Rev. B **76**, 233410 (2007)
42. B.F. Sørensen, S. Primdahl, J. Mater. Sci. **33**, 5291 (1998)
43. J. Johnson, J. Qu, J. Power Sources **181**, 85 (2008)
44. P. H. Larsen, S. Primdahl, M. Mogensen, In *High Temperature Electrochemistry: Ceramics and Metals*, ed. by F. W. Poulsen, N. Bonanos, S. Linderoth, M. Mogensen, B. Zachau-Christiansen, (Risø National Laboratory, Roskilde DK, Denmark 1996) p. 331
45. R. Gorte, J. Vohs. J. Catalysis **216**, 477 (2003)
46. E. Achenbach, E. Riensche, J. Power Sources **52**, 283–288 (1994)
47. E. Achenbach, J. Power Sources **49**, 333–348 (1994)
48. P. V. Hendriksen, In *SOFC V*, Eds. U. Stimming, S. C. Singhal, H. Tagawa, W. Lehnert, (The Electrochemical Society, Pennington, 1997) p. 1319
49. R. J. Aaberg, R. Tunold, S. Tjelle, R. Ødegaard, In *Proceedings of the 17th Risø International Symposium on Materials Science*, eds. F. W. Poulsen et al., Roskilde, Denmark, (1996), p. 511–516
50. P. Holtappels, L.G.J. de Haart, M. Mogensen, S. Stimming, I.C. Vinke, J. Appl. Elecrochem. **29**, 561 (1999)
51. S. Primdahl, M. Mogensen, In *Solid Oxide Fuel Cells (SOFC VI)*, Eds. S.C. Singhal and M. Dokiya, Proc. vol. 99–19, The Electrochemical Society, 1999, p. 530

52. K. Huang, J.B. Goodenough, *Solid Oxide Fuel Cell Technology: Principles, Performance and Operations* (Woodhead Publishing Ltd., UK, 2009)
53. J. Rostrup-Nielsen, J.B. Hansen, S. Helveg, N. Christiansen, A.K. Jannasch, Appl. Phys. A **85**, 427 (2006)
54. J.B. Hansen, Electrochem. Solid-State Lett. **11**, B178 (2008)
55. S. Zha, Z. Cheng, M. Liu, J. Electrochem. Soc. **154**, B201 (2007)
56. K. Sasaki et al., J. Electrochem. Soc, **153,** A2023
57. A. Hagen, J.F.B. Rasmussen, Proc. Eur. SOFC Forum **9**, 7–96 (2010)
58. H. He, R.J. Gorte, J.M. Vohs, Electrochem. Solid-State Lett. **8**, A279 (2005)
59. S.D. Ebbesen, M. Mogensen, pending patent application
60. S.D. Ebbesen, M. Mogensen, Electrochem. Solid-State Lett. **13**, B106 (2010)
61. T. Klemensø, C. Chung, P.H. Larsen, M. Mogensen, J. Electrochem. Soc. **152**, A2186 (2005)
62. M. Pihlatie, T. Ramos, A. Kaiser, J. Power Sources **193**, 322 (2009)
63. T. Klemensø, C. Chung, P.H. Larsen, M. Mogensen, J. Electrochem. Soc. **152**, A2186 (2005)
64. M. Pihlatie, A. Kaiser, P.H. Larsen, M. Mogensen, J. Electrochem. Soc. **156**, B322 (2009)
65. M.H. Pihlatie, H.L. Frandsen, A. Kaiser, M. Mogensen, J. Power Sources **195**, 2677 (2010)
66. D. Sarantaridis, R.A. Rudkin, A. Atkinson, J. Power Sources **180**, 704 (2008)
67. A.N. Busawon, D. Sarantaridis, A. Atkinson, Electrochem. Solid-State Lett. **11**, B186 (2008)
68. T. Klemensø, K. Thydén, M. Chen, H.-J. Wang, J. Power Sources **195**, 729 (2010)
69. V. Sonn, A. Leonide, E. Ivers-Tiffée, J. Electrochem. Soc. **155**, B675 (2008)
70. B. de Boer, *SOFC Anode, Hydrogen oxidation at porous nickel and nickel/yttria-stabilised zirconia cermet electrodes,* PhD Thesis, University of Twente, 1998
71. A. Bieberle, *The electrochemistry of solid oxide fuel cell anodes: experiments, modeling, and simulations*' PhD Thesis, Swiss Federal Institute of Technology, 2000
72. K.V. Jensen, *The nickel-YSZ interface. Structure, composition and electrochemical properties at 1000 °C,* Risø-R-1335(EN) (2002) 219 p. (PhD Thesis)
73. J. Guindet, C. Roux, A. Hammou, Proceeding 2nd International Symposium on SOFC (F. Grosz, P. Zegers, S.C. Singhal, O. Yamamoto, Hrsg.), Commission of the European Communities, Report EUR 13546 EN, Luxembourg (1991) 553 (see ref 39 for citation)
74. T. Norby, O. Velle, in *Proceedings 2. International Symposium on SOFC, The Electrochemical Soc Proc* vol 93–4 Pennington (1993) 473
75. T. Norby, In *Proceedings of the 2nd European Solid Oxide Fuel Cell Forum*, May 6–10, Oslo, p. 607, 1996
76. J. Mizusaki, H. Tagawa, T. Saito, K. Kamitani, T. Yamamura, K. Hirano, S. Ehara, T. Takagi, T. Hikita, M. Ippommatsu, S. Nakagawa, K. Hashimoto, J. Electrochem. Soc. **141**, 2129 (1994)
77. T. Yamamura, H. Yoshitake, H. Tagawa, N. Mori, K. Hirano, J. Mizusaki, S. Ehara, T. Takagi, M. Hishinuma, H. Sasaki, Y. Nakamura, K. Hashimoto, In *Proceedings of the 2nd European Solid Oxide Fuel Cell Forum*, May 6–10, Oslo, p. 617, 1996
78. J. Høgh, PhD Thesis, Technical University of Denmark, 2005
79. K. Thyden, PhD Thesis, Technical University of Denmark, 2008
80. R. Barfod, A. Hagen, S. Ramousse, P.V. Hendriksen, M. Mogensen, Fuel Cells **6**, 141 (2006)
81. K.V. Hansen, K. Norrman, M. Mogensen, J. Electrochem. Soc. **151**, A1436 (2004)
82. E.J. Schouler, M. Kleitz, J. Electrochem. Soc. **134**, 1045 (1987)
83. J. Mizusaki, H. Tagawa, T. Saito, T. Yamamura, K. Kamitani, K. Hirano, S. Ehara, T. Takagi, T. Hikita, M. Ippommatsu, S. Nakagawa, K. Hashimoto, Solid State Ionics **70/71,** 52 (1994)
84. M. Mogensen, S. Sunde and S. Primdahl, In *High Temperature Electrochemistry: ceramics and Metals*, ed. by F. W. Poulsen, N. Bonanos, S. Linderoth, M. Mogensen and B. Zachau-Christiansen (Risø National Laboratory, Roskilde DK 1996) p. 77
85. P. Holtappels, L.G.J. de Haart, U. Stimming, J. Electrochem. Soc. **146**, 1620 (1999)
86. P. Holtappels, I.C. Vinke, L.G.J. de Haart, U. Stimming, J. Electrochem. Soc. **146**, 2976 (1999)

87. S. Primdahl, *"Nickel/yttria-stabilised zirconia cermet anodes for solid oxide fuel cells"*, PhD-thesis, University of Twente, 1999, Risø-R-1137 (EN)
88. J.-H. Park, R.N. Blumenthal, J. Electrochem. Soc. **136**, 2867 (1989)
89. H.L. Tuller, in *Nonstoichiometric Oxides*, ed. by O. T. Sørensen (Academic Press, New York, 1981), p. 271
90. S. Linderoth, A. Kuzjukevics, in *Solid Oxide Fuel Cells V*, ed. by U. Stimming, S. C. Singhal, H. Tagawa, W. Lehnert. (The Electrochemical Society, Pennington, 1997) p. 1076
91. A. Kuzjukevics, S. Linderoth, Solid State Ionics **93**, 255 (1997)
92. M. Mogensen, S. Skaarup, Solid State Ionics **86–88**, 1151 (1996)
93. S. Skaarup, B. Zachau-Chistiansen, T. Jacobsen, in *High Temperature Electrochemistry: Ceramics and Metals*, ed. by F. W. Poulsen, N. Bonanos, S. Linderoth, M. Mogensen and B. Zachau-Christiansen, (Risø National Laboratory, Roskilde DK, Denmark 1996) p. 432
94. N. Nakagawa, H. Sakurai, K. Kondo, T. Morimoto, K. Hatanaka, K. Kato, J. Electrochem. Soc. **142**, 3474 (1995)
95. J. Mizusaki, T. Yamamura, N. Mori, H. Tagawa, K. Hirano, S. Ehara, T. Tagaki, M. Hishinuma, H.Sasaki, T. Sogi, Y. Nakamura, K. Hashimoto, in *High Temperature Electrochemistry: Ceramics and Metals*, ed. by F. W. Poulsen, N. Bonanos, S. Linderoth, M. Mogensen and B. Zachau-Christiansen, (Risø National Laboratory, Roskilde DK, Denmark 1996) p. 363
96. M. Mogensen, J. Høgh, K.V. Hansen, T. Jacobsen, ECS Trans. **7**(1), 1329 (2007)
97. S. Sunde, J. Electrochem. Soc. **143**, 1930 (1996)

Strategies for Perspective Cathode Materials for IT–SOFC

Janina Molenda and Konrad Świerczek

Abstract Lowering of the operation temperature of Solid Oxide Fuel Cells from present 1000 °C down to intermediate range 600–800 °C (so-called IT–SOFC), considered as a strategic goal in technological development, calls for application of suitable cathode material, which at that temperature range would possess high catalytic activity towards an oxygen reduction process. A crucial factor for IT–SOFC cells is the implementation of perovskite material, characterized by improved ionic–electronic transport for the cathode. In this chapter, we present studies of perovskites from group of $Ln_{1-x}A_xMO_3$ oxides, where: Ln: lanthanides, A: alkali–earth metals (e.g., Sr, Ba), M: composition of Fe, Co, and Mn. A challenging task was to assess interrelations between the types of point defects related to oxygen non-stoichiometry and foreign atoms of dopants, their concentration, crystal structure, electronic structure, transport, and catalytic properties.

1 Introduction

In classical SOFC with zirconia-based electrolyte, working at 1000 °C, $La_{1-x}Sr_xMnO_3$ oxide plays role of cathode. This compound exhibits perovskite structure and is usually referred to as LSM. LSM is characterized by high, practically purely electronic conductivity at high temperatures (> 900 °C). However, at lower temperatures, the efficiency of LSM cathode is significantly decreased [1, 2]. Therefore, a new class of the cathode materials, enabling reduction of SOFC

J. Molenda (✉) · K. Świerczek
Faculty of Energy and Fuels, AGH University of Science and Technology,
al. Mickiewicza 30, 30-059 Kraków, Poland
e-mail: molenda@agh.edu.pl

J. T. S. Irvine and P. Connor (eds.), *Solid Oxide Fuels Cells: Facts and Figures*,
Green Energy and Technology, DOI: 10.1007/978-1-4471-4456-4_3,

operating temperature down to 600 °C is at the focal point of vast number of researchers. Among numerous compounds considered for application in IT–SOFC, the majority crystallize in cubic or pseudo-cubic perovskite structure [3]. Such structure is typical for many complex oxides and in many cases is a building block for more complex structures (e.g., double perovskites, Ruddlesden–Popper type oxides) [4]. Therefore, good understanding of structural properties of the simple structure may greatly help with studies of more complex systems. The structural properties of majority of perovskites show presence of some type of unit cell distortion (e.g., rhombohedral, orthorhombic or tetragonal). This structure is also responsible for the effective transport of electrons, owing to the overlapping 3*d* orbitals of transition metal present in octahedral position with 2*p* orbitals of oxygen atoms. The ionic transport is also possible and it occurs via oxygen vacancies. The possibility of high, mixed ionic–electronic conductivity of some perovskites is highly desirable, because it may improve effectiveness of the cathode performance in SOFC in the intermediate and low temperature range [5].

There are many strict requirements for a candidate cathode material, among them the following seem to be particularly important [1, 2, 5, 6]:

- high temperature thermal and chemical stability in the oxidizing atmosphere,
- high chemical stability in relation to used electrolyte and interconnector,
- Cr tolerance in case of usage Cr containing interconnectors,
- CO_2 tolerance,
- high total electrical conductivity (>100 S cm^{-1}),
- if the material is used in IT–SOFC, high mixed ionic–electronic conductivity (MIEC) with as high as possible ionic transport,
- high catalytic activity for the oxygen reduction reaction,
- adequate thermomechanical properties, thermal expansion coefficient (TEC) matched with electrolyte and interconnector,
- possibility of preparation of cathode layers with appropriate porosity, adhesion and mechanical strength,
- as low as possible cost of substrates for synthesis and effective preparation method,
- environmental friendliness.

Oxides having perovskite structure have been recently the subject of extensive studies [7–13], but the main research activities concentrate on their functional properties, i.e., possible application of these materials as cathodes in IT–SOFC. Fundamental studies of perovskite oxides that might contribute to better understanding of relations between crystallographic structure, ionic and electronic defects resulting from oxygen non-stoichiometry, doping level, electronic structure, transport properties, and catalytic activity towards oxygen reduction are still lacking. These issues are of great importance. It has been found [2] that kinetics of the cathode reaction (oxygen reduction) may restrict the electrochemical effectiveness of SOFC. A microscopic mechanism of the catalytic process of the oxygen reduction has not been entirely explored. It has been established that the rate of the oxygen adsorption on the cathode material depends on the concentration

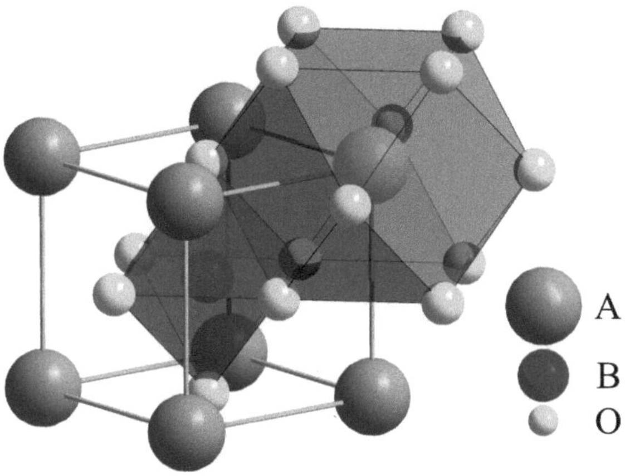

Fig. 1 Cubic structure of ABO_3 perovskite. A cations occupy Wyckoff 1a position (0,0,0), B cations 1b (½,½,½) position and O anions 3c (0,½,½) position

of oxygen vacancies and electrons in the cathode. A deviation from stoichiometry toward oxygen deficiency introduces donor centers, ionization of which leads to a change in the concentration of charge carriers. Therefore, the oxygen non-stoichiometry and the presence of dopants shift position of the Fermi level, which, according to electronic theory of catalysis is a vital factor for catalytic activity of the cathode material [5]. Magnitude of the oxygen non-stoichiometry is related to temperature, oxygen partial pressure pO_2 and type, and concentration of the dopant.

In order to examine the mechanism of the oxygen reduction in $La_{1-x}Sr_{x}Co_{1-y}Fe_{y}O_{3-\delta}$ (LSCF) in this paper, complex studies over the structure of the ionic and electronic defects, resulting from the deviation from stoichiometry or doping with different ions, were carried as a function of temperature and oxygen partial pressure. The results allowed for designing functional properties of perovskite-type oxides, perspective cathode materials for electrochemically effective IT–SOFC.

2 Perovskite Structure

Ideal, cubic ABO_3 perovskite structure belongs to $Pm\overline{3}m$ (no. 221) space group. It can be described as a 3-dimensional network of corner sharing BO_6 octahedra. In such a structure bigger, 12-fold A cations occupy all available positions. Alternatively, it can be described as a cubic close-packed (*ccp*) array of AO_3, in which one quarter of all octahedral positions is occupied by smaller B cations. Fig. 1 shows graphical representation of perovskite structure.

In many cases the nature of chemical bonds in ABO_3 is highly ionic, therefore a simple approximation, in which ions are approximated as spheres with fixed radius is valid. It allowed Goldschmidt to describe tolerance factor *t* using simple geometrical considerations:

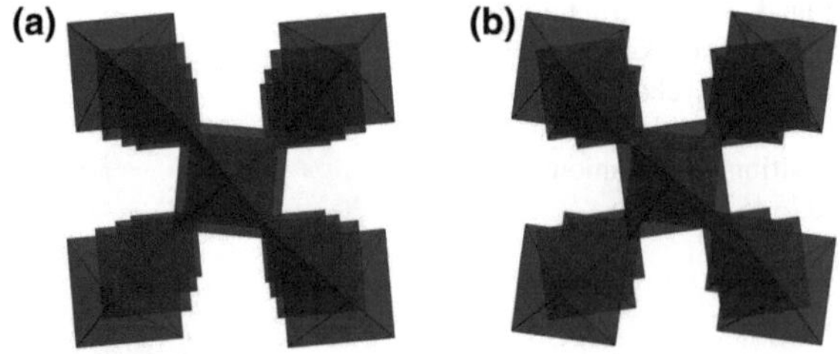

Fig. 2 Octahedra tilt systems: **a** $a^0a^0c^+$ and **b** $a^0a^0c^-$. Viewing direction along c axis

$$t = \frac{r_A + r_O}{\sqrt{2}(r_B + r_O)} \quad (1)$$

where: r_A, r_B, r_O represent appropriate ionic radii.

Perovskite structure is realized when $0.8 < t < 1.1$, while cubic Pm$\overline{3}$m symmetry is usually observed for materials, for which the tolerance factor is closer to 1 ($0.9 < t < 1.05$). However, it is important to realize that ionic approximation is not always valid and several other factors (e.g., Jahn–Teller effect, metal–metal interactions) may strongly affect crystal structure [4]. In other words, it is not possible to predict space group, in which considered material will crystallize only on the basis of calculated value of t. Additionally, there are other ways for calculating tolerance factor: one is based on Bond Valence method [14], while the other one uses real interatomic distances determined, for instance, from neutron diffraction studies. Values of t derived from the last method represent in fact the actual degree of distortion of the cell and are usually much closer to 1, comparing to values obtained from Shannon's ionic radii.

There are several other structural parameters commonly used for describing perovskite structure, among them: V_A/V_B ratio (V_A, V_B—volume of respectively A–site or B–site polyhedron) and free cell volume V_f (difference between unit cell volume a^3 and volume occupied by ions $V_f = a^3 - \Sigma_i m_i \frac{4\pi}{3} r_i^3$, m_i—stoichiometry factor, r_i—appropriate ionic radius). A more advanced description in form of global parameterization was proposed by Thomas [15]. Another approach, which is based on Bond Valence method and so called Global Instability Index GII was successfully applied in SPUDS software, which allows in many cases for accurate prediction of the actual space group of considered perovskite [16].

The Pm$\overline{3}$m aristotype is a base for a wide range of hettotypes, which are realized due to tilting or rotation of rigid BO_6 octahedra or distortion of octahedra or AO_{12} cubo-octahedra. Octahedral tilting and space group relationship in perovskites was studied by several authors, and nowadays Glazer notation of octahedral tilting is commonly used [17]. In Fig. 2 two typical examples: $a^0a^0c^+$ (in-phase tilting, space group P4/mbm) and $a^0a^0c^-$ (antiphase, I4/mcm) are shown. In case of $a^0a^0c^+$ system, consecutive octahedra layers possess the same tilt angle, while viewed along c axis. For $a^0a^0c^-$ system the following layer tilt angle is opposite.

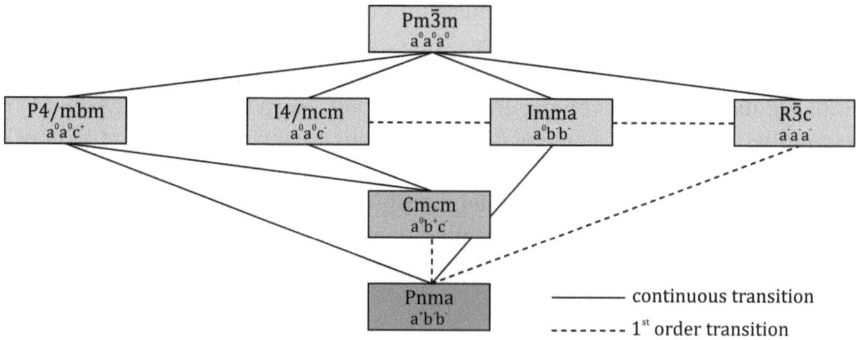

Fig. 3 Space group relationship after Thomas [15]

Group theory analysis allowed 15 tilt systems to be identified which should appear in real crystals, out of 23 originally proposed by Glazer [18]. Out of these, in 8 cases 3–tilt (along 3 axis) system is present, Four are 2–tilt, 2 are 1–tilt and the final one is no–tilt, Pm$\bar{3}$m space group. Moreover, space group relationship analysis allows phase transition diagrams to be proposed, which show permitted types of phase transition (first order or continuous), which will appear while a compound transforms from one to the other tilt system. This is particularly useful, because phase transitions in perovskites are common and often appear during heating or cooling. One such diagram is presented in Fig. 3. However, literature data on this subject is somewhat inconsistent, for instance in case of quite typical Pnma → R$\bar{3}$c transition and different authors propose slightly different space group relationships [18, 19].

It is necessary to point out that the presented short description of perovskite structure deals with simple perovskites only. In case of non-stoichiometric materials, double perovskites or materials with related, perovskite–based structure, structural analysis becomes very complicated and is not fully understood yet.

3 Oxygen Non-stoichiometry and Mixed Ionic-Electronic Transport in Perovskite Oxides

Due to a vast range of possible chemical compositions, perovskite oxides exhibit various physicochemical properties. Different combination of cation charges in A and B sublattice are realized, with typical being 2 + 4 or 3 + 3. Because both sublattices can be occupied by more than one type of cation, a fractional nominal charge distribution is often observed, e.g., $(La_{0.6}Sr_{0.4})^{2.6+}(Co_{0.2}Fe_{0.8})^{3.4+}O_3$. Appropriate selection of single–valence cations results in fixed level of oxygen nonstoichiometry and is widely used in designing perovskite–based solid state electrolytes, e.g., $La_{0.8}Sr_{0.2}Ga_{0.9}Mg_{0.1}O_{2.85}$ ($\delta = 0.2$) [20]. In these materials δ does not change in wide range of temperature and oxygen partial pressure (pO_2) and electronic conductivity is negligible. In the case of cations, which may possess two or more stable valences in octahedral coordination, a mixed-valence compounds can be obtained, e.g,. $La_{1-x}Sr_xMnO_3$, which are almost stoichiometric in a relatively wide range of oxygen partial pressures [21]. In such materials double exchange mechanism, involving electron transport via oxygen anion (M–O–M) is responsible for the electronic conductivity. At high temperatures and at high pO_2 $LaMnO_3$ and $La_{1-x}Sr_xMnO_3$ oxides can be hyperstoichiometric ($\delta > 0$), however various precise measurements unambiguously showed that the apparent excess of oxygen is in fact realized by a complex defect system present in both cationic sublattices. In case of undoped $LaMnO_3$, one of the proposed models can be written as [22]:

$$LaMnO_3 + \frac{\delta}{2}O_2 = \frac{3+\delta}{3}\left(La_{3/(3+\delta)}V_{\delta/(3+\delta)}\right)\left(Mn_{3/(3+\delta)}V_{\delta/(3+\delta)}\right)O_3 \quad (2)$$

where *V* stands for vacancy. The proposed non-stoichiometry model can be expected, as there is virtually no interstitial position available for the excessive oxygen in *ccp* array of AO_3. Between these two extremes, there are many examples of materials with temperature dependent equilibrium between non-stoichiometry level δ and mixed valence of 3*d* metals. This in turn allows for mixed ionic–electronic conductivity and such MIEC compounds are currently widely studied in terms of their possible application in IT–SOFC. Particularly, high mixed ionic–electronic conductivity exists in the case of $La_{1-x}Sr_xCo_{1-y}Fe_yO_{3-\delta}$ (LSCF) perovskites [8].

The general relationship (so called Brouwer diagram) between oxygen non-stoichiometry, oxygen vacancy concentration and electrical conductivity together with electroneutrality principle for A^{2+} doped $Ln_{1-x}A_xMeO_{3\pm\delta}$ is shown in Fig. 4.

4 Overview of Structural Properties of LSCF Perovskites

In $La_{1-x}Sr_xCo_{1-y}Fe_yO_{3-\delta}$ ($0 \leq x \leq 1$; $0 \leq y \leq 1$; $0 \leq \delta \leq 0, 5$) existence of single phase solid solutions is observed in whole chemical composition range (x, y) and in a relatively wide δ range [24]. At room temperature $LaCoO_3$ possesses rhombohedral symmetry with $R\bar{3}c$ space group. Strontium substituted compounds have identical structure for compositions with $x \leq 0.5$ in $La_{1-x}Sr_xCoO_3$ [25]. For higher Sr content these perovskites adopt regular symmetry. An especially broad range of oxygen non-stoichiometry, $2.29 < 3-\delta < 3.00$, can be observed for $SrCoO_{3-\delta}$. However synthesis of fully stoichiometric $SrCoO_3$ is difficult and so far the best results were obtained by electrochemical oxidation. In oxygen deficient samples there is a strong tendency for oxygen vacancy ordering. Materials (or regions of grains) with $Sr_nCo_nO_{3n-1}$ ($n = 2, 4, 5, 6, 7, 8$) composition were structurally identified [26]. In many cases coexistence of regions with different *n* was observed. For sample with $n = 2$ brownmillerite type phase exists, with alternative layers of vertex-connected CoO_4 tetrahedra and CoO_6 octahedra. On the other side of the

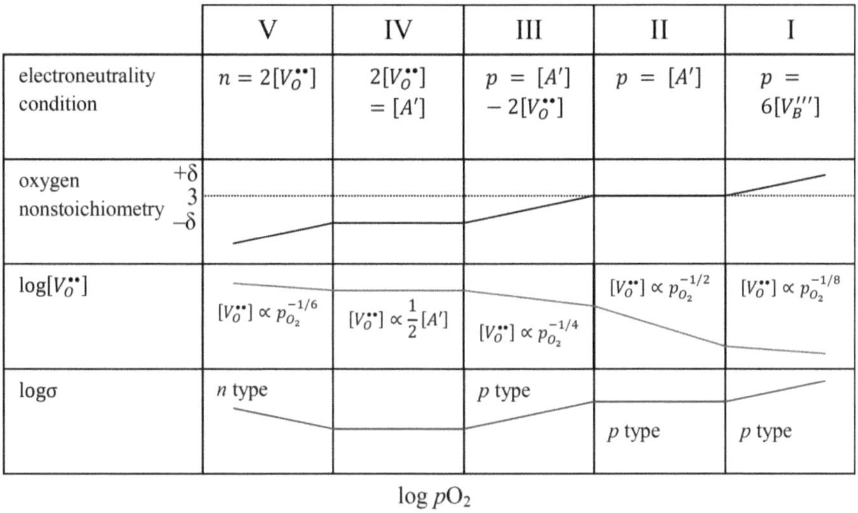

Fig. 4 Brouwer diagram for A^{2+} doped $Ln_{1-x}A_xMO_{3\pm\delta}$ [23]

chemical composition diagram $LaFeO_3$ possesses orthorhombic *Pnma* symmetry over a wide temperature range. With increasing x $La_{1-x}Sr_xFeO_3$ initial orthorhombic symmetry ($0 \leq x \leq 0.2$) is replaced by rhombohedral ($0.4 \leq x \leq 0.7$) and for even higher strontium content cubic structure is observed [27]. Synthesis of fully stoichiometric $SrFeO_3$ requires high oxygen pressure (~200 atm.), for samples with increasing y it is more and more difficult to obtain material with $\delta = 0$. Similarly to $Sr_nCo_nO_{3n-1}$, also in case of $Sr_nFe_nO_{3n-1}$ a homologous series was identified ($n = 2$, 4, 8) [28]. In all these materials a specific arrangement of ordered oxygen vacancies was observed. Oxygen vacancy ordered samples were observed in case of $LaCoO_{3-\delta}$, $La_{0.5}Sr_{0.5}CoO_{2.5}$, $SrCo_{1-y}Fe_yO_{3-\delta}$ and $La_{1-x}Sr_xFeO_{3-\delta}$. In contrast to this, there is no literature evidence of ordered oxygen sublattice in case of $La_{1-x}Sr_xCo_{1-y}Fe_yO_{3-\delta}$ ($0 < x < 1$; $0 < y < 1$; $0 < \delta < 0.5$). This might be explained by the existence of disorder in both cationic sublattices as cobalt/iron and lanthanum/strontium are statistically placed.

5 Influence of Sr Content in LSCF ($La_{1-x}Sr_xCo_{0.2}Fe_{0.8}O_{3-\delta}$) Perovskites on their Structural and Transport Properties

5.1 *Structural Properties*

The studied materials were obtained by ''soft chemistry'' EDTA method, details of the procedure can be found elsewhere [29]. In Fig. 5 an exemplary x–ray diffractogram of $La_{0.4}Sr_{0.6}Co_{0.2}Fe_{0.8}O_{3-\delta}$ is presented. While initial XRD studies

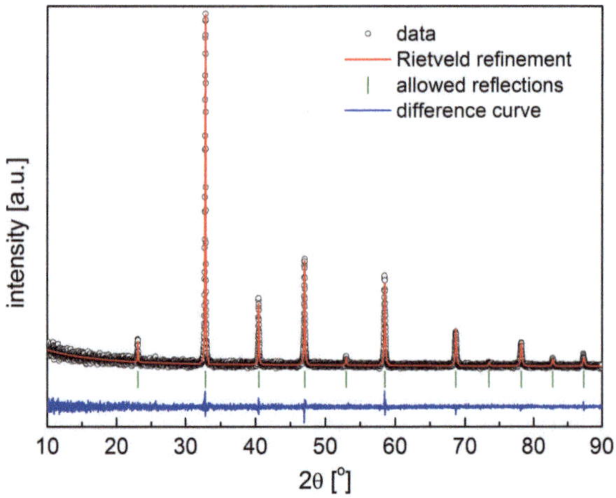

Fig. 5 Exemplary X-ray diffractogram of $La_{0.4}Sr_{0.6}Co_{0.2}Fe_{0.8}O_3$ perovskite

showed regular structure, precise neutron diffraction (ND) measurements (Fig. 6) revealed the presence of slight rhombohedral distortion of the unit cell. In this case, in $R\bar{3}c$ space group, characterized by $a^-a^-a^-$ octahedra tilt system, actually the oxygen sublattice is the one being distorted, while both cation sublattices are almost exactly regular. Therefore due to the nature of the XRD technique, such a distortion is poorly visible, contrary to the ND measurements, for which it can be easily seen. In the case of the first order phase transition, a non-monotonic evolution of unit cell parameters and volume is expected, which in turn may affect thermal compatibility between the cathode material and electrolyte. In the presented case, at high temperatures, the phase transition $R\bar{3}c \rightarrow Pm\bar{3}m$ might be expected, but it would be continuous (Fig. 3) and therefore should be less influential, additionally the degree of distortion is very small. Nevertheless, this result shows importance of precise structural studies of the cathode materials.

Unit cell parameters evolution for $La_{1-x}Sr_xCo_{0.2}Fe_{0.8}O_3$ ($0 \leq x \leq 0.8$) oxides is presented in Fig. 7. In order to directly compare materials with different structures (i.e. Pnma for $LaCo_{0.2}Fe_{0.8}O_3$, $R\bar{3}c$ for samples with $0.2 \leq x \leq 0.6$ and $Pm\bar{3}m$ for $La_{0.2}Sr_{0.8}Co_{0.2}Fe_{0.8}O_3$ a normalization of values was performed. In this process unit cell parameters or volume are recalculated to show a pseudo-cubic, elementary perovskite cell. Despite the rather complicated dependence seen in Fig. 7, with highest degree of distortion for samples with $x = 0.2$ and 0.4, unit cell volume behaves linearly with the increasing amount of Sr up to the $x = 0.6$ composition (Fig. 8). A deviation from the linear at high x dependence may be attributed to the existence of significant amount of oxygen vacancies present in the sample at room temperature, that lowers average valence of 3*d* metals. Higher than expected concentration of Co^{3+} and Fe^{3+} with bigger ionic radii contributes to larger unit cell volume. Two more effects are noteworthy, first because of *ccp* nature of the AO_3 array, empty places related to oxygen vacancies by themselves appear to have no significant influence on unit cell volume. Additionally, despite

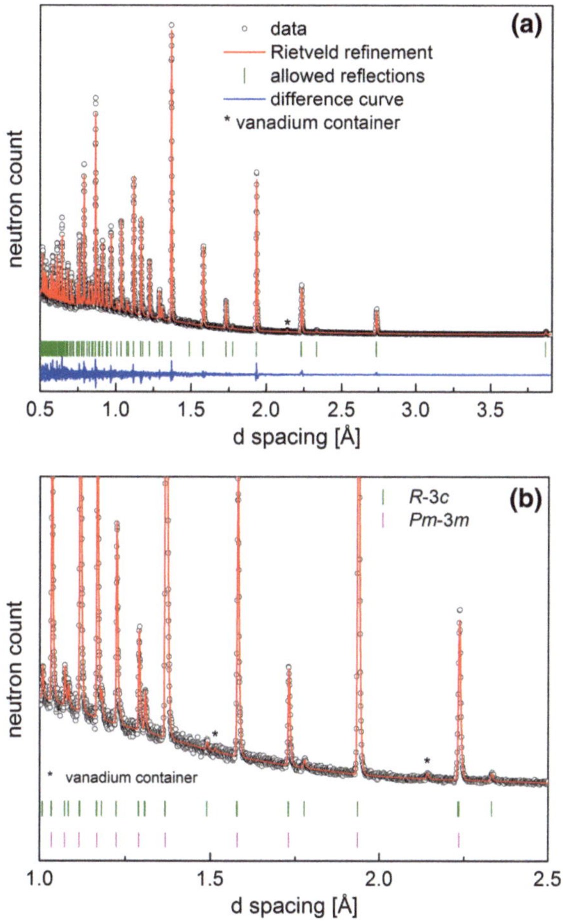

Fig. 6 Combined XRD and ND structural analysis of $La_{0.4}Sr_{0.6}Co_{0.2}Fe_{0.8}O_3$ **a** ND data, **b** magnification of ND diffractogram showing clearly visible $R\overline{3}c$ reflections. *Unit cell described in hexagonal convention

substitution of smaller La^{3+} (1.36 Å) ion by larger Sr^{2+} (1.44 Å) cations, a decrease of unit cell volume is observed as a function of x in $La_{1-x}Sr_xCo_{0.2}Fe_{0.8}O_3$. Finally, $SrCo_{0.2}Fe_{0.8}O_{3-\delta}$ material was found to have ordered oxygen sublattice similar to previously reported for $SrFeO_{2.75}$. High temperature structural studies performed for this compound revealed that above around 400 °C this ordering diminishes.

The tolerance factor t for $La_{1-x}Sr_xCo_{0.2}Fe_{0.8}O_3$ was calculated by two different methods. The obtained results are gathered in Fig. 9. As can be seen, t_S (calculated from ionic radii) and t_{BV} (calculated using SPUDS software [16]) dependences on composition are not parallel, nevertheless, both dependences are increasing with x. Using precise structural data obtained for $La_{0.4}Sr_{0.6}Co_{0.2}Fe_{0.8}O_3$ it was possible to calculate the actual value of t_g [30], which is equal to 0.9985, for $La_{0.2}Sr_{0.8}$-$Co_{0.2}Fe_{0.8}O_3$ composition this tolerance factor is defined to be 1.

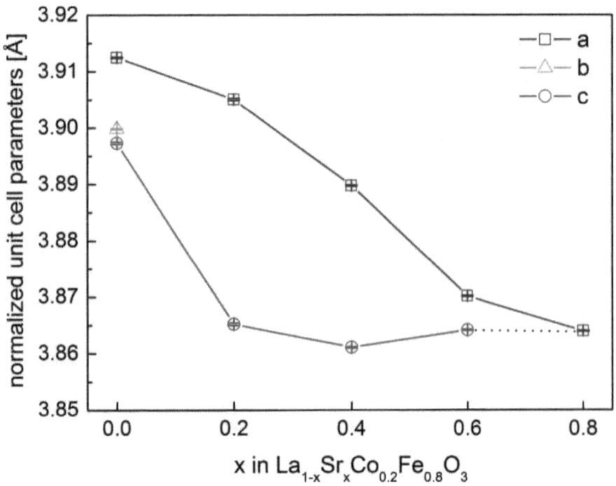

Fig. 7 Dependence of normalized unit cell parameters for $La_{1-x}Sr_xCo_{0.2}Fe_{0.8}O_3$ $(0 \leq x \leq 0.8)$ as a function of chemical composition

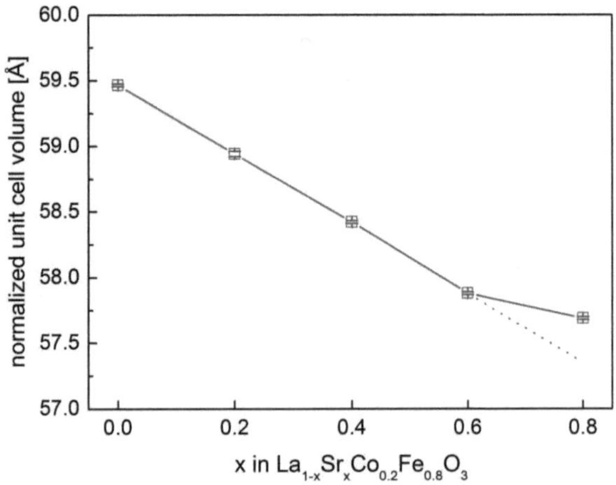

Fig. 8 Dependence of normalized unit cell volume for $La_{1-x}Sr_xCo_{0.2}Fe_{0.8}O_3$ $(0 \leq x \leq 0.8)$ as a function of chemical composition

6 Oxygen Non-stoichiometry

Thermogravimetric (TG) analysis was performed in air for as prepared $La_{1-x}Sr_xCo_{0.2}Fe_{0.8}O_3$ materials and is presented in Fig. 10. Due to the preparation technique an initial small decrease of mass of the samples may be attributed to the oxidation of residual carbon. Moreover, samples held for a prolonged time in air may have significant amount of CO_2 and other gases adsorbed. In any case, the second run in artificial air showed lack of these initial changes. As can be seen in Fig. 10, for $La_{0.4}Sr_{0.6}Co_{0.2}Fe_{0.8}O_3$ and $La_{0.2}Sr_{0.8}Co_{0.2}Fe_{0.8}O_3$ perovskites, above ~200 °C an increase of mass was recorded, with the maximum around 300 °C. In case of $La_{0.2}Sr_{0.8}Co_{0.2}Fe_{0.8}O_3$ the maximum mass at 330 °C was equal 100.26 %, while at 1000 °C it dropped down to 98.11 %. The observed

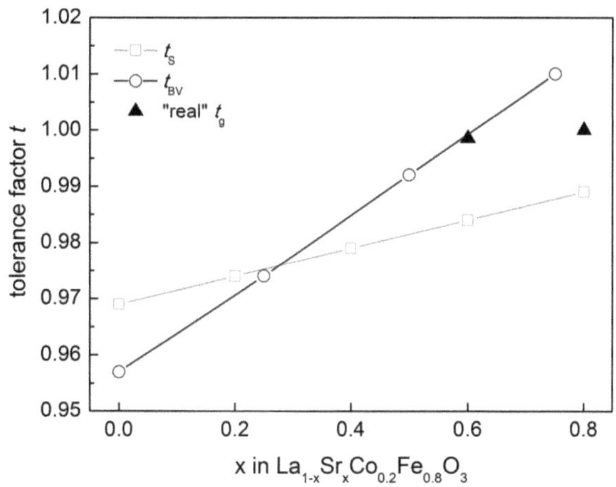

Fig. 9 Dependence of tolerance factor for $La_{1-x}Sr_xCo_{0.2}Fe_{0.8}O_3$ ($0 \leq x \leq 0.8$) as a function of chemical composition: t_S values were calculated from ionic radii, t_{BV} using SPUDS software (for slightly different composition $La_{1-x}Sr_xCo_{0.25}Fe_{0.75}O_3$ due to software limitation), t_g value for $La_{0.4}Sr_{0.6}Co_{0.2}Fe_{0.8}O_3$ was obtained as the geometric average of the actual interatomic distances

phenomena at high temperatures are related to incorporation/release of oxygen to/from the crystal lattice and disappearance/creation of oxygen vacancies $V_O^{\cdot\cdot}$. In case of $La_{0.2}Sr_{0.8}Co_{0.2}Fe_{0.8}O_3$ oxide the observed strongest dependence of mass(T) imply strong changes of $V_O^{\cdot\cdot}$ concentration. The observed dependences support known tight correlation of the amount of Sr in LSCF and high temperature concentration of the oxygen vacancies.

Additional more complex TG measurements were performed for the same samples, quenched from different temperatures (900 °C or 1000 °C). These results (Fig. 11, exemplary results for $La_{0.2}Sr_{0.8}Co_{0.2}Fe_{0.8}O_{3-\delta}$) yield valuable information on thermodynamic constraints in formation of the oxygen non-stoichiometry defects. The observed mass increase in the vicinity of 300 °C is bigger for samples quenched from higher temperature, and much bigger compared to the starting, not quenched samples. This points to more rapid oxygen uptake by samples frozen from high temperatures, with higher oxygen deficiency. Technologically important information may be derived from these results: when designing cathode materials for IT–SOFC, namely, it is not possible to profit from good catalytic and transport properties related to frozen oxygen non-stoichiometry in quenched samples, which exhibit very fast oxygen uptake, as at around 300 °C quenched in oxygen site vacancies are reoccupied.

On the basis of thermogravimetric measurements (performed in air as well as TG reduction studies) the equilibrium concentration of oxygen vacancies in air as a function of temperature was estimated in the studied perovskites. In Fig. 12 Arrhenius-type plot of such temperature dependence of $V_O^{\cdot\cdot}$ is presented. One can see that with higher concentration of strontium the formation energy of oxygen vacancies significantly decreases (from 1.01 eV for $La_{0.6}Sr_{0.4}Co_{0.2}Fe_{0.8}O_{3-\delta}$ to 0.30 eV for $La_{0.2}Sr_{0.8}Co_{0.2}Fe_{0.8}O_{3-\delta}$). This important conclusion will be discussed later, together with the considerations on the influence of metal transitions on the oxygen non-stoichiometry in LSCF perovskites.

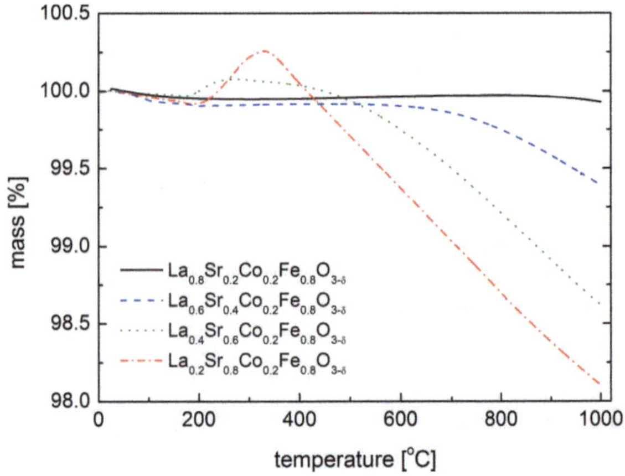

Fig. 10 Thermogravimetric measurements (relative changes of mass as a function of temperature) performed in air for as prepared $La_{1-x}Sr_xCo_{0.2}Fe_{0.8}O_3$ materials. Heating rate 2.5 deg min^{-1}

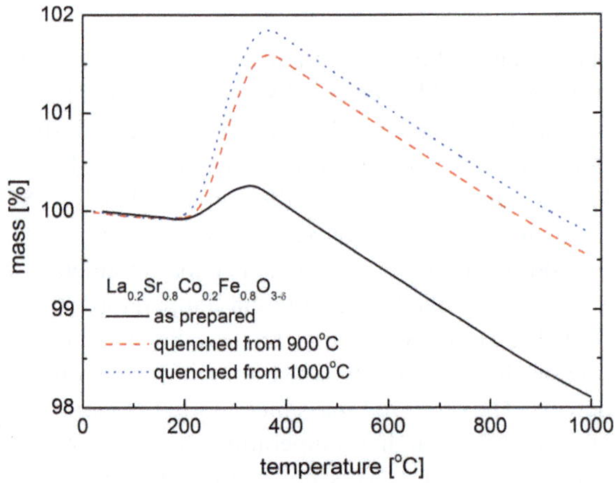

Fig. 11 TG studies of quenched $La_{0.2}Sr_{0.8}Co_{0.2}Fe_{0.8}O_{3-\delta}$. Heating rate 2.5 deg min^{-1} in air. Data from [31]

7 Electrical Properties

Figures 13 and 14 present results of electrical conductivity and thermoelectric power measurements of $La_{0.4}Sr_{0.6}Co_{0.2}Fe_{0.8}O_{3-\delta}$ exemplary sample as a function of oxygen partial pressure obtained in equilibrium conditions for selected temperatures. It is worth noting that the obtained character of these dependences at a constant temperature is uncharacteristic compared to simple non-stoichiometric oxide. For perovskite oxide, in pO_2 range, in which the prevailing defects are oxygen vacancies (Fig. 4, range III), according to theory of defects [23], lowering of oxygen partial pressure leads to the increase in oxygen vacancies concentration according to the relation $[V_O^{\cdot\cdot}] \propto p_{O_2}^{-1/4}$. Electrical conductivity as a rule follows this pressure dependence, however as seen in Fig. 13, the dependence is characterized with positive index n, $\log \sigma \propto p_{O_2}^{1/n}$. This apparent disagreement may be explained by

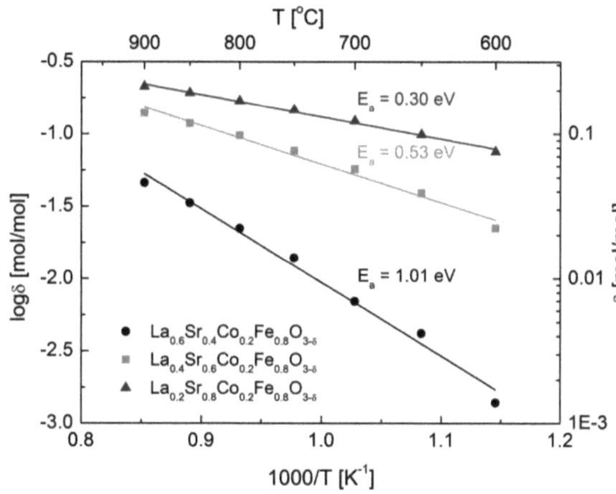

Fig. 12 Equilibrium concentration of oxygen vacancies for $La_{1-x}Sr_xCo_{0.2}Fe_{0.8}O_{3-\delta}$ as a function of temperature in air. E_a—enthalpy of formation of $V_O^{\cdot\cdot}$

electron–hole recombination mechanism: electrons originating from ionization of oxygen vacancies recombine with holes, which at high temperatures are effective charge carriers in LSCF perovskites, in wide range of chemical composition. This effect can be described by the following equations:

$$O_O = V_O^{\cdot\cdot} + 2e^- + 0.5O_2 \uparrow \tag{3}$$

$$2M(Co, Fe)^{4+} + 2e^- = 2M(Co, Fe)^{3+} \tag{4}$$

Furthermore, the analysis of the above oxygen partial pressure dependences (Figs. 13 and 14) points that $La_{0.4}Sr_{0.6}Co_{0.2}Fe_{0.8}O_{3-\delta}$ perovskite should be catalytically active already at 650 °C, because at this temperature the dependence of transport properties on pO_2 is observed. A conclusion may be drawn that this oxide will be useful as cathode material for SOFC working at intermediate temperatures.

Figures 15 and 16 present results of electrical conductivity and thermoelectric power measurements versus temperature for considered LSCF perovskites in wide temperature range up to 950 °C. In the lower temperature range (T < 500 °C) activated type of conductivity was observed for all studied LSCF compositions, with activation energies of the order of 0.05–0.1 eV. In this temperature range for samples with the highest strontium content, $x = 0.6$ and 0.8, values of thermoelectric power are negative and rather small in the order of 10 μV/K. It indicates that the effective transport takes place via electrons and Fermi level is situated at the bottom of $e_{g(\mathrm{Co,Fe})}$ band. Above 500 °C, for these compositions the release of lattice oxygen takes place, as seen on TG measurements (Fig. 10), and thermoelectric power changes its sign to positive (Fig. 16). The observed drop in conductivity and change of thermoelectric power sign is a consequence of oxygen vacancies creation. Oxygen atoms exiting perovskite lattice leave divalent vacancies and electrons, which fill $e_{g(\mathrm{Co,Fe})}$ band and shift position of Fermi level to the top of the band. This explains change of the thermoelectric power sign at high temperatures from negative to positive. For samples with lower concentration

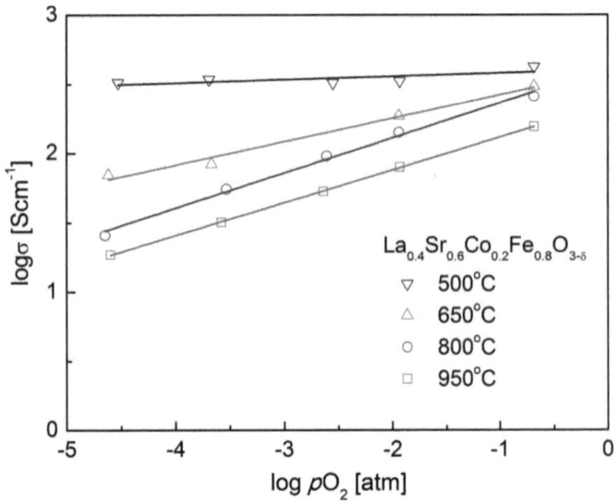

Fig. 13 Equilibrium electrical conductivity of $La_{0.4}Sr_{0.6}Co_{0.2}Fe_{0.8}O_{3-\delta}$ as a function of oxygen partial pressure. Data from [31]

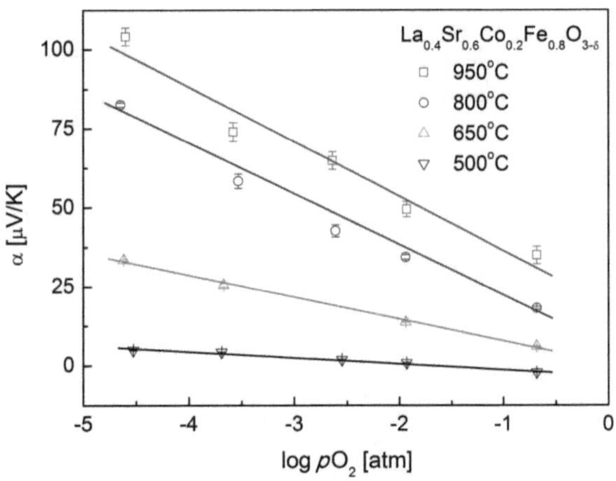

Fig. 14 Equilibrium thermoelectric power of $La_{0.4}Sr_{0.6}Co_{0.2}Fe_{0.8}O_{3-\delta}$ as a function of oxygen partial pressure. Data from [31]

of strontium, $x = 0.2$ and 0.4, Fermi level is situated at the top of $e_{g(Co,Fe)}$ band and holes are the effective charge carriers. In consequence, thermoelectric power presents positive sign in the whole temperature range.

As seen in Fig. 15, in the investigated perovskite oxides, with the increasing oxygen non-stoichiometry electrical conductivity significantly decreases, contrary to typically observed raise of σ with increasing nonstoichiometry. This leads to another conclusion that total conductivity of LSCF perovskite oxides is determined mainly by its electronic component and ionic component, associated with oxygen vacancy transport, is remarkably lower. The temperature at which characteristic maximum of electrical conductivity is observed has important technological meaning: it determines temperature of oxygen vacancies formation, which are catalytic centres for oxygen reduction. With higher strontium content this temperature decreases and higher catalytic activity may be expected.

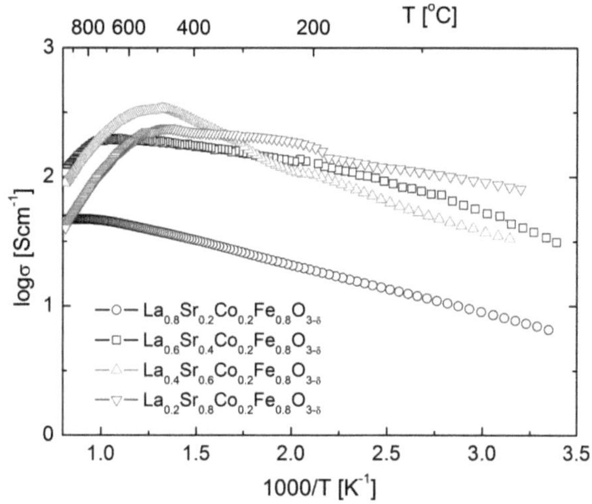

Fig. 15 High temperature electrical conductivity of $La_{1-x}Sr_xCo_{0.2}Fe_{0.8}O_{3-\delta}$ perovskites. Data from [5] and [29]

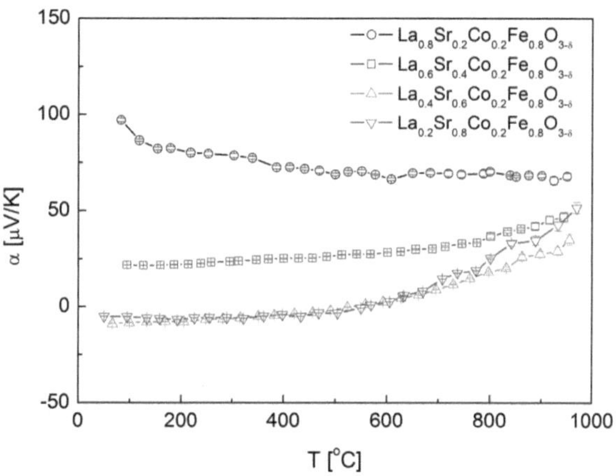

Fig. 16 High temperature thermoelectric power of $La_{1-x}Sr_xCo_{0.2}Fe_{0.8}O_{3-\delta}$ perovskites. Data from [5] and [29]

Results of electrical properties studies enable to propose a qualitative model of electronic structure of the studied LSCF materials (Fig. 17). According to the model, charge transport in considered perovskites takes place in a narrow $e_{g(\mathrm{Co,Fe})}$ band via localized states in the vicinity of Fermi level. The observed activation energy for T < 500 °C is related to activation of motion of electron carriers in $e_{g(\mathrm{Co,Fe})}$ band (Fig. 17a). At higher temperatures, change of electronic carriers concentration due to formation of oxygen vacancies rises Fermi level (Fig. 17b) and masks activated character of their motion. Further, with increasing non-stoichiometry, holes in $e_{g(\mathrm{Co,Fe})}$ band contribute remarkably to conductivity, what leads to change of sign of thermoelectric power from negative to positive. The observed change of α sign upon change of $V_O^{\cdot\cdot}$ concentration leads to conclusion that $e_{g(\mathrm{Co,Fe})}$ band is rather a narrow one.

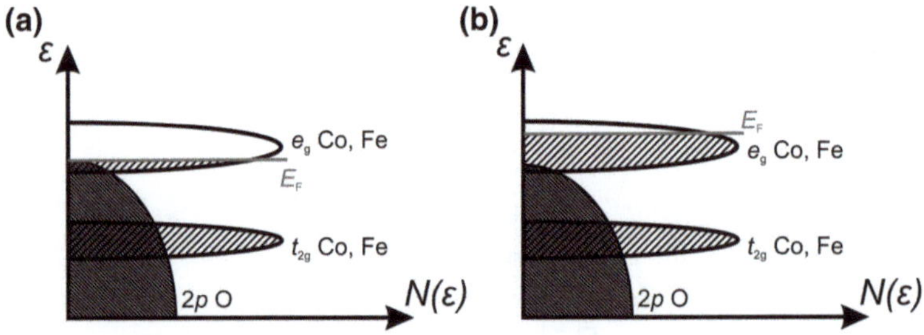

Fig. 17 Qualitative model of electronic structure of LSCF perovskites. Influence of transition metal content in LSCF perovskites on their structural and transport properties

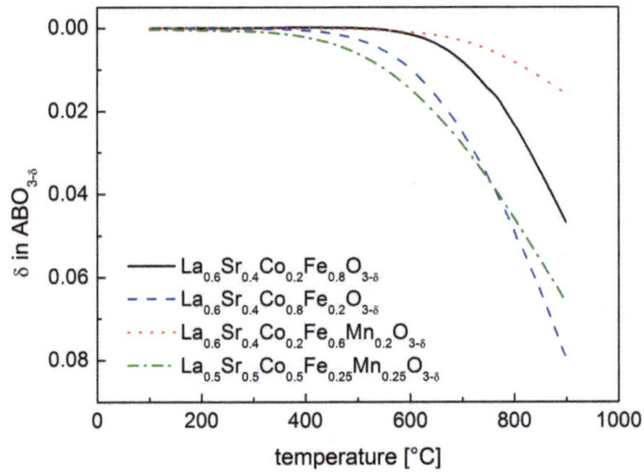

Fig. 18 Equilibrium oxygen nonstoichiometry as a function of temperature for considered perovskite oxides. Measurements performed in air

8 Influence of Transition Metal Content in LSCF Perovskites on their Structural and Transport Properties

8.1 Oxygen Non-stoichiometry

In order to analyze influence of transition metal content on oxygen nonstoichiometry δ in LSCF perovskites, materials with $La_{0.6}Sr_{0.4}Co_{0.2}Fe_{0.8}O_{3-\delta}$, $La_{0.6}Sr_{0.4}Co_{0.8}Fe_{0.2}O_{3-\delta}$, $La_{0.6}Sr_{0.4}Co_{0.2}Fe_{0.6}Mn_{0.2}O_{3-\delta}$ and $La_{0.5}Sr_{0.5}Co_{0.5}$-$Fe_{0.25}Mn_{0.25}O_{3-\delta}$ composition were measured and compared. The performed TG measurements in air allowed equilibrium values of δ as a function of temperature to be determined. The obtained results are gathered in Fig. 18. Depending on the composition, above 400–600 °C, a creation of oxygen vacancies is observed, which follows Eq. 3.

In Fig. 19 the Arrhenius-type plot of $V_O^{\cdot\cdot}$ concentration dependence on temperature is presented. A clear correlation can be seen: for $La_{0.6}Sr_{0.4}Co_{0.8}Fe_{0.2}O_{3-\delta}$ and $La_{0.5}Sr_{0.5}Co_{0.5}Fe_{0.25}Mn_{0.25}O_{3-\delta}$ materials, for which charge compensation is related to creation of a high concentration of oxygen vacancies, calculated

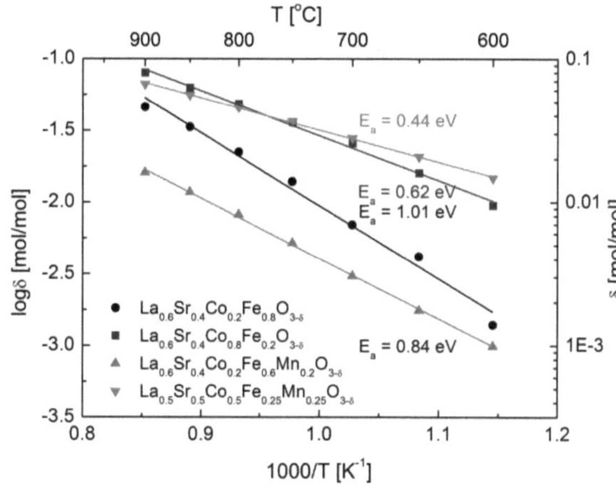

Fig. 19 Equilibrium concentration of oxygen vacancies for considered perovskite oxides as a function of temperature in air. E_a—enthalpy of formation of $V_O^{\cdot\cdot}$

enthalpies of formation are low, respectively 0.62 and 0.44 eV. The above results suggest that both Sr and Co concentration determine level of δ: substitution of La^{3+} by Sr^{2+} cations leads to effective negative charge in perovskite structure and increases average valence of Co and Fe and/or (depending on temperature) causes creation of positively charged $V_O^{\cdot\cdot}$ On the other hand, comparison of $La_{0.6}Sr_{0.4}$-$Co_{0.2}Fe_{0.8}O_{3-\delta}$ and $La_{0.6}Sr_{0.4}Co_{0.8}Fe_{0.2}O_{3-\delta}$ oxides, having identical Sr content, but different Co content, suggests that higher oxygen nonstoichiometry observed for Co-rich material (Fig. 18) is due to the fact that Co^{4+} configuration is energetically less favourable and negative charge compensation, introduced by Sr^{2+}, occurs by increase of iron valence $Fe^{3+} \rightarrow Fe^{4+}$ and creation of oxygen vacancies.

Interesting conclusions may be drawn from comparing behaviour of the Mn-containing perovskites: $La_{0.6}Sr_{0.4}Co_{0.2}Fe_{0.6}Mn_{0.2}O_{3-\delta}$ and $La_{0.5}Sr_{0.5}Co_{0.5}$-$Fe_{0.25}Mn_{0.25}O_{3-\delta}$. As manganese cations are in Mn^{4+} state, for the first perovskite, negative charge compensation occurs by increase of iron valence (Fe^{4+}), and therefore oxygen non-stoichiometry is small. For $La_{0.5}Sr_{0.5}Co_{0.5}Fe_{0.25}Mn_{0.25}O_{3-\delta}$ material, already all iron is in Fe^{4+} state and charge compensation is realized by increase of concentration of oxygen vacancies only. This analysis confirms that the +4 oxidation state for cobalt cations is unfavourable in perovskite structure. It is also in agreement with zero value of octahedral weak field stabilization energy for Co^{4+}. In such field Fe^{4+} and Fe^{5+} cations possess high value of stabilization energy, respectively $\Delta_{oct} = 0.6$ and 1.2. This suggests the possibility of a charge disproportionation reaction, which can be described by equation:

$$2Fe^{4+} \rightarrow Fe^{5+} + Fe^{4+} \qquad (5)$$

Mössbauer spectroscopy studies performed on LSCF materials confirmed crystal field theory conclusions. Results of such measurements, for the example of $La_{0.4}Sr_{0.6}Co_{0.2}Fe_{0.8}O_3$ were reported in [32]. For this material at room temperature a lack of magnetic ordering was observed with only one absorption line visible with relatively narrow FWHM = 0.24 mm/s. At 77 K studied material was magnetically

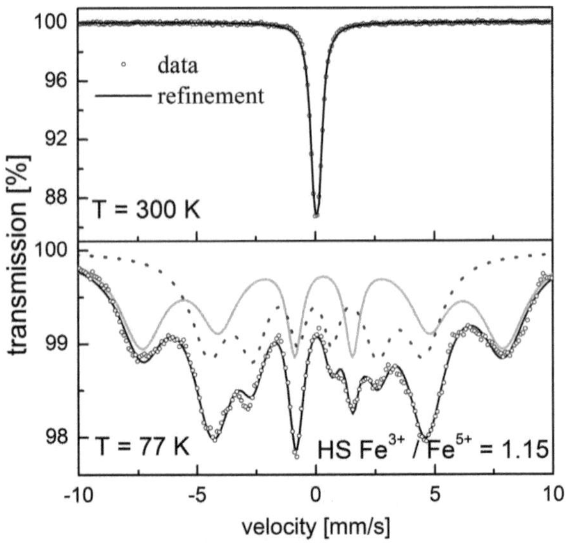

Fig. 20 Mössbauer spectra of $La_{0.4}Sr_{0.6}Co_{0.2}Fe_{0.8}O_3$ at 300 and 77 K. Data from [32]

ordered and the measured spectrum seems complicated (Fig. 20). This spectrum may be, however, fitted assuming only two magnetic components (sextets) arising from Fe^{3+} and Fe^{5+} having ratio equal 1.15. Evaluated hyperfine fields equal correspondingly 47 and 28 T and can be directly related to high-spin configuration of Fe^{3+} ($S_1 = 5/2$) and Fe^{5+} ($S_2 = 3/2$). Assuming that the value of hyperfine field is directly related to spin: $S_1/S_2 = 5/2:3/2 = 1.67$, which is in almost perfect agreement with: $47/28 = 1.68$. Additionally, measured values of isomeric shifts are typical of these valence states of iron. Performed measurements suggest instability of Fe^{4+} configuration in considered perovskite structures. Mössbauer spectra refinement gives Fe^{5+} concentration equal to 0.37 mol/mol for a total 0.8 mol/mol, which is slightly higher than value derived from stoichiometric calculations: 0.3 mol/mol. The above results imply that good catalytic properties of considered LSCF perovskites are also related to the existence of Fe^{5+}/Fe^{3+} redox pair.

8.2 Electrical Properties

Results of measurements of high temperature electrical conductivity and thermoelectric power for $La_{0.6}Sr_{0.4}Co_{0.8}Fe_{0.2}O_{3-\delta}$, $La_{0.6}Sr_{0.4}Co_{0.2}Fe_{0.6}Mn_{0.2}O_{3-\delta}$, $La_{0.5}Sr_{0.5}Co_{0.5}Fe_{0.25}Mn_{0.25}O_{3-\delta}$ and for comparison $La_{0.6}Sr_{0.4}Co_{0.2}Fe_{0.8}O_{3-\delta}$ perovskites are presented in Figs. 21 and 22.

The obtained results are similar to the ones presented for $La_{1-x}Sr_xCo_{0.2}Fe_{0.8}O_{3-\delta}$ oxides in Figs. 15 and 16. Qualitative model discussed above and shown in Fig. 17 fully explains the obtained characteristics. It is worth noting especially high electrical conductivity for $La_{0.6}Sr_{0.4}Co_{0.8}Fe_{0.2}O_{3-\delta}$ material, which reaches 800 S cm^{-1}.

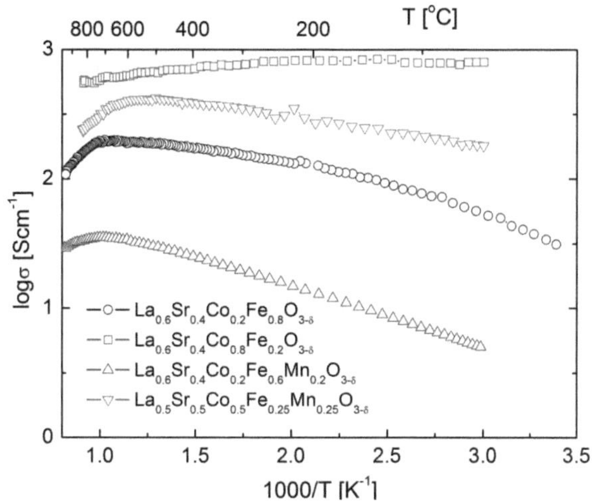

Fig. 21 High temperature electrical conductivity of considered perovskites

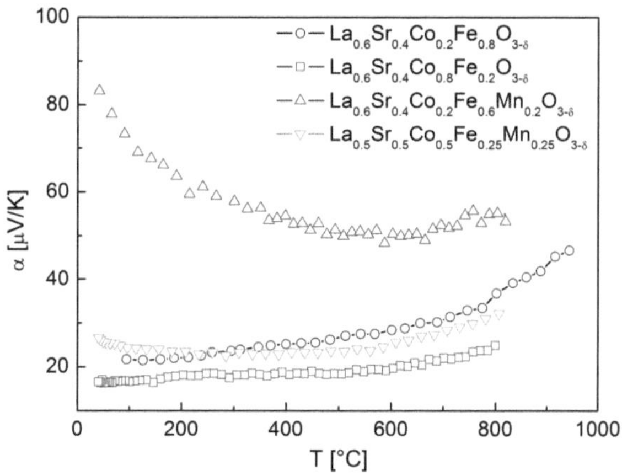

Fig. 22 High temperature thermoelectric power of considered perovskites

9 Properties of Highly Reduced Perovskites

Interesting behavior was found in the case of highly reduced perovskites. The reduction process was performed in 0.1 mol. % of H_2 in Ar atmosphere at 900 °C for 6 h. After reduction samples were quenched to room temperature. XRD measurements allowed modification of the crystal structure upon reduction to be determined. Except for $La_{0.5}Sr_{0.5}Co_{0.5}Fe_{0.25}Mn_{0.25}O_{3-\delta}$, in all other studied materials (Table 1) a transformation to cubic $Pm\overline{3}m$ symmetry was observed, clearly visible as a disappearance of (104) reflection as presented in Fig. 23. It is also worth noting a significant increase of unit cell volume, which may be associated with reactions 3 and 4 and, in consequence, with increase of 3*d* metals ionic radius. The magnitude of this

Table 1 Structural parameters before and after reduction of considered perovskites

	$La_{0.6}Sr_{0.4}Co_{0.2}Fe_{0.8}O_{3-\delta}$	$La_{0.6}Sr_{0.4}Co_{0.8}Fe_{0.2}O_{3-\delta}$	$La_{0.6}Sr_{0.4}Co_{0.2}Fe_{0.6}Mn_{0.2}O_{3-\delta}$
Before reduction	a = 5.5026(1) Å	a = 5.4444(1) Å	a = 5.4906(1) Å
$R\overline{3}c$	c = 13.3787(4) Å	c = 13.2515(2) Å	c = 13.3547(2) Å
	V/6 = 58.42(1) Å^3	V/6 = 56.70(1) Å^3	V/6 = 58.12(1) Å^3
After reduction	a = 3.8938(1) Å	a = 3.8582(1) Å	a = 3.8801(2) Å
$Pm\overline{3}m$	V = 59.04(1) Å	V = 57.43(1) Å	V = 58.42(1) Å
	δ = 0.15 mol/mol	δ = 0.18 mol/mol	δ = 0.08 mol/mol

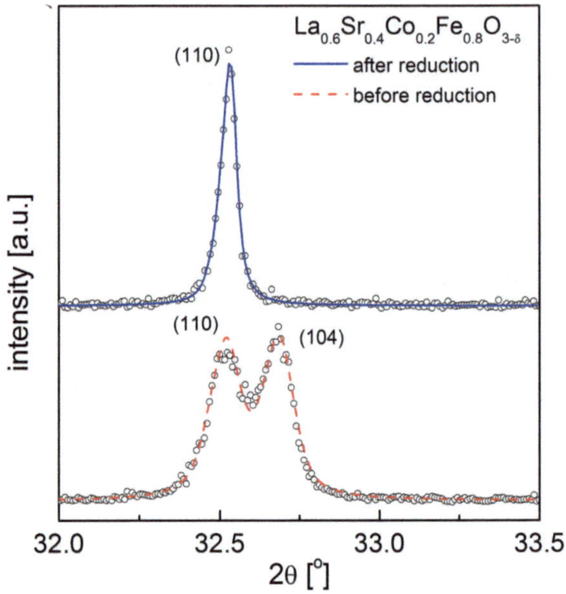

Fig. 23 Close-up of XRD diffraction pattern for $La_{0.6}Sr_{0.4}Co_{0.2}Fe_{0.8}O_{3-\delta}$ before and after reduction

increase is related to the oxygen non-stoichiometry (Fig. 18) and is highest for the $La_{0.6}Sr_{0.4}Co_{0.8}Fe_{0.2}O_{3-\delta}$ sample. $La_{0.5}Sr_{0.5}Co_{0.5}Fe_{0.25}Mn_{0.25}O_{3-\delta}$ perovskite decomposed with precipitation of layered Ruddlesden–Popper type oxide. This behavior is expected for over-reduced samples.

In Fig. 24 examples of thermogravimetric measurements of oxidation of reduced $La_{0.6}Sr_{0.4}Co_{0.2}Fe_{0.8}O_{3-\delta}$ are shown. The measurements were performed in air up to 900 °C with rather slow heating/cooling rate 1°/min. As can be seen in Fig. 24, there is a very strong tendency for oxygen uptake, which appears as low as at 200 °C and proceeds to the point where sample is exactly stoichiometric. For other materials from studied group similar behaviour was registered. Additionally, reduced samples with higher initial non-stoichiometry were bonding oxygen faster what could be seen as steeper mass change on heating. This effect can be also seen in case of LSCF samples quenched in air (Figs. 10 and 11). Exact mechanism of this process is yet unknown. Due to the fact that it occurs at very low temperatures and because kinetics of it are rather very fast it seems very unlikely that it proceeds

Fig. 24 Changes of oxygen content in reduced $La_{0.6}Sr_{0.4}Co_{0.2}Fe_{0.8}O_{3-\delta}$ upon heating in air

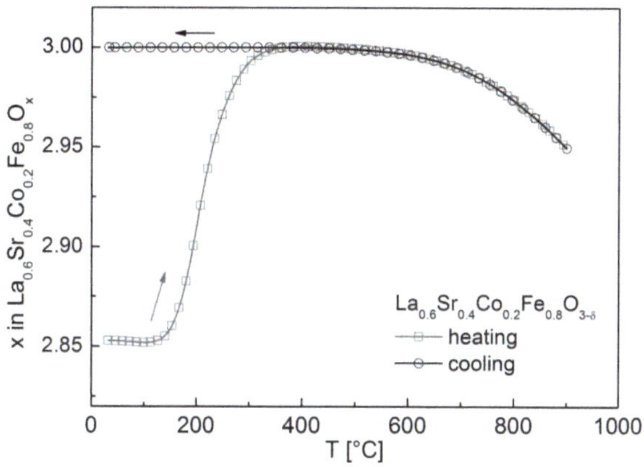

Fig. 25 Changes of electrical conductivity of reduced $La_{0.6}Sr_{0.4}Co_{0.2}Fe_{0.8}O_{3-\delta}$ upon heating in air

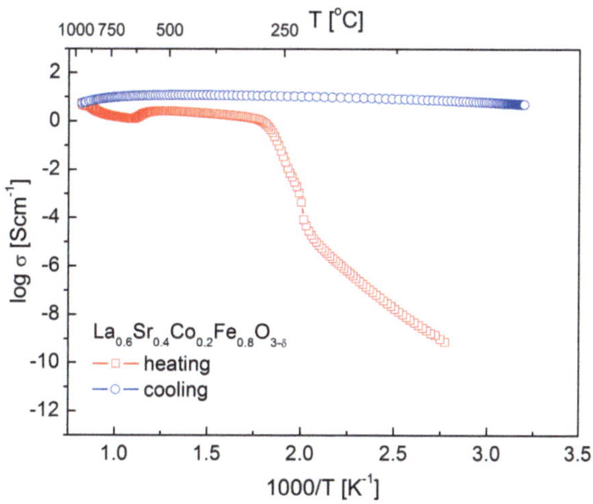

as a typical non-stoichiometry related reaction with oxygen anions occupying oxygen vacancy sites (reaction 3 backwards).

On the basis of electrical conductivity, thermoelectric power and thermogravimetric measurements, performed in thermodynamic equilibrium conditions it is known that oxygen diffusion in crystal lattice is possible only above 400–600 °C, depending on chemical composition of perovskite. In Fig. 24 above 600 °C oxygen removal from the sample can be observed. The first cycle irreversible change of the oxygen content has its counterpart on qualitatively identical characteristic of electrical conductivity (Fig. 25). Because σ is related to changes of transition metal valence it implies that the observed effect applies to whole volume of the sample. Possibly this rapid oxygen uptake, taking place at around 200 °C can be described as oxygen intercalation process into metastable structure of reduced perovskite. High temperature treatment of such structure brings back the high temperature equilibrium one. This quite untypical and interesting behavior is currently under detailed study.

10 Chemical Stability in Relation to Ceria-Based Electrolyte

Presented perovskite materials were tested for their chemical compatibility with $Ce_{0.85}Gd_{0.15}O_{1.925}$ solid-state electrolyte. In all cases a very good mutual stability was confirmed by XRD measurements performed on 50:50 mixtures consisting of perovskite and electrolyte powders, which were heated in 800–1000 °C range for up to 100 h. No secondary phases were observed and only a minimal changes of unite cell parameters, both for perovskite oxide and ceria electrolyte, less than about 0.1 % were recorded. These changes, which depended on temperature and heating time, might be attributed to a limited diffusion process occurring between both phases. Summarizing, all studied materials might be used in fuel cells, in which ceria-based electrolyte is used.

References

1. J.W. Fergus, R. Hui, X. Li, D.P. Wilkinson, J. Zhang (eds.), *Solid Oxide Fuel Cells Materials Properties and Performance* (CRC Press, Boca Raton, 2009)
2. S.C. Singhal, K. Kendall (eds.), *High Temperature Solid Oxide Fuel Cells: Fundamentals Design and Applications* (Elsevier, Oxford, 2003)
3. T. Ishihara (ed.), *Perovskite Oxide for Solid Oxide Fuel Cells* (Springer, Berlin, 2009)
4. R.H. Mitchell, *Perovskites Modern and Ancient* (Almaz Press Inc., Canada, 2002)
5. J. Molenda, K. Świerczek, W. Zając, J. Power Sources **173**, 657 (2007)
6. E. Ivers-Tiffée, A. Weber, D. Herbstritt, J. Eur. Ceram. Soc. **21**, 1805 (2001)
7. E.V. Tsipis, V.V. Kharton, J. Solid State Electrochem. **12**, 1039 (2008)
8. E.V. Tsipis, V.V. Kharton, J. Solid State Electrochem. **12**, 1367 (2008)
9. J. Richter, P. Holtappels, T. Graule, T. Nakamura, L.J. Gauckler, Monatsh. Chem. **140**, 985 (2009)
10. S.J. Skinner, Fuel Cells Bull. **339**, 6 (2001)
11. C. Sun, R. Hui, J. Roller, J. Solid State Electrochem. **14**, 1125 (2010)
12. W. Zhou, R. Ran, Z. Shao, J. Power Sources **192**, 231 (2009)
13. S.P. Jiang, J. Mater. Sci. **43**, 6799 (2008)
14. I.D. Brown, Phys. Chem. Minerals **15**, 30 (1987)
15. N.W. Thomas, Acta Cryst. **B54**, 585 (1998)
16. M.W. Lufaso, P.M. Woodward, Acta Cryst. **B57**, 725 (2001)
17. A.M. Glazer, Acta Cryst. **B28**, 3384 (1972)
18. C.J. Howard, H.T. Stokes, Acta Cryst. **B54**, 782 (1998)
19. K.S. Aleksandrov, Ferroelectrics **14**, 801 (1976)
20. P. Huang, A. Petric, J. Electrochem. Soc. **143**, 1644 (1996)
21. J. Mizusaki, Y. Yonemura, H. Kamata, K. Ohyama, N. Mori, H. Takai, H. Tagawa, M Dokiya, K. Naraya, T. Sasamoto, H. Inaba, T. Hashimoto, Solid State Ionics **132**, 167 (2000)
22. J.A.M. van Roosmalen, E.H.P. Cordfunke, J. Solid State Chem. **110**, 109 (1994)
23. F.W. Poulsen, J.J. Bentzen, T. Jacobsen, E. Skou, M.J.L. Østergård (eds.), in *Proceedings of the 14th Risø International Symposium on Materials Science*, Risø National Laboratory, Roskilde, 1993
24. K. Świerczek, B. Dabrowski, L. Suescun, S. Kolesnik, J. Solid State Chem. **182**, 280 (2009)
25. A. Mineshige, M. Inaba, T. Yao, Z. Ogumi, K. Kikuchi, M. Kawase, J. Solid State Chem. **121**, 423 (1996)
26. L. Karvonen, M. Valkeap, R.-S. Liu, J.-M. Chen, H. Yamauchi, M. Karppinen, Chem. Mater. **22**, 70 (2010)

27. S.E. Dann, D.B. Currie, M.T. Weller, M.F. Thomas, A.D. Al-Rawwas, J. Solid State Chem. **109**, 134 (1994)
28. J.P. Hodges, S. Short, J.D. Jorgensen, X. Xiong, B. Dabrowski, S.M. Mini, C.W. Kimball, J. Solid State Chem. **151**, 190 (2000)
29. K. Świerczek, M. Gozu, J. Power Sources **173**, 695 (2007)
30. B. Dabrowski, O. Chmaissem, J. Mais, S. Kolesnik, J.D. Jorgensen, S. Short, J. Solid State Chem. **170**, 154 (2003)
31. K. Świerczek, Solid State Ionics **179**, 126 (2008)
32. J. Marzec, J. Power Sources **173**, 671 (2007)

Critical Issues of Metal-Supported Fuel Cell

Yngve Larring and Marie-Laure Fontaine

Abstract Metal-Supported SOFCs (MS-SOFCs), wherein the supporting component of the cell is made of a porous alloy, are referred to as the third generation SOFC operating at temperature down to 500–650 °C. This technology is expected to decrease significantly capital and operational costs, while increasing the lifetime of cells due to lower operating temperature and higher redox stability. The chapter reviews MS-SOFC development with a focus given to main issues affecting the performance and longevity of single cells. It addresses critical issues for selection of alloy materials based on material cost, thermal expansion coefficient, corrosion rate, particle size, and Cr evaporation issues. Protective coatings, cell architecture, and advanced fabrication processes are then presented to illustrate the level of technical refinement currently achieved. Performance of produced MS-SOFCs is finally discussed to pinpoint factors contributing to major electrochemical losses and possible routes for improvement are reported.

1 Introduction

Solid oxide fuel cells (SOFCs) are an important class of environmentally friendly power generation devices. SOFCs can operate on a number of hydrocarbon containing fuels to directly convert chemical energy into electrical power at high efficiency. SOFC technology has been investigated worldwide for many decades. This has resulted in the development of two generations of cell designs with

Y. Larring (✉) · M.-L. Fontaine
SINTEF Materials and Chemistry, Forskningsveien 1,
0374 OSLO, Norway
e-mail: yngve.larring@sintef.no

J. T. S. Irvine and P. Connor (eds.), *Solid Oxide Fuels Cells: Facts and Figures*,
Green Energy and Technology, DOI: 10.1007/978-1-4471-4456-4_4,

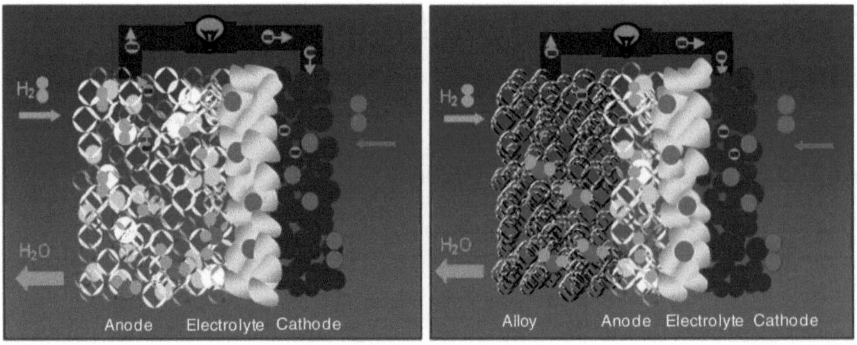

Fig. 1 Scheme of second generation SOFC (anode supported cell) and third generation MS-SOFC (alloy supported cell)

significant progress in cell performance. High cost of materials and cell processing is still hampering commercial deployment of this technology. Furthermore, long-term stability of anode supported SOFC is challenged by limited redox stability of nickel-based anodes. There are many strategies pursued by SOFC developers to overcome these challenges. Among them, introduction of metal support in the cell's architecture has recently offered new possibilities. Metal-Supported SOFCs (MS-SOFCs), wherein the supporting component of the cell is made of a porous alloy, are referred to as the third generation SOFC (see Fig. 1). In this design, the mechanical and the catalytic functions of the fuel electrode are separated: the thick porous alloy provides mechanical strength, electrical connection, and improved heat distribution, while the thin porous functional anode ensures electrocatalytic function.

MS-SOFCs can be operated at lower temperatures than the conventional SOFC, typically down to 500–650 °C. Long- term corrosion problems can thus be alleviated, depending on the type of metal support, fuel quality, and power load. Other advantages of this technology are lower materials cost and easier cells assembling (see Table 1). Besides the resulting decrease in costs, introduction of alloy support in the cell is expected to improve mechanical strength, tolerance to redox cycles, and ability to withstand repeated and rapid thermal cycles. The latter would increase the lifetime and enlarge applications windows of MS-SOFCs. New opportunities for portable applications and in the transport sector may be found, where the demand on robustness is severe and rapid thermal cycling desirable.

There are, however, major challenges to overcome for successful development of MS-SOFC. Among others, Cr poisoning of electrodes, which is often encountered in SOFCs, may also be considerable in MS-SOFCs. Several studies report on fast degradation of cell performance due to interdiffusion of Cr, Fe, and Ni in the metal-support and nickel containing anode. Besides, high polarization of the cathode is also lowering dramatically the overall cell performance. Many strategies are investigated worldwide to tackle these issues, including development of improved alloy compositions, protective coatings against corrosion and metal

Table 1 Advantages of MS-SOFC technology

	Advantages of MS-SOFC
Materials cost	Use of cheaper alloys in stack system; easier management of thermal insulation; ceramic layers are thinner and contain less expensive electrochemically active materials
Processing of cells	High ruggedness of cells enables fast and easier handling during manufacturing
Assembling of cells	Use of conventional metal joining techniques, such as welding or brazing for sealing

element diffusion, advanced manufacturing processes for low temperature fabrication of MS-SOFCs, etc.

This chapter reviews MS-SOFC development with a focus given to the main issues affecting performance and longevity of single cells. It addresses the selection of alloy materials on the basis of their composition, thermal expansion coefficient, costs, grain size, and corrosion rates. Protective coatings, new cell architectures, and advanced fabrication processes are then presented to illustrate the level of technical refinement currently achieved. Performance of produced MS-SOFCs is finally discussed to pinpoint factors contributing to major electrochemical losses and possible routes for improvement are reported.

2 Selection of Alloy for MS-SOFC

2.1 General Criteria

Porous alloy supports in MS-SOFCs must fulfill the following requirements. The alloy should have high electrical conductivity, high mechanical strength, and low corrosion rate. Low resistivity of oxide scale formed in oxidizing conditions during operation is also necessary (see "Future Fuel Cells" and references [1–6]). The metal support should have a thermal expansion coefficient (TEC) reasonably matching other cell elements. One should note that most alloys have TEC increasing with temperature, while ceramic materials have rather constant TEC (provided that no phase change or reduction occurs in the operating temperature range). few alloys have thermal expansion characteristics identical to those of well-known electrolyte materials (e.g., yttria stabilized zirconia YSZ and doped ceria) from room temperature to MS-SOFC operating temperatures. Alloy selection should thus be targeted toward specific application: different alloys might be used for long-term stationary application and auxiliary power units (APU), where cycling is required.

Metals properties are determined by their nominal composition and use of additives. Figure 2 shows the phase diagram of the Cr-Ni-Fe system. Reported phases have been evaluated by Yang et al. [7] and their properties are summarized

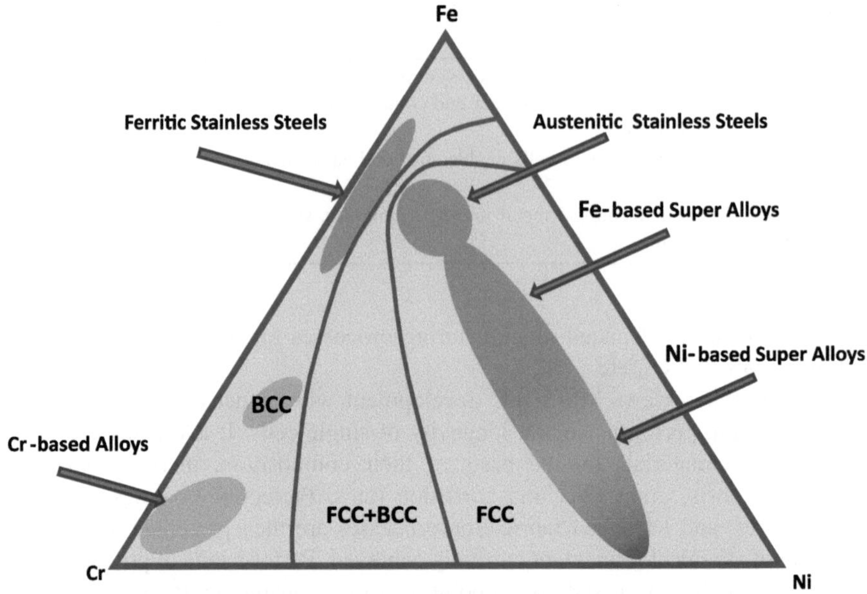

Fig. 2 Schematic phase diagram of Cr–Fe–Ni system [1, 7]

in Table 2. Properties of other metals currently investigated for MS-SOFC application are also reported in this table.

Ni and Ni–Fe supports are investigated as potential support candidates, as their established manufacturing technology alleviates considerably cell fabrication constraints. These materials have, however, poor redox resistance, low oxidation resistance, and relatively low mechanical strength. They are also more susceptible to coking under lower steam to CH_4 ratio and are expensive. Ferritic stainless steels (FSS) have an average TEC matching most of ceramic electrolytes investigated for SOFCs. They have good oxidation resistance, low cost, and their technology of production and forming is well developed. The drawback of FSS lies in their low mechanical strength. Austenitic SS have good oxidation resistance, high mechanical strength, good manufacturability, and low cost. A major drawback of these alloys is their high TEC. Ni–Cr containing alloys have high mechanical strength, excellent oxidation resistance, and good manufacturability. Their thermal expansion is, however, rather large compared to known electrolyte materials and they are expensive.

For these reasons, most SOFC developers favor ferritic stainless steels for interconnect materials. FSS steels also appear as the alloys of choice for MS-SOFC supports and will thus be discussed in detail in this chapter. For optimizing performance and stability of FSS candidates, two additional parameters should be considered, namely the particle size and the oxidation rate constant of the alloys, as discussed below.

Table 2 Comparison of key properties of selected metals [1, 7, 39–41]

Alloys	Main structure	TEC (RT-800 °C) [a]10^{-6} K^{-1}	Oxidation resistance	Mechanical strength	Manufacturability	Cost
Cr based	BCC	11.0–12.5	Good	High	Difficult	Very expensive
Ferritic SS	BCC	11.5–14.0	Good	Low	Fairly readily	Cheap
Austenitic SS	FCC	18.0–20.0	Good	Fairly high	Readily	Cheap
Fe based super alloys	FCC	15.0–20.0	Good	High	Readily	Fairly cheap
Ni based super alloys	FCC	14.0–19.0	Good	High	Readily	Expensive
Ni	FCC	16.0–17.0	Low	Low	Readily	Expensive
Ni–Fe (1:1)	FCC	13.0–14.0	Low	Low	Readily	Fairly expensive

[a] average between 25 and 900 °C

2.2 *Effect of Alloy Particle Size*

Particle size is a crucial parameter for manufacturability. In traditional powder metallurgy processes, large particles are favorable for forming porous interconnected networks enabling fast gas diffusion through the supports. Introduction of pyrolyzable fillers during processing also enables creation of connected pores and to some extent, alleviates the constraint on particle size selection. As the alloy will serve as mechanical support, smooth surface with small pores is desirable for further coating of the other functional layers. In this case, small alloy particles are desired.

Besides manufacturability issues, performance and stability of ferritic alloys are also dependent on their particle size. The particles should be large enough to provide sufficient Cr for forming a continuous Cr_2O_3 surface layer, while retaining sufficient Cr in the bulk [8].

It is possible to estimate a sensible particle size for an ideal support microstructure consisting of spherical spheres. Let us consider an alloy containing 22 wt % Cr and a critical thickness of the Cr_2O_3 scale around 1 μm, which is assumed to fully cover the sphere surface (± 0.5 μm of the original sphere radius). It is assumed that the necks between metal particles should have a thickness of least 4–5 μm to avoid complete corrosion of these conduction paths. Cr concentration needed to form this scale is calculated using the density of Cr_2O_3 scale. The remaining Cr concentration in the bulk can then be deduced, as illustrated in (Fig. 3).

Figure 3 shows that alloys with particle size below 10 μm do not contain sufficient Cr to form an oxide scale of 1 μm. Using a particle size of around 50 μm, Cr depletion from the bulk is less than 20 % yielding Cr content in the bulk above 17 %. This threshold enables to maintain the ferritic structure of the alloy

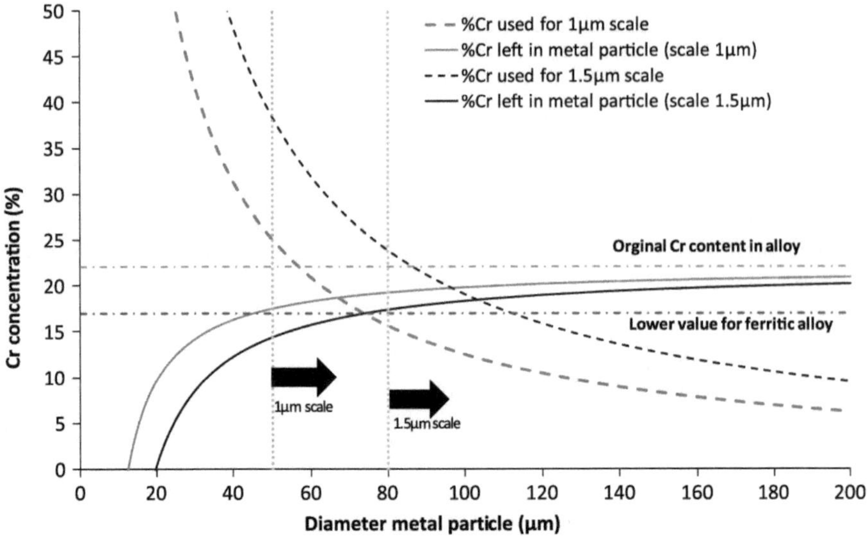

Fig. 3 Evolution of Cr concentration (in weight %) in the bulk and in the oxide scales (thickness: 1 and 1.5 μm) as a function of particle size. Selected alloy has 22 wt % Cr

and to avoid phase change, evolution of TEC, and to improve the alloy's ability to repair cracks formed in the oxide scale.

Similar calculations can be done for an oxide scale thickness of 1.5 μm. In this case, larger grains of around 80 μm size should be selected to reach a minimum of 17 wt % Cr in the bulk. It should be noted that chromia scale is the limiting conductive pathway between the functional anode and the power leads. Therefore, increasing the thickness of the oxide scale would contribute to increase cell resistance, and ultimately, to spallation of the layer.

2.3 Effect of Oxidation Rate Constant

The oxidation rate of the alloy must be low to limit the oxide growth, even if a protective coating is to be applied (see Sect. 3). This decreases the risks of failure if the protective coating is not continuous throughout the support. Growing of the oxide scale during operation may indeed restrict gas transport through the support. Besides, metallic bridges between the alloy grains should be maintained to ensure continuous electrical and thermal pathways, as illustrated in Fig. 4.

The thickness of the oxide scale can be calculated as a function of time, according to the parabolic growth rate law:

$$L^2 = \frac{K_P t}{(\rho\theta)^2} \tag{1}$$

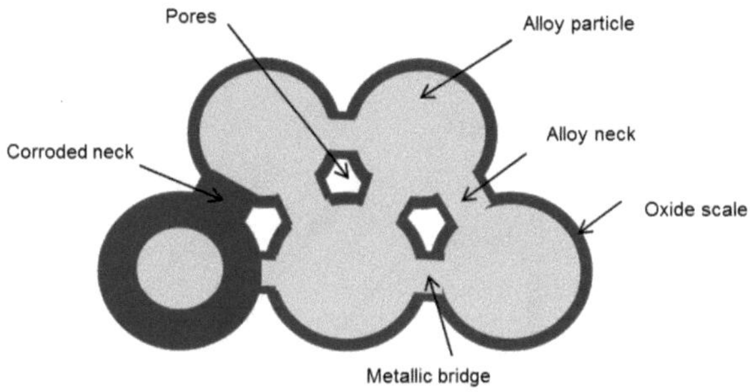

Fig. 4 Schematic ideal microstructure in cross-section view of porous supports of MS-SOFC showing metallic bridges between grains (necks) where the oxide scale Cr_2O_3 (in *blue*) does not disrupt the necks between the alloy particles. This enables fast thermal and electrical conductivity. Thick oxide scale contributes to corrode completely the necks and is thus not favorable

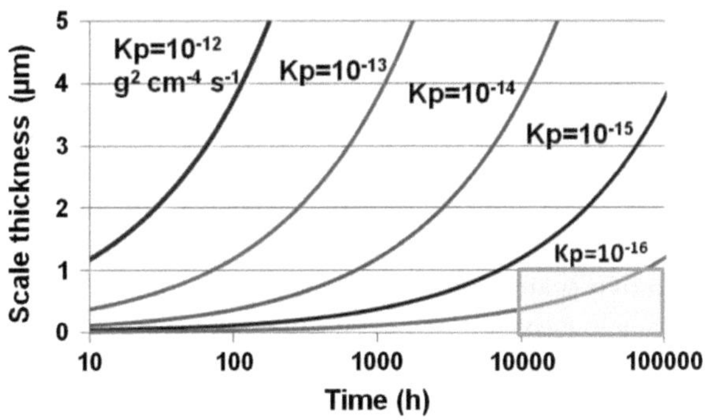

Fig. 5 Expected scale thickness as a function of oxidation time for various oxidation growth rate constants. If one aims at scale thickness below 1 µm after 10 kh, a parabolic rate constant less than 10^{-15} $g^2\ cm^{-4}s^{-1}$ is required (*grey* box) [9, 10]

Where L is the oxide scale thickness (in cm), k_p is the parabolic growth rate constant referring to as the oxygen mass uptake (in $g^2\ cm^{-4}\ s^{-1}$), t is time (in s), ρ is the density of the scale (in g cm^{-3}), and θ is the weight fraction of oxygen in the scale. A plot of the scale thickness as a function of the oxidation time for various parabolic growth rate constants is shown in Fig. 5.

Let us now consider a maximum acceptable scale thickness of 1 µm after 10 kh operation. This thickness is expected to avoid spallation of the oxide scale and to prevent corrosion of the metallic conduction paths between the grains. As shown in Fig. 5, the alloy support should have an oxidation rate constant lower than $10^{-15}\ g^2\ cm^{-4}s^{-1}$. The same plot can also be used to conversely predict the expected lifetime of the support by knowing the parabolic rate constant of the alloy. One should notice that these calculations are based on an ideal parabolic oxidation behavior of the alloy. Some deviations may be expected due to Cr evaporation in air from the scale, which is enhanced in the presence of steam. This linear evaporation term is not introduced in Eq. (1).

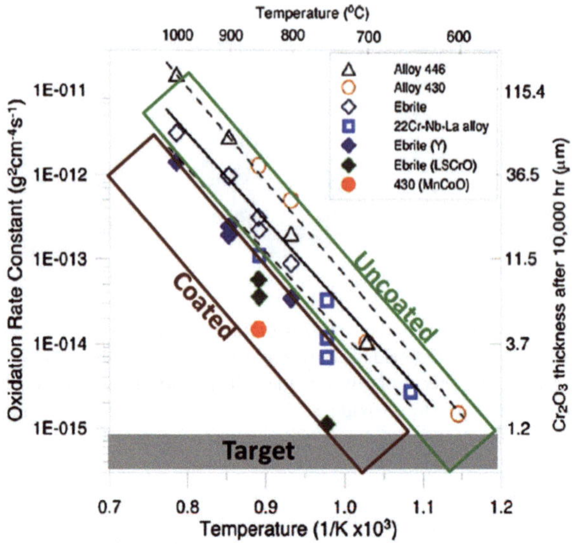

Fig. 6 Measured oxidation rate constants for selected uncoated and coated FSS [9, 10]. The *grey* square referred to as acceptable values set for the oxide scale thickness and parabolic oxidation rate constants after 10 kh operation. Reproduced here with kind permission from © The Electrochemical Society 2009

Figure 6 shows the parabolic oxidation rate constants for a variety of coated and uncoated FSS measured in oxidizing condition as a function of temperature. These alloys are specifically designed for high temperature application, and contain more than 17 wt % Cr with minor additions of Ti, Ce, Y, Nb, Ta, and Mo, to improve scale adhesion and to reduce oxidation rate. Accordingly, these alloys have a low oxidation rate constant between 700 and 850 °C in the range 1×10^{-14}–5×10^{-14} g^2 $cm^{-4}s^{-1}$. Deposition of a protective layer on Ebrite alloys decreases significantly the oxidation rate, as also observed for other alloys [4–6, 9, 10].

Based on these experimental data, setting an oxide scale around 1 μm after 10 kh operation for an alloy with a parabolic oxidation rate constant around 10^{-15} g^2 $cm^{-4}s^{-1}$ requires coated alloys for temperature above 750 °C. There are a few alloy candidates fulfilling all requirements to operate the cell at temperature below 700 °C, but it is anticipated that continuous progress in material design will enable to improve oxidation resistance of alloys.

2.4 Reported Performance of Selected Alloys

As discussed above, there are limited alloys with Cr content in the range of 17–30 wt % forming stable Cr_2O_3 layers and exhibiting low parabolic growth rate for low temperature operation. In order to improve properties of alloys, several additives are introduced in the metal composition. For instance, addition of rare earth elements in alloys allows improving oxide scale adherence, while reducing Cr transport through grain boundaries. Introduction of Ti and Mn in ferritic alloys is shown to improve conductivity of oxide scale. Additionally, introduction of Mn

Table 3 Nominal compositions and TEC of alloys (extracted from commercial data sheets)

Alloy	Composition (wt %)	Additives (wt %)	TEC[a] (ppm/K)
Hastelloy X	22Cr18Fe47Ni	9Mo0.6W1Si	16
SS430	17Cr-Bal.Fe	1(Mn,Si)0.3(P,S)	11.4
Sandvik sanergy	22Cr-Bal.Fe	1Mo0.5Mn0.3Si0.75Nb	12.0
Crofer 22APU	20-24Cr-Bal.Fe	0.5(Si, Cu, Al)0.2(Ti, La)	12.5
ITM	26Cr-Bal. Fe	Mo, Mn, Ti, Y2O3	11.3

[a] average between 25 and 900 °C

contributes to the formation of an additional protective layer $(Cr,Mn)_3O_4$, provided that Mn content is sufficient (ThyssenKrupp VDM GmbH, Crofer 22APU alloy and Plansee SE, ITM alloys). Table 3 reports compositions and TEC of selected alloys commonly investigated as interconnects of SOFCs. Some of these alloys are being tested as metal supports for MS-SOFCs (see ref. in Table 5).

Work on Crofer22APU alloy, which is one of the best alloy candidates for interconnects of SOFCs, indicates that critical oxidation of the porous support takes place after short operation time under oxidizing condition [8, 11]. An increased lifetime can, however, be obtained by appropriate selection of grain size. This failure for porous MS is expected to happen to other alloys traditionally used as dense interconnects. The increased surface area compared to bulk volume in porous metal supports creates new challenges for forming a continuous thin Cr_2O_3 layer. Failure might not only be due to depletion of Cr in the bulk, but can also result from poor oxide scale adherence. These alloys might require addition of rare earth elements to improve scale adherence and to restrict Cr diffusion [12].

3 Protective Coatings for MS-SOFC

Besides corrosion issues of metal substrates, the use of alloys as mechanical supports of the cells is subject to interdiffusion of iron, chromium, and nickel between ferritic steel and nickel-containing anodes during cells fabrication and operation. Diffusion of nickel into FSS substrates may cause austenitization of steels, which would result in TEC mismatch with other cell components. Diffusion of iron and chromium into Ni-based anodes may cause formation of oxide scales on nickel particles. This would result in fast degradation of cell performance during operation, as the electrochemically active surface is passivated. In order to overcome these issues, one possibility investigated by MS-SOFC developers is to use protective coatings [1–6, 13].

Extensive work has been dedicated to the development of protective coatings for SOFC interconnects (see "Future Fuel Cells"). Mitigation of Cr evaporation and excessive growth of oxide scale can be achieved with the use of a protective layer and a contact layer deposited on the dense alloy. The protective layer is obtained with the deposition of Co-Mn oxides to form $(Mn,Co,Cr)_3O_4$ spinel layer [4–6].

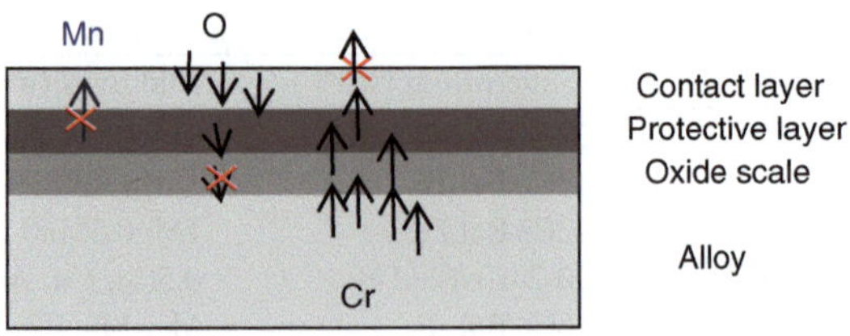

Fig. 7 Schematic representation of the effects of the protective layer, e.g., $(Cr,Mn)_3O_4$ spinel, and the contact layer on ions diffusion/evaporation for SOFC interconnects

Spinel layer shows high performance in preventing oxygen inward diffusion and Cr outward diffusion and exhibits good corrosion resistance and low electrical resistance. As both Cr and Mn are to some degree volatile, an additional contact layer of perovskite material (e.g., LSM or LSC) is used to reduce Cr evaporation (see Fig. 7) [1, 10]. Several techniques are investigated to form these layers, such as slurry coating, aerosol deposition, screen-printing, PVD, electroplating, etc.

It is challenging to successfully apply this strategy to MS-SOFCs as porous alloy supports have high surface area compared to bulk volume (see Fig. 8a). Fabrication of fully dense and uniform coatings throughout the porous metal supports raises strong technical issues.

Another strategy is to make use of electrically conducting porous diffusion barrier layers (DBLs) coated on the surface of the alloy in order to suppress metal element diffusion between the electrode and the substrate (Fig. 8b). Manufacturability of these layers is easier to achieve. The challenges are to make the layers porous enough to ensure gas transport from the metal substrate towards the anode and to cover all contact points between electrode particles and substrate particles for efficient inhibition of mass transport [14, 15].

DLR is developing DBLs of perovskite materials, such as doped $LaCrO_3$-type and $LaMnO_3$-type, by atmospheric plasma spray (APS), vacuum plasma spray (VPS), or pulse laser deposition (PLD) [16–21]. Using doped $LaCrO_3$ layer deposited by VPS on FSS alloy [15], they showed that after 230 h of cell operation, both Ni in the substrate and Fe in the anode replace less than 0.1 % of the primary metal atoms. This is about hundred times less than the 17 % of Ni atoms in grain substituted by Fe and Cr in an anode of a cell prepared without DBL.

Brandner et al. [11] have tested two DBLS based on doped and undoped CeO_2 on dense Crofer22APU supports. They measured the Cr diffusion profile in Ni-based anodes after co-sintering of the alloy/DBL/anode assembly in argon at 1100 °C for 3 h. No interdiffusion was observed when using CeO_2 DBL. The latter was also integrated in a single cell, which was run at 800 °C for 165 h in air/H_2. It was shown that the DBL effectively prevents metal element diffusion between the anode and the steel support (see Fig. 9a).

Use of DBLs is effective to prevent metal element diffusion between Ni-based anode and steel. Nevertheless, as the DBLs are porous, they will unlikely contribute to prevent Cr evaporation and poisoning of electrodes in the long run. Besides, it was also shown that high depletion of Cr content in the bulk of

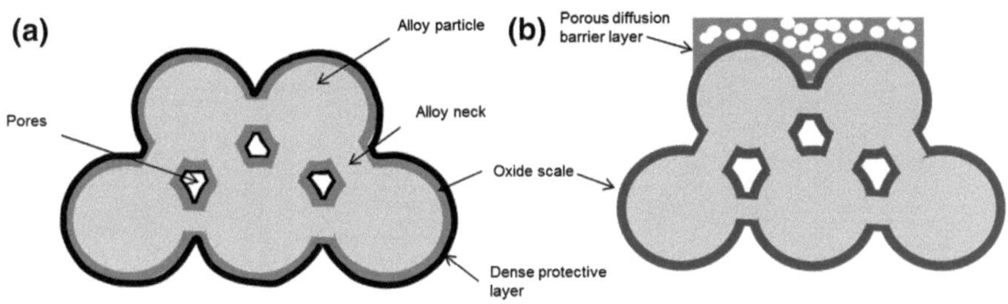

Fig. 8 Schematic representations of porous metallic supports with a protective layer coated on the grains of the alloy (**a**) and a porous electrically conducting diffusion barrier layer coated on the surface of the alloy (**b**)

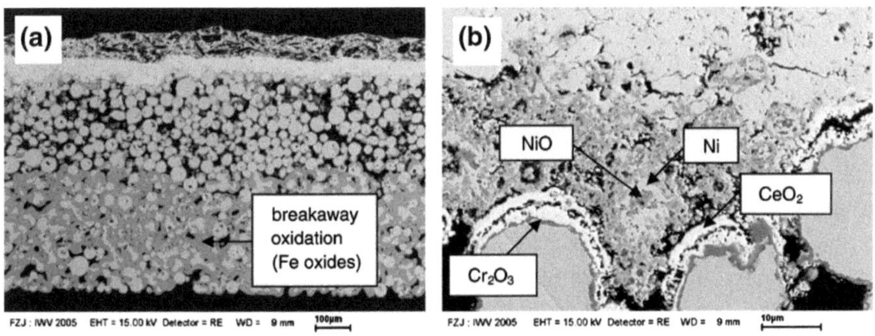

Fig. 9 (**a**), (**b**) MS-SOFC with CeO_2 DBL between a CroFer22APU substrate and a sintered Ni/YSZ anode after 165 h of electrochemical testing at 800 °C [11]. Reproduced here with kind permission from © Elsevier 2008

un-coated alloy particles yielded non-protective breakaway oxidation after operation of the cell (Fig. 9b). Clearly, further alloy design efforts are needed to improve the overall performance of metallic interconnect alloys.

4 Fabrication of MS-SOFC

4.1 Fabrication Requirements

Introduction of stainless steel support in cell's architecture increases dramatically the demands on the fabrication processes. Two critical issues must be considered: (i) in order to avoid alloy oxidation and meltiong during fabrication, annealing at reasonably low temperature in reducing atmosphere is required; (ii) alloy support experiences none or low sintering shrinkage during fabrication. These considerations have drastic consequences on the selection of suitable deposition techniques

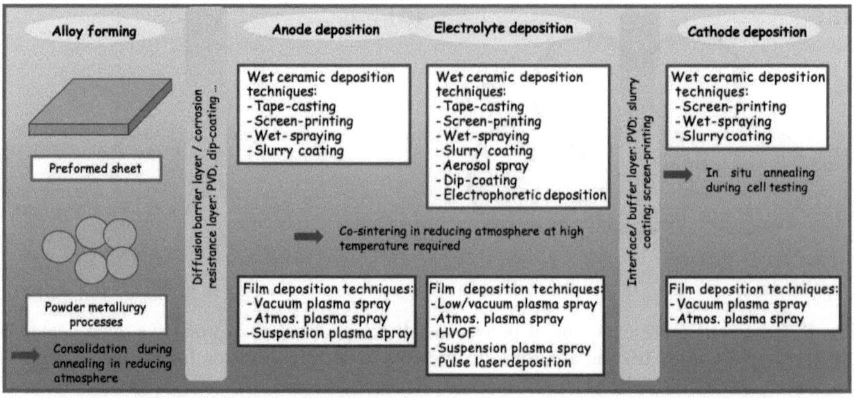

Fig. 10 Techniques currently used for the preparation of MS-SOFCs

and (co)-sintering methods. Hence, there has been a large body of work undertaken to accommodate these limitations, as illustrated by the number of techniques investigated shown in Fig. 10. At present, one may distinguish three main fabrication routes:

- MS-SOFCs are prepared by *wet ceramic deposition techniques*. Densification of the electrolyte requires at least one firing step in reducing atmosphere at high temperature. This route is often referred to as "high temperature process".
- MS-SOFCs are prepared by *film deposition techniques* encompassing physical/vapor/plasma deposition methods. This route is often referred to as "low temperature process", as the layers are produced at lower temperature. Post annealing of the films may only be required to release strains or to improve crystallinity.
- MS-SOFCs are prepared by *a "mixed process"* combining wet ceramic deposition techniques for producing alloy and anode layers and film deposition techniques for coating of the electrolyte.

Advantages and drawbacks of deposition techniques currently used for the production of MS-SOFCs are summarized in Table 4. Details about each route are given in the next paragraph.

4.2 Fabrication Achievements

4.2.1 Fabrication of Alloy Support

The alloy support should exhibit sufficient porosity for gas diffusion. Appropriate procedures to achieve the desired porosity must thus be taken, or else the support

Table 4 Advantages and drawbacks of techniques investigated for MS-SOFCs fabrication

Technique	Estimated cost	Advantages	Drawbacks
PLD	High	Enables deposition of layers with controllable microstructure at intermediate temperature; can produce multi-layered films of different materials by sequential ablation of assorted targets	Deposition in vacuum; low pO_2 during deposition results in strain relaxation in post annealing and formation of micro-cracks; uneven deposition reported for large samples; post annealing may be required to obtain crystalline materials; requires preparation of dense targets
EPD	Moderate	Less restriction of substrate shape and roughness; high powder packing density of deposited layers favorable for lowering sintering temperature; manufacturability is easy; fast deposition rates; thickness of films variable	Crack formation reported for layer with thickness above 50 μm; substrate should be conductor or a conductive backing such as carbon or platinum should be applied; post sintering required to achieve full densification of electrolyte; control of porosity in electrodes challenging
PS, SPS, VPS, HVOF	Moderate to high	Possibility of multi-layer deposition; fast deposition rates of PS and short deposition time avoiding chemical reaction between different layers; flexibility for automated production; VPS has higher particles velocities than PS yielding higher density of films	Low pO_2 during deposition results in strain relaxation in post annealing and formation of micro-cracks; electrodes with low porosity <15 vol. %; requires electrolyte thickness >30 μm to be gas tight; high cost of VPS; SPS needs to be optimized to improve quality of coating; anisotropy in electrolyte layer and thickness variations
Tape-casting	Low	Robust multi-layer deposition technique; continuous process; easy manufacturability; reproducibility; scalable	Bending; high temperature needed for densification of electrolyte; difficult to prepare layers with thickness below 30 μm; burnout of organic compounds preferably done in air to avoid carbides or nitrides formation in reducing atmosphere
Screen-printing	Low	Robust multi-layer deposition technique; dense or porous layers prepared with thickness down to 5 μm and good uniformity; porosity is easily controlled; reproducibility; scalable	Minimum coated layer >5 μm; requires evenness of the substrate for deposition; size of the screen and resolution given by the mesh opening; high sintering step needed for densification of electrolyte

(continued)

Table 4 (continued)

Technique	Estimated cost	Advantages	Drawbacks
Slurry coating (spray-, dip-, spin-)	Low	Multi-layer deposition technique; dense or porous layers prepared with thickness down to hundred nanometers; flexibility of size and shape of supports to be coated; scalable	High temperature needed for densification of electrolyte; multiple steps of coating and annealing often required to achieve sufficient density
Wet powder spraying	Low	Multi-layer deposition technique; thin films of a few microns can be produced; flexibility of size and shape of samples to be coated; scalable	High temperature needed for densification of electrolyte; several deposition steps required to achieve sufficient thickness; overspray

microstructure can impose severe limitation of current density due to concentration polarization.

Alloy supports may be prepared from commercially available preformed sheets, which are found in a wide range of compositions and microstructures. The sheets can be wire cut to form flow channels or laser drilled to create regularly spaced transversal open pores. Laser drilling a preformed sheet provides a smooth, well-defined surface, which facilitates the deposition of thin inorganic layers. On the other hand, prefabricated sheets will not shrink during MS-SOFC fabrication. This implies that the electrolyte material should be selected so as to tolerate constrained-sintering (such as ceria based oxide) or that film deposition techniques should be used for electrolyte fabrication. Wet ceramic deposition techniques, which require high temperature co-sintering steps, are thus not favorable for the fabrication of dense electrolytes.

Alloy supports can also be engineered by traditional powder metallurgy methods involving molding, pressing and sintering. Extrusion, free-sintering, tape-casting, and isostatic powder pressing-sintering are mainly employed. These methods require sintering shrinkage for consolidation of the support during annealing. Shrinkage can be monitored for further processing of functional layers. Formation of a porous open network can be obtained by addition of pyrolyzable fillers and organic additives to the alloy powders. Final support porosity will depend on a complex balance between sintering temperature, particle size, particle size distribution, alloy composition, particle morphology, pore former, organic additives, and shrinkage of other layers.

4.2.2 Fabrication by Wet Deposition Techniques

In this processes, the alloy support, the AFL, and the electrolyte are prepared by wet ceramic deposition techniques (see Fig. 10). The assembly is subjected to one

Fig. 11 Cross-sectional SEM micrograph of MS-SOFC fabricated by Ceres: FSS/Ni-CGO/CGO/CGO-Lanthanum ferrite [25]. Dense CGO is obtained after firing at 1000 °C. Reproduced here with kind permission from © A S M International 2004

or several co-sintering steps at high temperature in reducing atmosphere to densify the electrolyte and consolidate the AFL and the alloy support. The cathode layer is typically screen-printed and usually sintered in situ during testing of the cells to avoid decomposition of the oxide. This route is flexible, easy to upscale, and relatively economic provided that limited numbers of annealing steps are carried out. It is actively developed e.g. by Ceres Power Ltd. and Imperial College in England using wet spraying for the AFL and EPD for the electrolyte (Fig. 11) [22–25].

This route presents, however, several drawbacks. It is difficult to establish a heating schedule, which meets the conflicting requirements of densifying the electrolyte, enabling sufficient shrinkage of the alloy substrate and not causing excessive oxidation of the alloy. Co-sintering of the AFL and the electrolyte layer in reducing atmosphere tends also to promote diffusion of iron, chromium, and nickel between the ferritic FeCr steel and nickel containing AFL, which is detrimental for the cell (see Sect. 3) [11, 13]. Furthermore, Ni particles are easily coarsened in reducing atmosphere, which reduces both surface area and triple-phase boundaries.

To overcome these problems, Lawrence Berkeley National Laboratory in USA (LBNL) has developed a process based on the infiltration of electrochemically active catalysts in the cathode (LSM) and in the anode (Ni) after high temperature co-sintering of the electrolyte [10, 26, 27]. Another benefit of this approach is the higher mechanical integrity of the cell, which only relies on the steel and the ceramic electrolyte, as LSM and Ni are not structural elements of the cell. This route is also investigated by Risø National Laboratory in Denmark with Karlsruhe Institut für Technologie (KIT) in Germany [28, 29] (Fig. 12). A drawback of this process is the need for multiple infiltration-annealing cycles to achieve sufficient connectivity between catalyst particles, thereby making the process time and energy consuming.

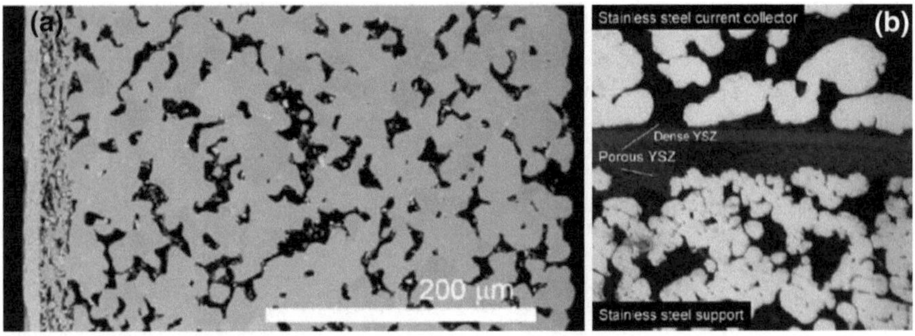

Fig. 12 SEM micrographs of (**a**) MS-SOFC of 25 cm^2 developed by Risø DTU- KIT [28], (**b**) MS-SOFC developed by LBNL [10]. Reproduced here with kind permission from © Elsevier 2010

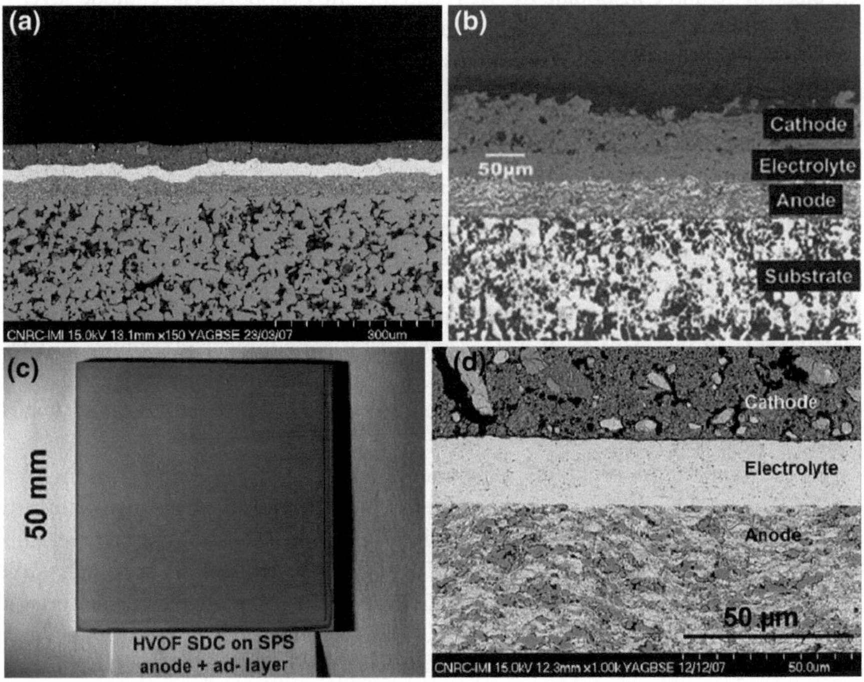

Fig. 13 Cross-sectional SEM micrographs of MS-SOFC fabricated (**a**) by suspension plasma spraying at the NRC: Hastelloy X/Ni-SDC/SDC/SSCo-SDC [30, 31]. Reproduced here with kind permission from © Elsevier 2007. (**b**) DLR concept [19, 37]. Reproduced here from [44] with kind permission from © John Wiley and Sons 2005. (**c**) and (**d**) Picture and SEM micrograph of MS-SOFC developed at NRC using HVOF for electrolyte deposition [32]. Reproduced here with kind permission from © A S M International 2008

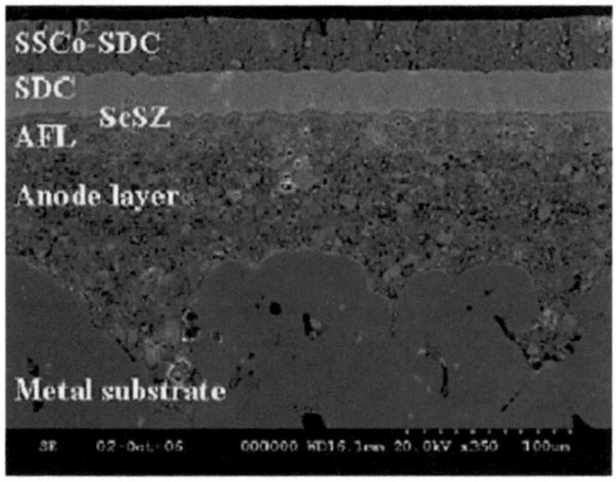

Fig. 14 Cross-sectional SEM micrograph of MS-SOFC developed by NRC [31, 38]. Reproduced here with kind permission from © Elsevier 2008

4.2.3 Fabrication by Film Deposition Techniques

Low temperature film deposition techniques are investigated for producing MS-SOFCs without the need for post sintering treatment at high temperature (Fig. 13). These processes are actively developed e.g. by the National Research Council Canada (NRC) using PLD [30, 31] and HVOF [32], and the Aerospace Research Center and Space Agency (DLR) in Germany using APS and VPS [19, 21, 33]. DLR fabrication processes have been scaled up to square cells of $20 \times 20\ cm^2$. These processes offer the possibility to produce MS-SOFCs in a single step by consecutive coatings of all the functional layers on the alloy substrate. Other advantages are the limited coarsening of nickel particles and interdiffusion of chromium, iron, and nickel in the ALF and steel support during fabrication.

Drawbacks of these processes are capital cost and limited flexibility. Use of plasma spray deposition techniques requires sufficiently smooth surface of the underlying layer to be coated. Preformed perforated alloy sheets are thus preferably used for these processes. If the alloy is to be processed by powder metallurgy methods, fine-grained alloy powder is more favorable to produce substrate with small pore size and surface roughness. On the other hand, the resulting enhanced specific surface area may yield, after a certain operating time to Cr depletion in the porous substrate. This might induce non-protective breakaway oxidation if the critical Cr threshold is reached and subsequent breakdown in electrochemical performance.

4.2.4 Fabrication by Mixed Processes

In mixed processes, electrodes are deposited by wet ceramic deposition techniques, while the electrolyte is deposited by film deposition techniques in order to avoid high temperature sintering. This route is actively developed e.g. by the Research Center Jülich using screen-printing for AFL coating and atmospheric plasma spray for electrolyte deposition [11] and NRC with Huazhong University of Science and Technology using screen-printing for AFL coating and suspension plasma spray for electrolyte deposition (Fig. 14) [30, 31].

Table 5 Examples of performance of MS-SOFCs integrating ferritic alloy substrates measured at 0.7 V in air/H_2

	Component	Substrate	Anode	Electrolyte	Cathode	Other layers	Power density at 0.7 V in W/cm^2	Remark and references
Wet ceramic deposition processes	Materials	Ametek ferritic steel	NiO-YSZ	YSZ	YSZ-LSM	–	0.23 at 650 °C	In situ annealing of electrodes [27]
	Processing	Tubular support	Infiltrated NiO	Aerosol spray	Infiltrated LSM			
	Thickness	Not reported	15 µm	15 µm	Not reported			
	Materials	Ferritic 22 % Cr	Fe22Cr-YSZ	ScYSZ	LSCF-CGO		0.40 650 °C	Infiltration of NiO-CGO, Second cathode layer LSC [28]
	Processing	Tape casting	Tape casted	Tape casted	Screen printed			
	Thickness	400 µm	20 µm	10–15 µm				
	Materials	Ti-Nb (Fe-17Cr)	NiO-GDC	GDC	Doped $LaFeO_3$/GDC	Contact layer not defined 30 µm	0.22 at 600 °C	In situ sintering of cathode [25]
	Processing	Sheet	Wet spraying	EPD	Screen-printing			
	Thickness	Not reported	10 µm	15 µm	15 µm			
	Materials	ANSI-430 (Fe-17Cr)	NiO-$Ce_{0.8}Y_{0.2}O_{1.9}$	GDC	LSCF	–	0.11 at 600 °C	In situ sintering of cathode [42]
	Processing	Powder pressing/sheet	Wet powder spraying	Wet powder spraying	Spraying			
	Thickness	Not reported	26 µm	12 µm	30 µm			
Film deposition processes	Materials	Hastelloy X	NiO-SDC	SDC	SSCo-SDC	–	0.11 at 600 °C	In situ sintering of cathode [31]
	Processing	Sheet	SPS	SPS	Screen-printing			
	Thickness	Not reported	30 µm	30 µm	45 µm			
	Materials	Hastelloy X	NiO-SDC	SDC	SSCo-SDC	–	0.10 at 650 °C	In situ sintering of cathode [30]
	Processing	Sheet	SPS	SPS	Screen-printing		0.08 at 600 °C	
	Thickness	1.25 mm	20–25 µm	30 µm	45 µm			
	Materials	Hastelloy X	NiO-SDC	SDC	SDC-SSCo	–	0.53 at 650 °C	In situ sintering of cathode [43]
	Processing	Sheet	SPS	HVOF	Screen-printing		0.35 at 600 °C	
	Thickness	1.25 mm	70 µm	20 µm	55 µm			
	Materials	ITM	NiO-YSZ	YSZ	LSM/LSCF	$La_{0.6}Sr_{0.2}Ca_{0.2}O_3$	0.38 at 800 °C	[21]
	Processing	Sheet	APS	VPS/LPPS	APS	APS	Reformate NG;	
	Thickness	0.95–1.05 mm	40–60 µm	35–50 µm	20–30 µm	10–30 µm	0.61 at 800 °C	

(continued)

Table 5 (continued)

	Component	Substrate	Anode	Electrolyte	Cathode	Other layers	Power density at 0.7 V in W/ cm^2	Remark and references
Mixed processes	Materials	SS430	NiO-SDC	ScSZ/SDC	SSCo-SDC	–	0.14 at 600 °C	In situ sintering of cathode [31]
	Processing	Sheet	Screen-printing	PLD	Screen printing			
	Thickness	Not reported	90 μm	2/20 μm	25 μm			
	Materials	CroFer22APU	NiO-YSZ	YSZ	LSCF	CeO_2 barrier layer on the alloy by PVD—5 μm thick	0.43 at 800 °C 0.34 after 165 h	In situ sintering of cathode [11]
	Processing	Tape-casting	Screen-printing	VPS	Screen-printing			
	Thickness	1 mm	10–15 μm	60 μm	146 μm			

In order to achieve gas tight electrolytes, the underlying AFL should have a smooth surface with small pores, which can be obtained by screen-printing or spray-coating. This process requires multiple steps of coating and annealing, which makes it time and energy consuming.

5 MS-SOFC Performance

Several demonstrations with metal-supported cells using CGO or YSZ electrolyte are reported in the recent review of Tucker [10], which shows encouraging cell performance and progress in MS-SOFC development. Power density of MS-SOFCs measured at 0.7 V in air/H_2 is reported in Table 5 for selected references, which illustrate technical refinement of fabrication progress and cell architectures. The best power density reaches around 0.5 W/cm^2 at 650 °C for an SDC electrolyte of 20 μm thickness. Two main factors are contributing to lowering power density and reducing longevity of the cells, as discussed below.

5.1 Impact of Cathodic Electrode on Overall Cell Performance

Cathode catalysts, such as LSM and LSCF decompose in reducing atmosphere at elevated temperature, precluding them for being co-fired with the alloy support. For these reasons, wet coated electrodes are usually sintered in situ during electrochemical testing of cells at temperature below 1100 °C. Experimental and numerical analyses showed that limitations from oxygen reduction reaction of the cathode dominate the overall cell performance. Besides, significantly larger polarization resistances of cathodes in MS-SOFCs are also observed than similar cathodes used in SOFCs [34, 35]. This can be explained by better electrolyte/electrode interface and microstructure of the electrodes in SOFCs due to higher annealing temperature. Clearly, work is needed to improve the microstructure of the cathode using low temperature processes.

Use of film deposition techniques, such as plasma spray processes is an alternative to produce cathode layers at low temperature. One of the main challenges for these processes is to improve the open porosity of plasma sprayed cathodic layer, which remains too low [36].

5.2 Impact of Anodic Electrode on Overall Cell Performance

As discussed in Sect. 3, interdiffusion of Ni, Fe, Cr in steel and Ni-based anode can damage the cell. Coarsening of Ni particles during manufacturing will also reduce significantly the electrochemical performance. LBNL, and more recently

Risø, address these issues by using infiltration techniques to incorporate Ni catalyst in the anode after sintering of YSZ electrolyte coated on porous YSZ anode backbone. The anode catalyst is never exposed to high-temperature reducing sintering conditions, solving the issues of interdiffusion and coarsening during cell fabrication. The anode catalyst also coats the metal support, so inclusion of rare earth salts in the catalyst precursor is expected to enhance oxidation resistance of the metal support. The catalyst is not part of the mechanical structure of the cell, but rather forms a thin coating on YSZ and steel backbone. Hence, expansion and contraction due to thermal or redox cycling do not cause stress in the electrolyte layer. Favorable redox and thermal cycling tolerance of cells with infiltrated anode design were reported [26, 28]. A similar strategy is also applied with LSM catalyst infiltrated in the cathodic electrode, and shows promising results.

Replacement of most or all of the Ni is also investigated using an anode catalyst, such as doped ceria, that is stable in contact with YSZ in SOFC fuel conditions [10]. Ceria offers improved stability, though its electronic conductivity is much less than that of Ni. As the electronic current passing from the metal support to the active electrode area is carried by the ceria catalyst coating, cell performance may be limited by ceria electronic conductivity. This situation can be improved by addition of a conductive component to the anode backbone, such as copper or stainless steel particles. Risø has demonstrated the latter choice and promising durability of the cell is reported.

6 Conclusions

MS-SOFCs are expected to decrease significantly capital and operational costs, and increase the lifetime of the cells due to lower operating temperature and higher redox stability. This chapter reviews technology development with a focus given to the main issues affecting performance and longevity of single cells. Criteria for selection of alloy materials were discussed based on material cost, thermal expansion coefficient, corrosion rate, and particle size. Ferritic stainless steels appeared as promising candidates for MS-SOFC supports, though research for improving design of these materials is still needed to further improve their oxidation resistance. Surface coatings are of great potential to mitigate Cr volatility and interdiffusion of metal element between Ni-based anode and steel, though manufacturability of these layers is to be further improved. New cell architectures and advanced fabrication processes were presented illustrating the level of technical refinement currently achieved. One of the main issues remaining is the fabrication of the cathode. Possible alternatives may be found with the infiltration of electrode catalysts. Improvements in catalyst composition and current collection schemes are needed, and this is a fruitful area for further research. Performance of the produced MS-SOFCs was discussed highlighting significant progress in the field. MS-SOFCs are thus expected to fulfill expectations in the coming years.

Acknowledgments The authors acknowledge FP7 EFFIPRO project funded by the European Union (FP7 Project—Grant Agreement 3227560) EU—FCH JU RAMSES Project Under Grant agreement Number 256768 and StackPro project (number 18532/530) funded by the Norwegian Research Council (Renergi Program).

References

1. J. Wu, X. Liu, Recent development of SOFC metallic interconnect. J. Mater. Sci. Technol. **26**(4), 293–305 (2010)
2. S. Megel, E. Girdauskaite et al., Area specific resistance of oxide scales grown on ferritic alloys for solid oxide fuel cell interconnects. J. Power Sources (2010). doi:10.1016/2010.09.003
3. J. Froitzheim, G.H. Meier et al., Development of high strength ferritic steel for interconnect application in SOFCs. J. Power Sources **178**, 163–173 (2008)
4. N. Shaigan, W. Qu et al., A review of recent progress in coatings, surface modifications and alloy developments for solid oxide fuel cell ferritic stainless steel interconnects. J. Power Sources **195**, 1529–1542 (2010)
5. Z. Yang, G.G. Xia et al., (Mn, Co)3O4 spinel coatings on ferritic stainless steels for SOFC interconnect applications. Int. J. Hydrogen Energ. **32**(16), 3648–3654 (2007)
6. X. Montero, F. Tietz et al., $MnCo_{1.9}Fe_{0.1}O_4$ spinel protection layer on commercial ferritic steels for interconnect applications in solid oxide fuel cells. J. Power Sources **184**, 172–179 (2008)
7. Z.G. Yang, J.W. Stevenson et al., Material properties database for selection of high-temperature alloys and concepts of alloy design for SOFC applications. PNNL-14116. Prepared for the U.S. Department of Energy under Contract DE-AC06-76RL01830 (2002)
8. M. Brandner, Herstellung einer Metall/keramic-Verbundstruktur für Hochtemperatur-brennstoffzellen in mobilen Anwendungen. Berichte des Forschungzentrum Jülich-4238, (2006) ISSN 0944-2952
9. M.C. Tucker, T.Z. Sholklapper et al., Progress in metal-supported SOFCs. ECS Trans. **25**(2), 673–680 (2009)
10. M.C. Tucker, In metal-supported solid oxide fuel cells: a review. J. Power Sources **195**(15), 4570–4582 (2010)
11. M. Brandner, M. Bram et al., Electrically conductive diffusion barrier layers for metal-supported SOFC. Solid State Ionics **179**, 1501–1504 (2008)
12. S. Molin, M. Gazda et al., High temperature oxidation of porous alloys for solid oxide fuel cell applications. Solid State Ionics **181**, 1214–1220 (2010)
13. Y. Liu, D.Y. Chen, Protective coatings for Cr_2O_3—forming interconnects of solid oxide fuel cells. Int. J. Hydrogen Energ. **34**, 9220–9226 (2009)
14. I. Villarreal, M. Rivas et al., Characterization of metal-supported axial injection plasma sprayed solid oxide fuel cells with aqueous suspension plasma sprayed electrolyte layers. ECS Trans. **25**(2), 689–694 (2009)
15. D. Wiedenmann, U.F. Vogt et al., WDX studies on ceramic diffusion barrier layers of metal supported SOECs. Fuel Cells **9**(6), 861–866 (2009)
16. R.H. Henne, T. Franco et al., High-velocity DC-VPS for diffusion and protecting barrier layers in solid oxide fuel cells (SOFCs). J. Therm. Spray Technol. **15**(4), 695–700 (2006)
17. T. Franco, Z. HoshiarDin et al., Plasma sprayed diffusion barrier layers based on doped perovskite-type $LaCrO_3$ at substrate-anode interface in solid oxide fuel cells. J Fuel Cell Sci. Tech. **4**(4), 406–412 (2007)
18. T. Franco, K. Schibinger et al., Ceramic diffusion barrier layers for metal supported SOFCs. ECS Trans. **7**(1), 771–780 (2007)
19. G. Schiller, A. Ansar et al., High temperature water electrolysis using metal supported solid oxide electrolyser cells (SOEC). J. Appl. Electrochem. **39**, 293–301 (2009)

20. T. Franco, M. Brandner et al., Recent development aspects of metal supported thin-film SOFC. ECS Trans. **25**(2), 681–688 (2009)
21. P. Szabo, J. Arnold et al., Progress in metal supported solid oxide fuel cells and stacks for APU. ECS Trans. **25**(2), 175–185 (2009)
22. Y.B. Matus, L.C. De Jonghe et al., Metal-supported solid oxide fuel cell membranes for rapid thermal cycling. Solid State Ionics **176**(5–6), 443–449 (2005)
23. R.T. Leah, N.P. Brandon, Modelling of cells, stacks and systems based around metal-supported planar IT-SOFC cells with CGO electrolytes operating at 500–600 °C. J. Power Sources **145**, 336–352 (2005)
24. H. Kurokawa, G.Y. Lau, Water-based binder system for SOFC porous steel substrate. J. Process. Technol. **182**, 469–476 (2007)
25. N.P. Brandon, D. Corcoran et al., Development of metal supported solid oxide fuel cells for operation at 500–600 °C. J. Mater. Eng. Performance **13**(3), 253–256 (2004)
26. M.C. Tucker, G.Y. Grace, Stability and robustness of metal-supported SOFCs. J. Power Sources **175**, 447–451 (2008)
27. M.C. Tucker, G.Y. Lau et al., Performance of metal-supported SOFCs with infiltrated electrodes. J. Power Sources **171**(2), 477–482 (2007)
28. P. Blennow, J. Hjelm et al., Manufacturing and characterization of metal-supported solid oxide fuel cells. J. Power Sources (2010). doi:10.1016/2010.08.088
29. P. Blennow, J. Hjelm et al., Development of planar metal supported SOFC with novel cermet anode. ECS Trans. **25**(2), 701–710 (2009)
30. Z. Wang, J.O. Berghaus et al., Dynamic evaluation of low-temperature metal-supported solid oxide fuel cell oriented to auxiliary power units. J. Power Sources **176**(1), 90–95 (2008)
31. Q.A. Huang, J. Oberste-Berghaus et al., Polarization analysis for metal-supported SOFCs from different fabrication processes. J. Power Sources **177**(2), 339–347 (2008)
32. J.O. Berghaus, J.G. Legoux et al., Suspension HVOF spraying of reduced temperature solid oxide fuel cell electrolytes. J. Therm. Spray Technol. **17**, 700–707 (2008)
33. G. Schiller, R.H. Henne et al., Development of vacuum plasma sprayed thin-film SOFC for reduced operating temperature. Fuel Cells Bull. **21**, 7–12 (2000)
34. Y.M. Kim, P. Kim-Lohsoontorn et al., Effect of unsintered gadolinium-doped ceria buffer layer on performance of metal-supported solid oxide fuel cells using unsintered barium strontium cobalt ferrite cathode. J. Power Sources **195**(19), 6420–6427 (2010)
35. Q. Huang, B. Wang et al., Impedance diagnosis of metal-supported SOFCs with SDC as electrolyte. J. Power Sources **191**, 297–303 (2009)
36. D. Waldbillig, O. Kesler, Characterization of metal-supported axial injection plasma sprayed solid oxide fuel cells with aqueous suspension plasma sprayed electrolyte layers. J. Power Sources **191**, 320–329 (2009)
37. L.R. Pederson, P. Singh et al., Review: application of vacuum deposition methods to solid oxide fuel cells. Vaccum **80**, 1066–1083 (2006)
38. S.R. Hui, D. Yang et al., Metal-supported solid oxide fuel cell operated at 400–600 °C. J. Power Sources **167**, 336–339 (2007)
39. Z. Yang, G.G. Xia et al., Evaluation of Ni-Cr base alloys for SOFC interconnect applications. J. Power Sources **160**, 1104–1110 (2006)
40. Y. Kong, B. Hua et al., A cost-effective process for fabrication of metal-supported solid oxide fuel cells. Int. J. Hydrogen Energ. **35**, 4592–4596 (2010)
41. K.H. Kim, Y.M. Park et al., Fabrication and evaluation of the thin NiFe supported solid oxide fuel cell by co-firing method. Energy (2010). doi:10.1016/.2010.07.018
42. N. Oishi, Y. Yoo, Fabrication of cerium oxide based SOFC having a porous stainless steel support. ECS Trans. **25**(2), 739–744 (2009)
43. R. Hui, J.O. Berghaus et al., High performance metal-supported solid oxide fuel cells fabricated by thermal spray. J. Power Sources **191**(2), 371–376 (2009)
44. M. Lang, A. Dresel, T. Franco, Z. Ilhan, A. Nestle, G. Schiller, P. Szabo, Electrochemical characterization of vacuum plasma sprayed planar solid oxide fuel cells and short stacks for mobile application. Ceram. Eng. Sci. Proc. **26**(4), 67–74 (2005)

Challenges Imposed by Thermochemical Expansion of Solid State Electrochemical Materials

Jorge R. Frade

Abstract Thermal expansion compatibility is usually considered a *sine qua non* condition for selection of prospective materials for SOFC and other solid state electrochemical systems due to fragile behaviour of ceramic components, and their limited ability to withstand significant strain. However, chemical expansion may also add non-negligible strain contributions on materials with variable oxygen stoichiometry. Thus, one measured the chemical expansion of representative materials and compiled information on a variety of other electronic or mixed conductors proposed for electrode, membrane or interconnector applications. Selected materials were used to identify trends and guidelines for effects of composition and for structural effects. Dependence on working conditions was focused on temperature and oxygen partial pressure. It was also extended to predict effects exerted by overpotential or changes in gas composition, and dependence on fuel conversion when membrane materials are exposed to high chemical potential gradients under fuel/membrane/air conditions. Redox conditions were predicted by thermodynamic analysis of fuel conversion, for hydrogen and methane-based fuels. These thermodynamic calculations were combined with thermochemical expansion data for representative materials, to predict strain induced on: (1) solid electrolytes or mixed conducting membranes exposed to high chemical potential gradients, (2) cathode materials under combined effects of changes from processing to operation temperatures and polarisation, (3) anode materials on changing from high temperature processing in air to reduce fuel atmospheres. Differences between thermochemical effects on anode supported configurations and electrolyte supported cells were analysed, including lab scale conditions based on thick solid electrolyte cells with thin electrodes.

J. R. Frade (✉)
University of Aveiro, Aveiro, Portugal
e-mail: jfrade@ua.pt

J. T. S. Irvine and P. Connor (eds.), *Solid Oxide Fuels Cells: Facts and Figures*,
Green Energy and Technology, DOI: 10.1007/978-1-4471-4456-4_5,

1 Introduction

High temperature or intermediate temperature electrochemical systems have been extensively studied for their potential advantages in terms of clean energy conversion with enhanced efficiency. However, these systems are mainly based on fragile ceramic materials, including solid electrolytes, electrodes and interconnector materials for solid oxide fuel cells, and other prospective solid state electrochemical systems, such as steam electrolysers, electrochemical oxygen pumps, etc. These ceramic-based systems are conditioned by mechanical limitations of cell materials, and mostly intended for stationary applications.

Although thermal expansion compatibility [1–6] is usually assessed by the differences between thermal expansion coefficients, the magnitude of stresses also depends on elastic constants and on the amplitude of temperature changes, from sintering or deposition/bonding conditions to operation temperatures, and down to room temperature due to discontinuous operation or for maintenance. Excessive thermal expansion of individual materials also implies greater risks of thermal shock or failure under temperature gradients.

Recent finding of unusual elastic behaviour of highly defective materials may contribute to counter the destructive effects of thermal expansion mismatch, as reported for ceria-based materials [7–9]. Kossoy and coauthors [8, 9] demonstrated important changes in lattice parameters of CGO in strained conditions, resembling the dependence on the contents of Gd, and corresponding changes in point defect chemistry. Thus, this unique response may enhance tolerance to high thermo-chemical strain without mechanical failure.

Chemical incompatibility in firing conditions may also exclude selection of prospective materials with excellent electrochemical or electrocatalytic properties [10–14], especially if one attempts to co-fire all the cell components [15]. This shortcoming may be minimised by avoiding co-firing at excessive temperatures, developing alternative processing technologies at lower temperatures [16], and using buffer layers or bi-layered electrolytes [17–22]. Other interactions between cell materials may involve long-term interdiffusion, often associated with redox changes, as reported for diffusion of chromium from interconnecting materials [23–25], nickel from the Ni–YSZ cermet anodes into other cell materials [26–29] or diffusion of Mn from $LaMnO_3$-based cathodes [11, 14, 30–34]. Buffer layers [29, 35, 36] or microstructural optimisation [37] may contribute to countering some of these effects.

Optimised processing may minimise the impact of thermal expansion mismatch, by lowering temperature differences between processing and operation conditions. Processing is also crucial for enhanced redox tolerance of Ni/NiO in cermet anodes, which also depends on contents of Ni, porosity, original grain size of NiO/YSZ precursors and on the kinetics of reduction to metallic Ni [38, 39]. Double or multilayer anodes are also commonly used [40, 41]. However, optimised processing may still be insufficient to avoid stresses induced by thermal shock or relatively fast thermal cycling from operation conditions to room temperature. The greatest concern about cermet anodes is usually related to microstructural changes upon redox cycling, due to discontinuous operation, maintenance, repairing etc. and corresponding stresses imposed on other cell materials (e.g. [6, 37, 42–47]). Note that, redox induced strain of Ni/YSZ cermet electrodes may be as much as a few percent, thus imposing excessive stresses on thinner components of anode supported cells, and eventually causing fracture or delamination (e.g. [44]).

Many recent studies and developments on SOFC anodes have been dedicated to the search for alternative non-metallic anode materials [20, 48–51]. Though this was originally intended mainly to avoid carbon deposition in hydrocarbon fuelled cells, alternative ceramic anodes may also be essential to overcome risks associated with poor redox tolerance of Ni/YSZ cermet anodes. Other authors lowered the fraction of metallic Ni to a minimum by adding electronic or mixed conducting ceramic phases [52]. Yet, redox-induced changes also affect the potential applicability of ceramic fuel electrode materials. For example, important changes in oxygen stoichiometry and transport properties have also been reported for representative materials such as $(La_{0.75}Sr_{0.25})_{0.95}Cr_{0.5}Mn_{0.5}O_{3-\delta}$ [53]. In general, degradation or even collapse of ceramic electrode materials may be caused by a combination of excessive concentrations of point defects and lattice expansion upon partial reduction of transition metal cations to a lower valence state. However, good performance of anode material may also be related to the onset of a significant ionic conductivity contribution on approaching the redox stability limit. For example, some titanates show outstanding redox tolerance [54] and enhanced performance after a pre-reduction [55] or by sintering under reducing conditions [56]; this enhances the conductivity of strontium titanate-based materials with rare earth or yttria additions on A-site [57]. $Sr_{1-x}Y_xTiO_{3\pm\delta}$ combines ability to attain good redox tuned properties [58] and compatibility with the state-of-the-art YSZ electrolyte [59].

A variety of prospective ceramic components of solid state electrochemical systems also undergo significant chemical expansion upon being exposed to changes in atmosphere, as found on exposing air fired anode materials to reducing fuel atmospheres. In this case, the true compatibility between different cell materials must combine the thermal and chemical strain of bonded materials. Chemical expansion may also be the main cause of strain and corresponding thermochemical stresses under high gradients of chemical potential, due to reducibility on the fuel side, as emphasised for ceria-based materials [60–63]. Risks of fracture and/or delamination are also often found at electrolyte/cathode interfaces in oxidising conditions [64].

Chemical expansion may be related to gradual changes in point defect concentrations, with emphasis on oxygen ion vacancies, and/or increase in cationic radii upon gradual reduction to a lower valence state. This is quite different from first-order redox reactions (e.g. Ni/YSZ cermet anodes) or phase transformations (e.g. brownmillerite to perovskite transformation $Sr(Fe,Co)O_{2.5} \rightarrow Sr(Fe,Co)O_{3-\delta}$), with discontinuous structural changes and more or less abrupt property changes [65–67], except possibly for strong kinetic limitations [68] or 2-phase intermediate conditions

Fig. 1 Combined thermal and chemical expansion of $La_{0.3}Sr_{0.7}Fe_{0.6}Ga_{0.4}O_{3-\delta}$ in pure O_2 (1 atm) and in inert atmosphere with $pO_2 = 10^{-4}$ atm (redrawn from [80])

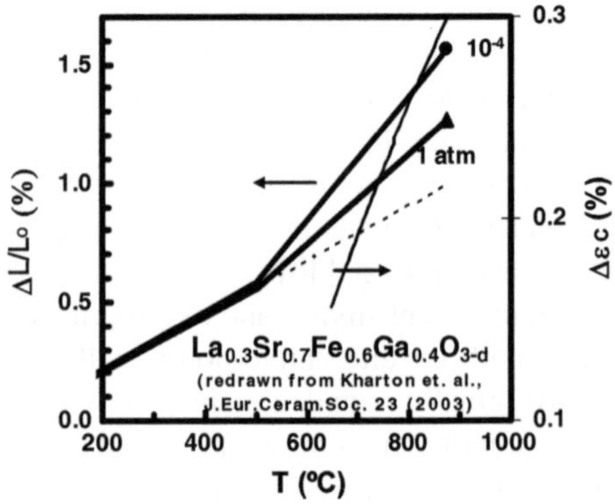

on varying temperature or oxygen partial pressure [69]. In other cases, it might be difficult to establish clear borderlines between chemical expansion and subtle structural changes, such as formation of nanodomains, defect ordering or formation of extended defects [70–73].

Clear evidence of chemical expansion is revealed by gradual dependence of lattice parameters on oxygen partial pressure obtained by XRD or neutron diffraction [74–76]. Often these studies are based on a combination of results obtained in air, pure oxygen and/or in inert atmospheres of Ar or N_2. Others (e.g., [77]) analysed the chemical expansion of samples after quenching under different atmospheres. However, dependence of lattice parameters on oxygen partial pressure does not necessarily imply significant cell volume expansion, as reported for Ln_2NiO_4-based materials with Ln = La, Nd [78, 79]. Therefore, isothermal dilatometry, performed with variable oxygen partial pressure, is still often used to assess chemical expansion [80–86]. Though one would expect reasonable consistency between data obtained by XRD and dilatometry [80], this is not always the case [75].

2 Relations Between Chemical Expansion and Point Defect Chemistry

Many mixed conductors show combined contributions of chemical expansion and thermal expansion on heating in fixed atmospheres (e.g. in air or Ar). Differences between high-and low temperature regimes (e.g. Fig. 1) are also ascribed to gradual chemical expansion contributions, and can be related to increasing oxygen sub-stoichiometry. A variety of other ferrite-based materials also show major increase in expansion coefficient at high temperatures [87], most probably due to oxygen loss and corresponding chemical expansion contribution.

(La, Sr)(Fe, Ga)$O_{3-\delta}$ materials represent an interesting study case for chemical expansion [80]. Though partial substitution of Fe by redox stable Ga resulted in

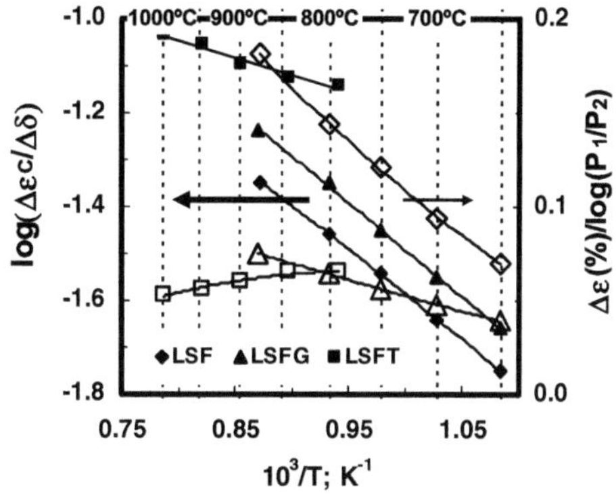

Fig. 2 Chemical expansion data of $La_{0.3}Sr_{0.7}Fe_{0.6}Ga_{0.4}O_{3-\delta}$ (LSFG) and $La_{0.3}Sr_{0.7}FeO_{3-\delta}$ (LSF) on changing from $pO_2 = 1\text{–}10^{-4}$ atm [80], and for $La_{0.2}Sr_{0.8}Fe_{0.55}Ti_{0.45}O_{3-\delta}$ (LSFT) on changing from $pO_2 = 0.21\text{–}10^{-4}$ atm [86]

lower thermochemical expansion of $La_{0.3}Sr_{0.7}Fe_{1-x}Ga_xO_{3-\delta}$, these LSFG compositions retain higher oxygen sub-stoichiometry than $La_{0.3}Sr_{0.7}FeO_{3-\delta}$. Yet, the oxygen sub-stoichiometry of LSFG is less sensitive to changes in temperature or oxygen partial pressure. For example, changes in the temperature range of 923–1,123 K, in air, are $\Delta\delta = 0.078$ for LSF and $\Delta\delta = 0.033$ for LSFG. Similarly, changes on lowering pO_2 from 0.21 to 10^{-4} atm, at 1,123 K, were $\Delta\delta = 0.127$ for LSF and $\Delta\delta = 0.039$ for LSFG. However, chemical expansion of LSFG is more sensitive to changes of oxygen sub-stoichiometry, as given by the so-called chemical expansion coefficient $\Delta\varepsilon_C/\Delta\delta$ (Fig. 2). Note that the ionic conductivity of LSFG is higher than for LSF, probably due to a combination of higher concentration of charge carrier (oxygen ion vacancy) and its mobility.

Point defect chemistry can assist the interpretation of the main factors governing chemical expansion, with emphasis on temperature and oxygen partial pressure. For example, if one considers generic $La_{1-x}A_xFe_{1-y}B_yO_{3-\delta}$ materials with redox stable divalent A-site additive and trivalent B-site additive, one can analyse redox changes and corresponding change in oxygen sub-stoichiometry as follows:

$$2\mathrm{Fe}_{\mathrm{Fe}}^{\cdot} + \mathrm{O}_O^x \Leftrightarrow 2\mathrm{Fe}_{\mathrm{Fe}}^x + V_O^{\cdot\cdot} + 0.5\mathrm{O}_2. \quad (1)$$

On expressing concentrations of point defects per unit formula one obtains a simple correspondence with oxygen sub-stoichiometry:

$$\left[V_O^{\cdot\cdot}\right] = \delta. \quad (2)$$

In addition, most mixed conductors have prevailing p-type contribution in oxidising conditions, and on assuming a simplified neutrality condition $2\delta + \left[\mathrm{Fe}_{\mathrm{Fe}}^{\cdot}\right] \approx x$ one obtains:

$$\left[\mathrm{Fe}_{\mathrm{Fe}}^{\cdot}\right] \approx x - 2\delta. \quad (3)$$

On inserting both concentrations of point defects in the mass action constant of Eq. 1:

$$\log(\mathrm{pO_2}) + 2\log\left\{\frac{(1-y+2\delta-x)^2\delta}{(x-2\delta)^2(3-\delta)}\right\} = 2\log\{k_o\} - \frac{2\Delta H}{2.3R}\mathrm{T}^{-1} \qquad (4)$$

where k_o is the pre-exponential factor of mass action constant and ΔH is the corresponding enthalpy. The enthalpy change and pre-exponential factor can be extracted from plots of $\log(\mathrm{pO_2})$ versus 1/T, at constant oxygen sub-stoichiometry [80].

Chemical expansion may involve contributions of different point defects, as predicted on assuming additive contributions of oxygen vacancies and fraction of Fe reduced to the trivalent state:

$$\varepsilon_C = \varepsilon^V\delta + \varepsilon^{\mathrm{red}}(1-y+2\delta-x) \qquad (5)$$

where ε^V and $\varepsilon^{\mathrm{red}}$ correspond to unit contributions of oxygen vacancies and reduced ions. On combining this with Eqs. 2and 3 one obtains:

$$\frac{\varepsilon_C}{\delta} = \varepsilon^V + 2\varepsilon^{\mathrm{red}} + \frac{\varepsilon^{\mathrm{red}}(1-x-y)}{\delta} \qquad (6)$$

and the corresponding chemical expansion coefficient:

$$\frac{\Delta\varepsilon_C}{\Delta\delta} = \varepsilon^V + 2\varepsilon^{\mathrm{red}}. \qquad (7)$$

Thus, Eq. 7 shows that $\Delta\varepsilon_C/\Delta\delta$ is a suitable criterion for the relevance of chemical expansion of mixed conductors even if this involves complex dependence on defect chemistry. It also emphasises that the chemical expansion coefficient $\Delta\varepsilon_C/\Delta\delta$ includes contributions of oxygen ion vacancies as well as changes in ionic radii of cations undergoing gradual reduction to a lower valence state, at least for this class of mixed ionic and p-type conductors.

Figure 2 also shows results reported for $La_{0.2}Sr_{0.8}Fe_{0.55}Ti_{0.45}O_{3-\delta}$ (LSFT) [86]. These results confirm that suitable B-site partial substitution tends to lower the dependence of oxygen stoichiometry on oxygen partial pressure. However, this is somewhat countered by increase in chemical expansion coefficient $\Delta\varepsilon_C/\Delta\delta$ which is lower for LSF than for modified LSFT compositions.

For $La_{1-x}Sr_xCoO_{3-\delta}$ (LSC) and mixed compositions $La_{1-x}Sr_xCo_{1-y}Fe_yO_{3-\delta}$ (LSCF), one finds values of $\Delta\varepsilon_C/\Delta\delta$ in the same order of magnitude [75]. Other source [81] reported an overall trend for the dependence of chemical expansion coefficient of LSC materials on the fraction of oxygen vacancies x_V, $\beta_C/3 = 0.061 + 0.584x_V$. Since the fraction of vacancies is related to oxygen sub-stoichiometry $\delta/3 = x_V$:

$$\frac{\partial\varepsilon_C}{\partial\delta} = 0.061 + 0.195\delta. \qquad (8)$$

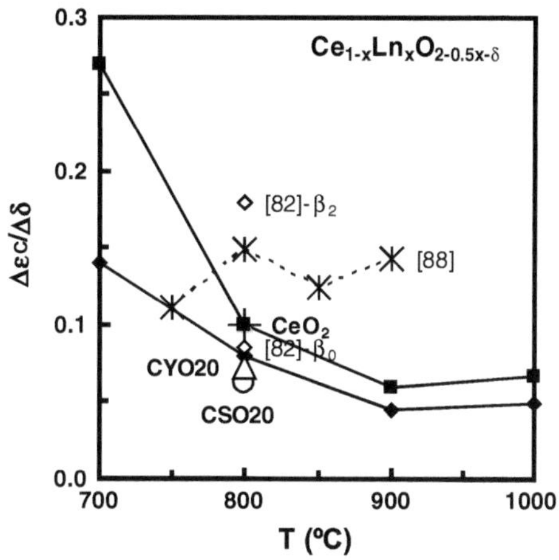

Fig. 3 Representative results of chemical expansion of ceria-based materials, with emphasis on $Ce_{0.9}Gd_{0.1}O_{1.95-\delta}$ (◆—[89], *—[91], and ◇—[85]) and $Ce_{0.8}Gd_{0.2}O_{1.9-\delta}$ (■—[89]). Additional results are shown for undoped ceria (+[85]), $Ce_{0.8}Y_{0.2}O_{1.9}$ (Δ—[90]) and $Ce_{0.8}Sm_{0.2}O_{1.9}$ (O—[90])

Note that this trend was obtained on combining data for a variety of LSC compositions and also for relatively wide temperature ranges.

Ceria-based materials $Ce_{1-x}Ln_xO_{2-0.5x-\delta}$, with a trivalent lanthanide additive Ln, also undergo significant oxygen stoichiometry changes and chemical expansion under strongly reducing conditions [88]. The chemical expansion coefficient of these ceria-based materials (Fig. 3) is usually far greater than for perovskite materials with prevailing p-type electronic conductivity [89–91]. Some values of chemical expansion coefficient $\Delta\varepsilon_C/\Delta\delta$ reported for $Ce_{1-x}Ln_xO_{2-0.5x-\delta}$ are higher than for undoped ceria [85], and tend to increase with the contents of trivalent lanthanide and with decreasing operation temperatures [89]. However, other results [91] do not show clear trends for the temperature dependence, and suggest mainly significant deviations from linear dependence on oxygen sub-stoichiometry.

The simplest point defect model to describe expansion upon reduction of Ce^{4+} to Ce^{3+} and oxygen stoichiometry changes under reducing conditions can be described by:

$$2\mathrm{Ce}_{\mathrm{Ce}}^{x} + \mathrm{O}_{O}^{x} \overset{K_0}{\Leftrightarrow} 2\mathrm{Ce}_{\mathrm{Ce}}^{\prime} + V_{O}^{\cdot\cdot} + 0.5\mathrm{O}_2. \tag{9}$$

and neutrality establishes a corresponding relation between point defect concentrations, and thus:

$$\left[\mathrm{Ce}_{\mathrm{Ce}}^{\prime}\right] = 2\left[V_{O}^{\cdot\cdot}\right] - x = 2\delta \tag{10}$$

when concentrations are expressed per formula unit. On assuming isolated point defects this yields $\left[V_{O}^{\cdot\cdot}\right] = 0.5x + \delta$, $\left[\mathrm{Ce}_{\mathrm{Ce}}^{\prime}\right] = 2\delta$, and by taking additive chemical expansion contributions of isolated vacancies and trivalent Ce^{3+} one also obtains:

$$\frac{\Delta\varepsilon_C}{\Delta\delta} = \varepsilon_V + 2\varepsilon_{\mathrm{red}}. \tag{11}$$

Actually, the relevant literature shows far from negligible dispersion of oxygen sub-stoichiometry data and chemical expansion [85]. In addition, the previous description is probably an oversimplified model, mainly because there is convincing evidence that point defects interact to form dimers ($(\mathrm{Ce}'_{\mathrm{Ce}}V_O^{\cdot\cdot})$, $(\mathrm{Ln}'_{\mathrm{Ce}}V_O^{\cdot\cdot})$), or trimers ($(\mathrm{Ce}'_{\mathrm{Ce}}V_O^{\cdot\cdot}\mathrm{Ce}'_{\mathrm{Ce}})$, $(\mathrm{Ln}_{\mathrm{Ce}}V_O^{\cdot\cdot}\mathrm{Ln}'_{\mathrm{Ce}})$, $(\mathrm{Ce}'_{\mathrm{Ce}}V_O^{\cdot\cdot}\mathrm{Ln}'_{\mathrm{Ce}})$); this may explain the increase in chemical expansion coefficient with decreasing temperature. Bishop and co-authors [85] reported an intermediate range with dlog(δ)/dlog(pO_2) $\approx -1/2$, mainly for undoped ceria, and interpreted this by assuming combination of isolated defects and intrinsic trimers formed as follows:

$$2\mathrm{Ce}^x_{\mathrm{Ce}} + \mathrm{O}^x_O \overset{K_2}{\Leftrightarrow} (\mathrm{Ce}'_{\mathrm{Ce}}V_O^{\cdot\cdot}\mathrm{Ce}'_{\mathrm{Ce}}) + 0.5\mathrm{O}_2. \tag{12}$$

Note that preference for intrinsic dimers$(\mathrm{Ce}'_{\mathrm{Ce}}V_O^{\cdot\cdot})$ or trimers $(\mathrm{Ce}'_{\mathrm{Ce}}V_O^{\cdot\cdot}\mathrm{Ce}'_{\mathrm{Ce}})$ may be expected for $\mathrm{Ce}_{1-x}\mathrm{Gd}_x\mathrm{O}_{2-0.5x-\delta}$ by taking into account the dependence of ionic radius on coordination; this is 0.1143, 0.107 and 0.101 nm for Ce^{3+} with CN = 8, 7 and 6, respectively. Corresponding changes in ionic radius of Gd^{3+} are smaller (i.e. 0.1053, 0.100 and 0.0938 nm for CN = 8, 7 and 6). Unfortunately, the dependence of oxygen stoichiometry on oxygen partial pressure has limited capability to quantify the contributions of different types of defect associations. In addition, the fitting reported in [85]. still shows clear deviations from the expected trend for very reducing conditions.

3 Dependence of Chemical Expansion on Changes in Working Conditions

Though the values of $\Delta\varepsilon_C/\Delta\delta$ or the corresponding chemical expansion coefficient [81] are often closely related to point defect chemistry and may be guidelines to understand relevant factors, useful information should rely mainly on the dependence of actual working conditions, such as temperature and chemical potential changes relative to a reference condition $\Delta\mu_{\mathrm{O}_2} = \mathrm{RT} \times \Delta\ln(\mathrm{pO}_2)$, or corresponding potential gradients across membranes. Thus, Fig. 2 also shows results expressed as:

$$\frac{\Delta\varepsilon_C}{\Delta\log(\mathrm{pO}_2)} = 2.3\mathrm{RT}\frac{\Delta\varepsilon_C}{\Delta\mu_{\mathrm{O}_2}}. \tag{13}$$

This is a suitable description for the dependence of chemical expansion caused by changes in chemical potential or under potential gradients. It also allows one to obtain the dependence of chemical expansion of electrode materials on overpotential, by combining Eq. 13 with the Nernst law:

$$\frac{\Delta\varepsilon_C}{\eta} = \frac{4F}{2.3\mathrm{RT}}\left(\frac{\Delta\varepsilon_C}{\Delta\log(\mathrm{pO}_2)}\right). \tag{14}$$

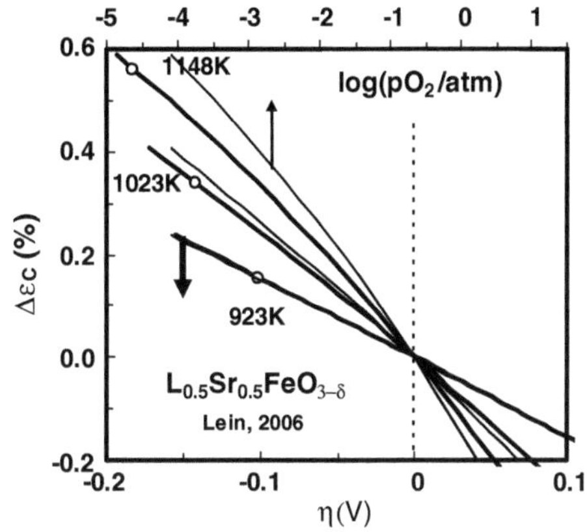

Fig. 4 Chemical expansion of $La_{0.5}Sr_{0.5}FeO_3$ versus working conditions, redrawn with data from [75]

This relative effect of electrode overpotential should not be underestimated, at least for materials with relatively large chemical expansion coefficient, as observed in Fig. 4 for $La_{0.5}Sr_{0.5}FeO_3$ [75].

The dependence of oxygen stoichiometry of some $La_{1-x}Sr_xCoO_{3-\delta}$ (LSC) materials with high contents of Sr (e.g. $La_{0.3}Sr_{0.7}CoO_{3-\delta}$) follows almost linear relations of δ versus $\log(pO_2)$ [81]. On combining this with Eq. 8 and Eqs. 13–14 one obtained predictions for the dependence of chemical expansion on oxygen partial pressure and on electrode overpotential (Fig. 5a). Corresponding results for LSC with low contents of Sr (e.g. $La_{0.9}Sr_{0.1}CoO_{3-\delta}$) show stronger dependence on temperature (Fig. 5b).

One may also obtain the dependence of chemical expansion of ceria-based materials on combining the dependence of oxygen stoichiometry on oxygen partial pressure (e.g., [92]) with typical values of chemical expansion coefficient (Fig. 6). Predictions are plotted versus oxygen partial pressure and also versus $\log(pH_2O{:}pH_2)$, assuming equilibrium in the gas phase ($H_2 + 0.5O_2 \Leftrightarrow H_2O$); this emphasises dependence on fuel conversion, and shows that chemical expansion increases in contact with fuel rich atmospheres and drops in fuel lean conditions.

4 Structural Guidelines for Redox Effects on Chemical Expansion

4.1 Perovskite Materials with Mixed Ionic and p-Type Conduction

Chemical expansion studies were reported for a variety of other perovskite materials with different B-site redox cations such as Cr [93], Co [81], Mn [94] or Ni [82] and combinations of different redox cations [75, 76, 95]. A summary of chemical expansion results ($\Delta\varepsilon_C/\Delta\delta$) is shown in Fig. 7. Differences in this variety of perovskite materials are far from trivial and should not be underestimated,

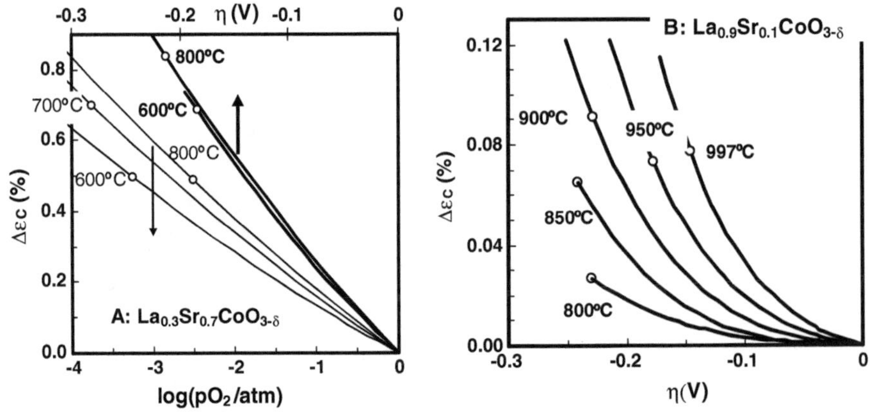

Fig. 5 Dependence of chemical expansion on working conditions for $La_{0.3}Sr_{0.7}CoO_{3-\delta}$ (*left*) and $La_{0.9}Sr_{0.1}CoO_{3-\delta}$ (*right*), redrawn with data from [101]

Fig. 6 Dependence of chemical expansion CGO20 on oxygen partial pressure (*closed symbols*) and on the H_2O:H_2 ratio (*open symbols*), based on combining oxygen sub-stoichiometry from [92]. and chemical expansion coefficients from [89]

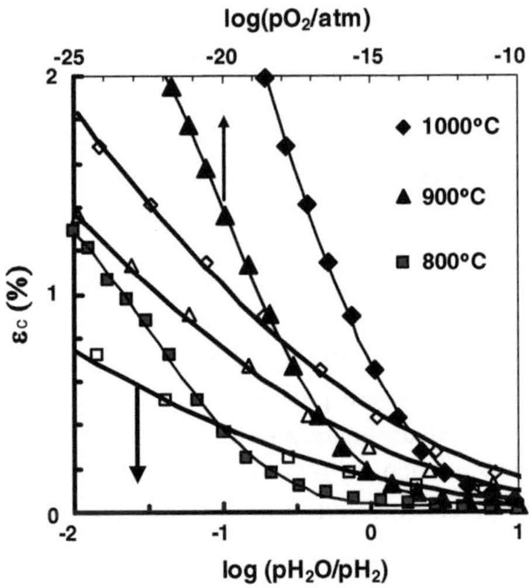

mainly because this may correspond to significant stress contributions under expected working conditions.

Another important conclusion from Fig. 7 is that chemical expansion coefficient can be related to the predicted stability of perovskites after reduction of the B-site species, as given by the so-called tolerance factor:

$$\tau = \frac{r_A + r_O}{(r_B + r_O)\sqrt{2}}. \tag{15}$$

This was obtained with ionic radii r_A, r_B, r_O [95], mainly for trivalent Cr^{3+}, Mn^{3+}, Fe^{3+} (high spin) and Co^{3+} (high spin), i.e., on assuming that chemical

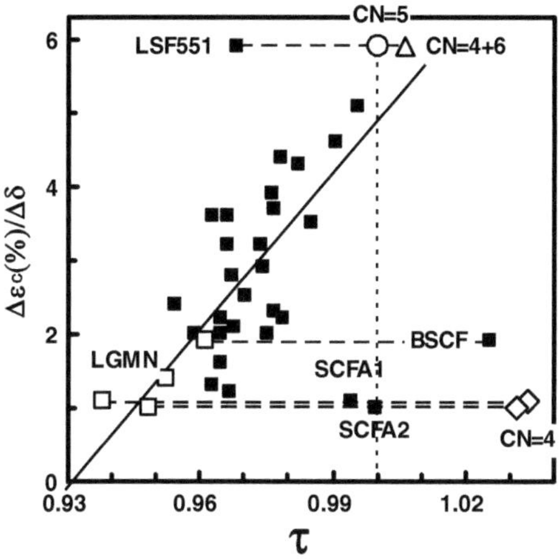

Fig. 7 Correlation between the chemical expansion coefficient and tolerance factor of representative ABO_3 pevoskites containing redox susceptible B-site species (Cr, Mn, Fe, Co, Ni) (from [75, 76, 81, 82, 93–96]). *Closed squares* correspond to tolerance factor based on 3+ B-site redox cations with CN = 6, and open squares for 2+ B-site redox cations with CN = 6. Other open symbols are for Fe^{3+} with CN = 4 (◇), CN = 5 (o) or equal fractions of CN = 4 and CN = 6 (Δ)

expansion corresponds to reductions $Cr^{4+} \rightarrow Cr^{3+}$, $Fe^{4+} \rightarrow Fe^{3+}$ and $Co^{4+} \rightarrow Co^{3+}$. Open squares show exceptional cases for $Fe^{3+} \rightarrow Fe^{2+}$ and/or $Co^{3+} \rightarrow Co^{2+}$ in SCFA = $SrCo_{0.4}Fe_{0.6-x}Al_xO_{3-\delta}$ [96] and BSCF = $Ba_{0.5}Sr_{0.5}C_{0.8}Fe_{0.2}O_{3-\delta}$ [76] or $Ni^{3+} \rightarrow Ni^{2+}$ in $LaGa_{0.65}Mg_{0.15}Ni_{0.2}O_{3-\delta}$ [82]. Coordination for B-site species was taken as CN = 6, except for exceptional cases indicated in the figure.

Most results in Fig. 7 are nearly described by a linear dependence on tolerance factor:

$$\frac{\Delta\varepsilon_C}{\Delta\delta} = 70.8\tau - 65.9. \tag{16}$$

However, one clear exception to this trend is the chemical expansion coefficient of $La_{0.5}Sr_{0.5}FeO_{3-\delta}$ (LSF551), from [75]. In this case, one can obtain closer matching to the general trend by assuming either Fe^{3+} with coordination CN = 5 (O) or equal fractions of CN = 4 and CN = 6 (Δ), rather than coordination CN = 6 (closed symbol). In both cases one may expect stabilisation of the perovskite phase upon reduction from Fe^{4+} to Fe^{3+}, provided that this implies decrease in coordination, as emphasised for CN = 5 ($\tau \approx 1.00$). The phase stability of $La_{0.5}Sr_{0.5}FeO_3$ is also demonstrated by ability to retain the cubic structure (space group Pm3m), down to about 300 °C. Other $La_{1-x}Sr_xFeO_{3-\delta}$ are rhombohedral (space group R$\bar{3}$c), for intermediate Sr contents or orthorhombic (Pnma) for even lower Sr [75]. The enhanced mixed conductivity of $La_{0.5}Sr_{0.5}FeO_3$ also distinguishes this composition relative to other LSF materials [97]. However, this seems to contradict the hypothesis of lower coordination for Fe^{3+} ions, which often implies trapping of oxygen ion vacancies and decrease in ionic conductivity, as found for other Fe-based mixed conducting perovskites [98]. Thus, detailed information on structural effects is needed to seek minimisation of chemical expansion and optimisation of mixed transport properties of these materials.

$Ba_{0.5}Sr_{0.5}Co_{0.8}Fe_{0.2}O_{3-\delta}$ (BSCF) also deviates from the main trend in Fig. 7. In this case, one may interpret the deviations by taking into account that both prevailing 3+ cations or prevailing 2+ cations imply important deviations from the ideal tolerance factor. Yet, the most likely hypothesis is that chemical expansion is related to reduction to Co^{2+} and Fe^{2+}, due to effects of large A-site cations (Ba^{2+}), promoting increase in average Co–O and Fe–O bonding and preventing oxidation to tetravalent B-site species, even in the most oxidising conditions. This is consistent with the oxygen stoichiometry of BSCF, which remains well below the value for prevailing trivalent B-site cations (i.e., $3-\delta < 2.5$) [76].

Figure 7 also shows corresponding results for tolerance factor of $SrCo_{0.4}$-$Fe_{0.6-x}Al_xO_{3-\delta}$ with $x = 0.1$ (SCFA1) or $x = 0.2$ (SCFA2). In these cases, one may assume two alternative explanations for low chemical expansion coefficients based on the relatively poor tolerance factor of the perovskite phase after either reduction to divalent cations (Co^{2+} and/or Fe^{2}), or reduction of Fe^{4+} to Fe^{3+} with lower coordination (CN = 4). Note that oxygen sub-stoichiometry is still $\delta < 0.5$ [96], indicating that the average valence of B-site species is likely to remain >3+, at least under oxidising conditions.

4.2 Chemical Expansion of Perovskite Materials with Coexisting Redox Pairs

Mixed transport and electrocatalytic properties of electrode or membrane materials may also involve interactions between co-existing redox pairs, whilst minimising oxygen stoichiometry changes and corresponding chemical expansion. For example, different interactions between redox pairs were suggested based upon Mn–Cr interactions [53] for $(La_{0.75}Sr_{0.25})_{0.95}Cr_{0.5}Mn_{0.5}O_{3-\delta}$; however, in view of XRD results indicating that only Mn changes oxidation state/coordination on redox, it seems more likely that Mn^{3+} disproportionation is more likely [99]:

$$Mn^{4+} + Cr^{3+} \Leftrightarrow Mn^{3+} + Cr^{4+}; \quad 2Mn^{3+} \Leftrightarrow Mn^{2+} + Mn^{4+}. \tag{17}$$

Such interactions are likely to give rise to the p-type contribution, without significant chemical expansion. The low chemical expansion obtained for oxidising conditions (Fig. 8) is likely to extend to relatively wide chemical potential ranges down to quite reducing conditions [53]. Higher chemical expansion was only observed for sufficiently reducing conditions, due to reduction to Mn^{2+} becoming dominant, as revealed by $p \rightarrow n$ transition.

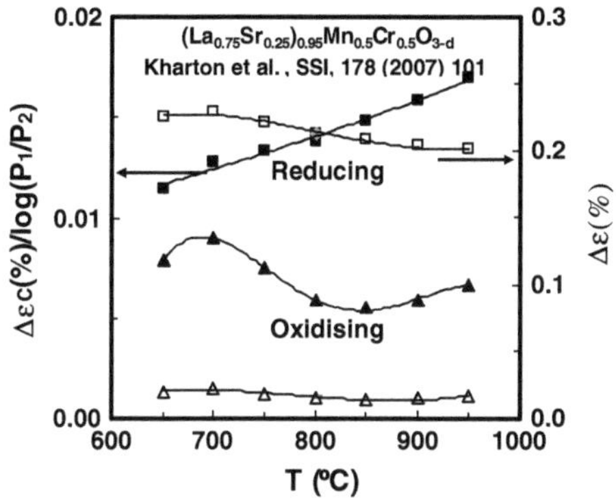

Fig. 8 Chemical expansion of $(La_{0.75}Sr_{0.25})_{0.95}Cr_{0.5}Mn_{0.5}O_{3-\delta}$ in oxidising conditions (from air to $pO_2 = 6 \times 10^{-4}$ atm), and upon exposition to reducing conditions (pO_2 = in the range $3 \times 10^{-13} - 5 \times 10^{-21}$ atm) [53]

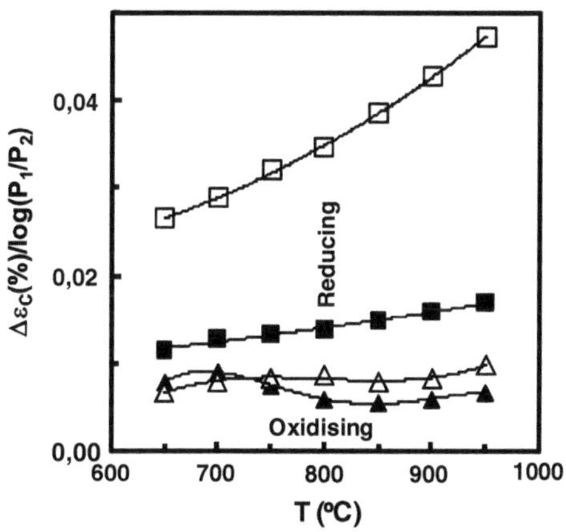

Fig. 9 Comparison of chemical expansion results obtained for $(La_{0.75}Sr_{0.25})_{0.95}Cr_{0.5}Mn_{0.5}O_{3-\delta}$ (*closed symbols*) [53] and $(La_{0.75}Sr_{0.25})_{0.95}Ti_{0.5}Mn_{0.5}O_{3-\delta}$ (*open symbols*) [100]. *Triangles* correspond to results obtained in oxidising conditions (from air to $pO_2 = 6 \times 10^{-4}$ atm), and *squares* for reducing conditions (pO_2 in the range 3×10^{-13}–5×10^{-21} atm)

Gradual substitution of Cr by Ti in $(La_{0.75}Sr_{0.25})_{0.95}Cr_{0.5-y}Ti_yMn_{0.5}O_{3-\delta}$ shows significant effects on chemical expansion, both for oxidising and reducing conditions [100] (see Fig. 9). Changes in transport properties, including the location of *p–n* transition, also confirm that these effects are related to differences in redox changes involving co-existing redox pairs. Redox interactions between Mn and Cr species may also account for the contributions of chemical expansion on heating $Sr_{0.7}Ce_{0.3}Mn_{1-y}Cr_yO_{3-\delta}$ materials (SCMC), and its dependence on the contents of Cr [84]. Note that the conductivity and oxygen permeability of these SCMC

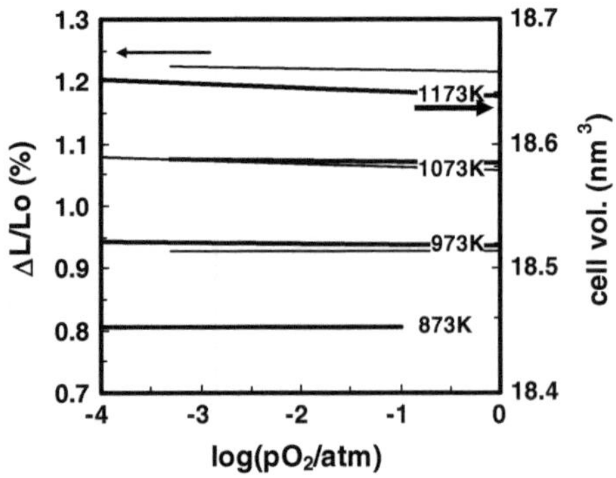

Fig. 10 Dependence of expansion of $La_2NiO_{4+\delta}$ on oxygen partial pressure (*thin lines*), relative to reference conditions in air at room temperature and corresponding changes in unit cell volume of Nd_2NiO_{4+d} (*thick lines*), (redrawn from [78, 79])

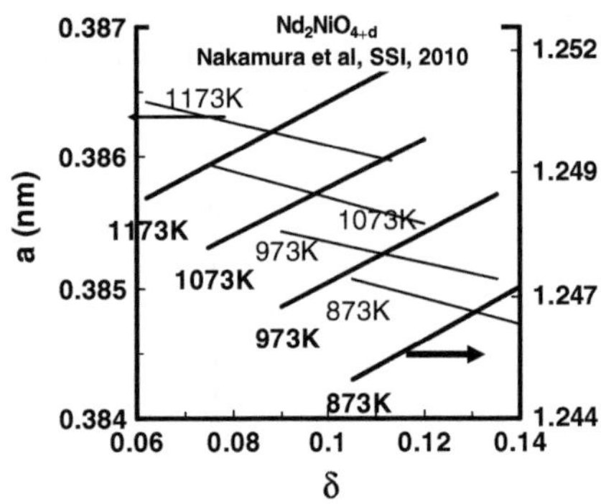

Fig. 11 Dependence of lattice parameters of $Nd_2NiO_{4+\delta}$ on oxygen hyperstoichiometry, at representative temperatures (redrawn from [79])

materials are superior to those reported for $(La_{0.75}Sr_{0.25})_{0.95}Cr_{0.5}Mn_{0.5}O_{3-\delta}$ [53], in similar oxidising conditions.

4.3 HyperStoichiometric Materials with Low Chemical Expansion

Hyperstoichiometric nickelates, $La_2NiO_{4+\delta}$ and $Nd_2NiO_{4+\delta}$, are interesting model materials for good mixed conductors with low chemical expansion [78, 79], as shown in Fig. 10. The behaviour of these materials can be understood by taking into account that the *a* and *c* lattice parameters show opposite trends for dependence on oxygen excess (Fig. 11), with minor impact on unit cell volume, at least for conditions when these materials retain tetragonal symmetry (space group 14/mmm or F4/mmm). Indeed, this unique chemical expansion behaviour may change for excessive hyperstoichiometry, due to partial or complete transformation into orthorhombic phase (space group Fmmm or Cmca) [79].

5 Thermochemically Induced Stresses

5.1 Stresses Under Thermochemical Gradients

Solid electrolytes in solid state electrochemical applications (SOFC, SOEC, etc.) are exposed to high gradients of chemical potential and similar conditions are imposed to mixed conducting membranes used for partial oxidation of hydrocarbons, etc. The resulting stresses across CGO electrolytes were analysed by Atkinson [60] for typical gradients of oxygen partial pressures. On assuming flat constrained conditions and without additional lateral constraint, total strain will combine chemical expansion across the membrane and a stress-related contribution as follows:

$$\Delta\varepsilon_r = \sigma(x)(1-v)/E + \Delta\varepsilon_C(x) \tag{18}$$

where $\Delta\varepsilon_C(x)$ denotes chemically induced strain, relative to a reference condition, E is Young modulus and v is the Poisson ratio. The resulting total strain will not change across the membrane if the membrane remains flat constrained, and thus:

$$\sigma(x) = \frac{E}{(1-v)}[\Delta\varepsilon_r - \Delta\varepsilon_C(x)]. \tag{19}$$

In addition, for externally unconstrained conditions:

$$\int_0^X \sigma \times \mathrm{d}x \approx \frac{X}{\Delta\mu O_2}\int_{\mu O_2'}^{\mu O_2''} \sigma \times \mathrm{d}\mu O_2 = 0 \tag{20}$$

and, since $\Delta\mu O_2 = \mathrm{RT} \times \ln\left(pO_2''/pO_2'\right)$, Eqs. 18–20 yield:

$$\sigma(0) = \frac{E \times \Delta\varepsilon_r}{(1-v)} \tag{21}$$

and

$$\sigma(X) = \frac{E\left(\Delta\varepsilon_r - \Delta\varepsilon_C(pO_2'')\right)}{(1-v)} \tag{22}$$

with

$$\Delta\varepsilon_r = (\Delta\mu O_2)^{-1}\int_{\mu O_2'}^{\mu O_2''} \varepsilon_C \mathrm{d}\mu O_2 = \left(\log(pO_2''/pO_2')\right)^{-1}\int_{pO_2'}^{pO_2''} \varepsilon_C \mathrm{d}\log(\mathrm{pO_2}). \tag{23}$$

In this analysis, one assumed a reference at the oxidising side (i.e. $\Delta\varepsilon_C(pO_2') = 0$).

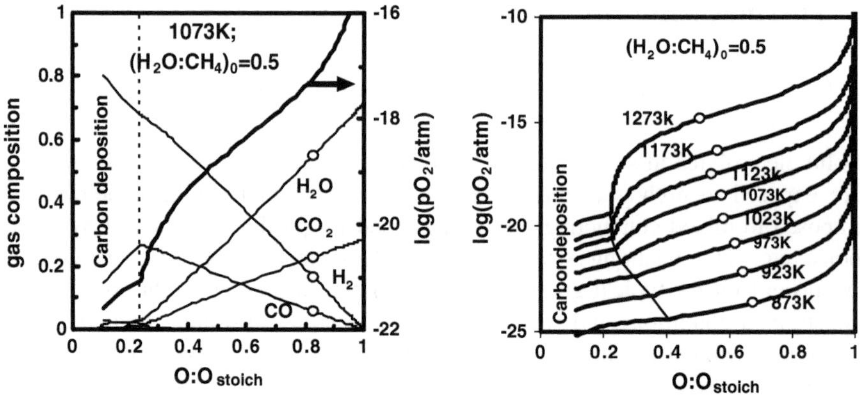

Fig. 12 Thermodynamic predictions of gas composition and oxygen partial pressure on relative oxygen content for the starting condition $H_2O{:}CH_4 = 0.5$ in gas feed

Fig. 13 Predictions of limiting stresses, and overall strain in a CGO20 membrane under air/fuel gradients versus fuel conversion, where $\sigma(0)$ refers to the air side and $\sigma(X)$ is for the fuel side

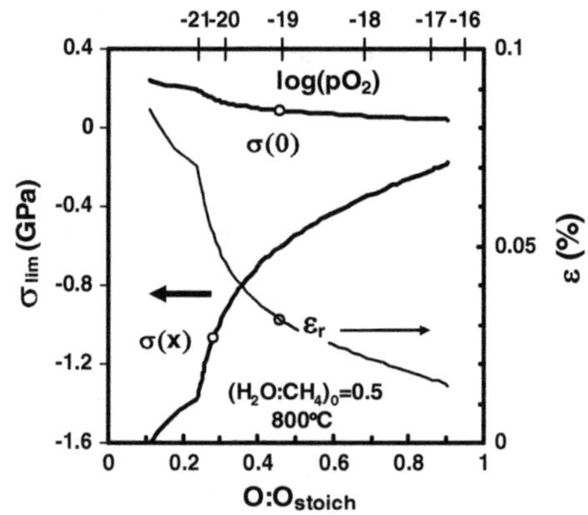

This dependence on oxygen partial pressure is easily transformed into the corresponding conversion of H_2 in fuel cells or steam electrolysis, based on the thermodynamics of the relevant reaction $H_2 + 0.5O_2 \Leftrightarrow H_2O$. For conversion of methane or other hydrocarbons, one may use suitable codes to predict composition changes with gradual conversion and corresponding dependence on equilibrium oxygen partial pressure (e.g. [101]). Representative examples in Fig. 12 show the dependence of composition (at 800 °C) and equilibrium oxygen partial pressure on relative degree of fuel oxidation $O{:}O_{stoich}$, where O_{stoich} denotes the oxygen contents for complete conversion to fully oxidised species (H_2O and CO_2). Note that discontinuities in Fig. 12 show the transition from carbon deposition into carbon free conditions, i.e., the value of relative oxygen content needed to avoid risks of carbon deposition.

The effects of fuel conversion on thermochemical stresses imposed on ceria-based membranes (Figs. 13 and 14) were evaluated by combining this dependence of equilibrium oxygen partial pressure on fuel conversion with the dependence of

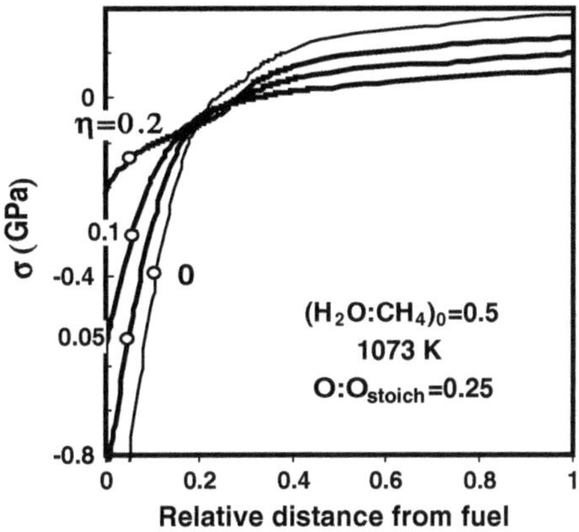

Fig. 14 Predictions of stresses across a CGO20 membrane versus relative distance to fuel, for $(H_2O:CH_4)_0 = 0.5$, $O:O_{stoich} = 0.25$, and $\eta = 0$, 0.05, 0.1, and 0.2 V

chemical expansion on oxygen partial pressure, and corresponding chemically induced stresses (Eqs. 18, 21–23). Figure 13 shows that the values of tensile stresses at the air side $\sigma(0)$ are about one order of magnitude lower than the absolute value of compressive stresses at the fuel side $\sigma(X)$. Both values depend on the relative degree of fuel oxidation (lower horizontal scale), with major risks of failure for very low fuel conversion. Note that, even the tensile stresses at the air side are far from negligible and reach about 200 MPa at the onset of carbon deposition. The axis on the right side of Fig. 13 shows the overall resulting strain (ε_r).

The predictions in Fig. 14 show more detailed stress profiles across the membrane under air/membrane/fuel gradients with initial steam/methane ratio = 0.5, relative degree of oxidation $O:O_{stoich} = 0.25$, and different values of overpotential ($\eta = 0$, 0.05, 0.1, 0.2 V). These predictions were obtained on combining Eqs. 19 and 23, for linear potential distance dependence across the membrane. The dependence of chemical expansion on oxygen partial pressure was taken from Fig. 6, and effects of anodic polarisation given by Eq. 14. Relevant elastic properties $E = 187$ GPa and $\nu = 0.334$ were taken from ref [102]. These predictions also show that overpotential imposes a gap $\eta = 0.25(\mathrm{RT}/F) \times \ln\left(\mathrm{pO}_2''/\mathrm{pO}_{2,\mathrm{fuel}}\right)$ between very reducing conditions in the fuel $(\mathrm{pO}_{2,\mathrm{fuel}})$ and less aggressive conditions at the surface of the membrane (pO_2''), with corresponding decrease in chemical expansion and also decrease in corresponding stresses under high chemical potential differences.

5.2 *Stresses in Multilayer Systems*

Thermochemical stresses in SOFC and other energy conversion systems are also due to differences between thermal and chemical expansion of component materials, as well as differences between processing, prospective operation conditions and room temperature. These changes will impose different strain contributions, including the effects of constrained conditions imposed by bonding of different cell components. On assuming isotropic behaviour for every cell component, with nearly flat configuration:

$$\varepsilon_i = \varepsilon_i^0 + \bar{\alpha}_i(T - T_0) + \sigma_i(1 - \nu_i)/E_i + \Delta\varepsilon'_{C,i} \tag{24}$$

where $\bar{\alpha}_i$ is the average thermal expansion coefficient, T_0 is the reference temperature, σ_i is a stress contribution imposed externally or due to bonding to other cell components, and $\Delta\varepsilon'_{C,i}$ represents combined contributions of chemical expansion due to changes in oxygen stoichiometry and additional effects of phase transformations, redox reactions, etc. i.e.:

$$\Delta\varepsilon'_{C,i} = \beta_{C,i}\Delta\delta + \varepsilon_{\mathrm{tr},i} + \ldots \tag{25}$$

The initial strain at reference conditions will be negligible ($\varepsilon_i^0 \approx 0$) if one assumes stress free conditions.

One may assume the following conditions for flat constrained multilayer systems without additional lateral constrains:

$$\varepsilon_1 = \varepsilon_2 = \ldots = \varepsilon_n \tag{26}$$

and

$$\sum_{i=1}^{n}(\sigma_i X_i) = 0. \tag{27}$$

thus, on combining the flat constrained conditions for two generic layers (Eqs. 24 and 26) one obtains:

$$\sigma_i = \frac{\sigma_j(1 - \nu_j)}{Ej}\frac{E_i}{(1 - \nu_i)} - \frac{E_i}{(1 - \nu_i)}\left[(\bar{\alpha}_i - \bar{\alpha}_j)(T - T_0) + \varepsilon_i^0 - \varepsilon_j^0 + \Delta\varepsilon'_{C,i} - \Delta\varepsilon'_{C,j}\right] \tag{28}$$

and on combining with laterally unconstrained conditions (Eq. 21):

$$\sigma_j = \left[\frac{E_j}{(1 - \nu_j)X_j}\right]\frac{(T - T_0)\sum_{i=1}^{n}\frac{(\bar{\alpha}_i - \bar{\alpha}_j)E_i X_i}{(1 - \nu_i)} + \sum_{i=1}^{n}\frac{(\varepsilon_i^0 - \varepsilon_j^0 + \Delta\varepsilon'_{C,i} - \Delta\varepsilon'_{C,j})E_i X_i}{(1 - \nu_i)}}{\sum_{i=1}^{n}\left[\frac{E_i}{(1 - \nu_i)}\frac{X_i}{X_j}\right]}. \tag{29}$$

This equation emphasises that a simple criterion of compatibility based on thermal expansion mismatch may be misleading. It also shows that the layers at greater risk are the thinner and/or most rigid ones. Simulated examples are shown in the next figures to illustrate expected trends for representative combinations of materials.

Table 1 Representative parameters of $Sr_{0.7}Ce_{0.3}Mn_{0.9}Cr_{0.1}O_{3-\delta}$, $(La_{0.75}Sr_{0.25})_{0.95}Cr_{0.5}Mn_{0.5}O_{3-\delta}$ and 8YSZ used to simulate thermochemical stresses

Material	SCMC [84]	LSMC [53]	8YSZ [103]
$10^6 \times \bar{\alpha}$ (K^{-1})	15.9	15.0	10.5
ΔT	(800–1,300 °C)	(800–1,300 °C)	
$\Delta\varepsilon_C$(%)	0.14	0.212	–
(ΔP/atm)	(7×10^{-4}–0.21)	(10^{-16}–0.21)	
$E/(1-v)$(GPa)	43 (Est.)	43 (Est.)	313

In real electrochemical systems, the reference temperature T_0 may differ from true stress free conditions, because different firing schedules are often used for different cell components. For example, processing of anode supported SOFCs may comprise separate stages for deposition and sintering of the thin dense electrolyte layer on the anode substrate and then deposition and firing of the thin porous cathode. Constrained sintering may also impose initial strain contributions. There are also uncertainties concerning stress/strain changes upon reduction of the NiO/YSZ precursor to the resulting Ni/YSZ cermet, especially after further redox cycling. The relevant literature also shows inconsistencies between predictions of thermal stresses and experimental validation (e.g. [103]). Thus, the actual predictions are exclusively based on simulated all ceramic cells. One combined effects of thermal contraction upon cooling from high temperature stress free conditions to prospective operation temperatures, with additional chemical expansion effects for the air electrode under cathodic polarisation, and for the fuel electrode under reducing conditions and anodic polarisation.

For the purpose of illustrating trends, one selected $(La_{0.75}Sr_{0.25})_{0.95}Cr_{0.5}Mn_{0.5}O_{3-\delta}$ (LSCM) [53] as prospective anode, and $Sr_{0.7}Ce_{0.3}Mn_{0.9}Cr_{0.1}O_{3-\delta}$ (SCMC) [84], as prospective cathode, with 8YSZ electrolyte. The relevant properties are listed in Table 1. Thermal and chemical expansion data can be taken from [53] for LSCM [53] and [84] for SCMC. Since the elastic properties of these porous LSMC and SCMC layers are unknown, one assumed $E/(1-v) \approx 43$ GPa, which is close to typical elastic properties for other porous ceramic electrodes (e.g. porous LSM [104]).

Simulations are shown in Fig. 15, for electrolyte supported cells, with typical thickness 20 μm for LSMC, 200 μm for 8YSZ and 20 μm for SCMC. One simulated the effects of stress free temperature on thermal stresses developed on cooling to 800 °C, in air (thin lines) and for H_2O,H_2/LSCM/8YSZ/SCMC/air cells with fuel conditions $H_2O{:}H_2 = 1{:}1$, and electrode polarisations of $\eta_a = 0.1$ V for LSCM and $\eta_c = -0.1$ V for the SCMC electrode. The dependence of chemical expansion on oxygen partial pressure [53, 84] was transformed to dependence on gas composition and/or overpotential as described above (Eq. 14).

For these electrolyte supported cells, most constrain is imposed by the thicker and denser electrolyte layer; this hinders contraction of other cell components with higher thermal expansion, and gives rise to tensile stresses. The magnitude of these stresses (Fig. 15) decreases on lowering the gap between the effective stress free

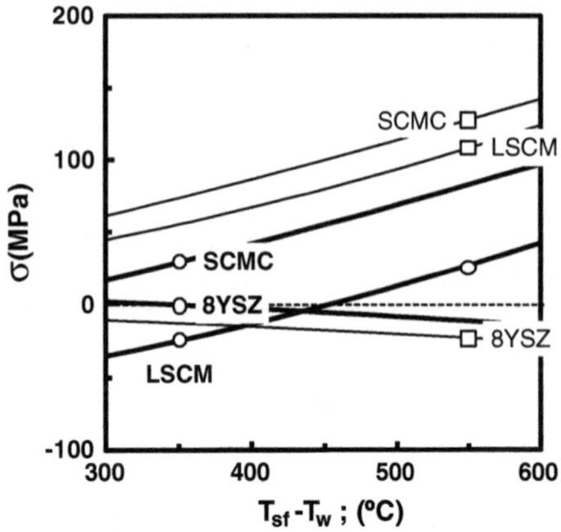

Fig. 15 Predictions of dependence of thermochemical stresses on the temperature gap between stress free temperature (T_{sf}) and working temperature, for electrolyte supported LSCM/8YSZ/SCMC cells, at 800 °C, in air (*thin lines*) and for prospective operation of H_2O,H_2/LSCM/8YSZ/SCMC/air cells with $H_2O:H_2 = 1:1$, and polarizations $\eta_a = 0.1$ V for LSCM and $\eta_c = -0.1$ V for SCMC

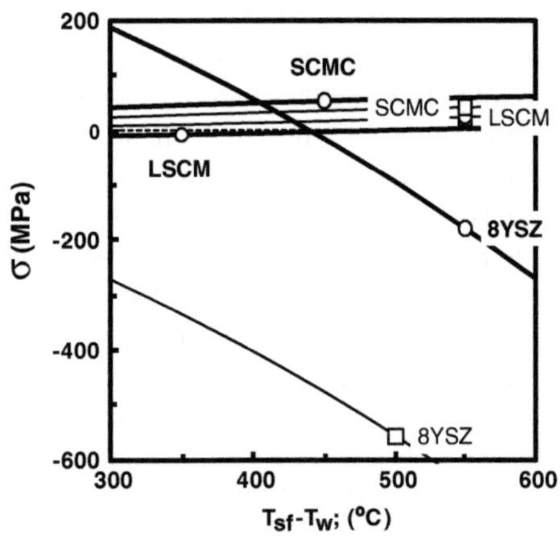

Fig. 16 Predictions of dependence of thermochemical stresses on the temperature gap between stress free temperature (T_{sf}) and working temperature, for anode supported LSCM/8YSZ/SCMC cells, at 800 °C, in air (*thin lines*) and for prospective operation of H_2O,H_2/LSCM/8YSZ/SCMC/air cells with $H_2O:H_2 = 1:1$, and polarisations $\eta_a = 0.1$ V for LSCM and $\eta_c = -0.1$ V for SCMC

temperature and operation conditions. In addition, reducing conditions imposed by the fuel and cathodic polarisation of the air electrode superimposes positive strain contributions on these electrode layers and lowers tensile stresses. These simulations indicate that chemically induced strain may even revert differences between the thermal expansion contributions of electrode and electrolyte layers for conditions when the gap between stress free temperature and working temperature is small, as shown for the LSCM layer. Unfortunately, one hardly expects significant beneficial effects of chemical expansion on cooling to room temperature, when the gap between stress free temperature and room temperature is likely to exceed 1,000 °C. In addition, porous ceramic layers readily uptake oxygen on cooling or on re-oxidising, with corresponding suppression of chemical expansion (e.g. [105]).

Figure 16 shows similar predictions of thermochemical stresses in electrode supported cells, assuming typical thickness of 500 μm for LSMC, 20 μm for 8YSZ and 20 μm for SCMC. In this case, the thin solid electrolyte layer is under high compressive stresses. Thermal stresses developed on cooling from the stress free temperature to 800 °C, in air, are in the same order of magnitude as reported for anode supported cells with Ni/YSZ cermets (e.g. [106]). However, compressive stresses imposed on the YSZ electrolyte may be substantially relaxed by inducing chemical expansion of the supporting LSCM layer in contact with fuels. Stresses in the thin electrolyte layer may even be reverted from compressive to tensile when the temperature gap is relatively low, with negligible stresses for intermediate temperature gaps.

6 Conclusions

Chemical expansion adds significant contributions to total strain of ceramic components of solid state electrochemical systems under expected working conditions. Chemical expansion can be related to point defect chemistry and corresponding relations to changes in oxygen stoichiometry, as emphasised for the most relevant types of materials with perovskite and fluorite structures. A compilation of data for a variety of perovskites with electronic or mixed conduction shows that chemical expansion upon partial reduction is dependent on the relative stability of the resulting phase. The chemical expansion coefficient of ceria-based materials is much higher than for perovskite materials, probably because differences between ionic radii of reduced and oxidised cations (i.e. Ce^{3+} and Ce^{4+}) have a greater influence on the unit cell volume in fluorite materials. The low chemical expansion of Ln_2NiO_4 materials (with Ln = La, Nd, …) can be ascribed to opposing effects of increasing oxygen stoichiometry on the *a* and *c* cell parameters. This type of materials may provide guidelines for development of other mixed conductors without undue chemical expansion effects.

The dependence of chemical expansion on oxygen chemical potential can be used to predict the corresponding dependence on fuel conversion and/or overpotential under typical operation conditions expected for SOFC and other electrochemical systems. These results allow one to evaluate stresses in mixed conducting membranes under high chemical potential gradients, as found for fuel/ceria/air configurations, and to assess the dependence on fuel conversion and overpotential. In addition, one can evaluate the impact of combined thermal and chemical expansion effects on stress profiles for electrochemical cells with mixed conducting electrodes. These predictions indicate that chemical expansion of fuel electrodes under reducing conditions and cathodic polarisation of the air electrode may counter the thermal expansion mismatch between electrodes and solid electrolytes. These simulations also reveal major differences between stress profiles in the most common electrode supported cells and previous electrolyte supported cells. Similarly, the ability to perform lab scale experiments with thick electrolyte samples and thin electrodes is insufficient to ensure that identical materials will keep thermochemical stresses within safe levels for electrode supported cells.

Acknowledgments This work was supported by the FCT, Portugal (Project REEQ/710/CTM/2005) and the European Commission (Project Matsilc-(STRP 033410).

References

1. H.U. Anderson, Solid State Ionics **52**, 33 (1992)
2. K.L. Ley, M. Krumpelt, R. Kumar, J. Mat. Res. **11**, 1489 (1996)
3. C.S. Montross, H. Yokokawa, M. Dokiya, Brit. Ceram. Trans. **101**, 85 (2002)
4. H. Ullmann, N. Trofimenko, F. Tietz, D. Stover, A. Ahmad-Khanlou, Solid State Ionics **138**, 79 (2000)
5. A. Petric, P. Huang, F. Tietz, Solid State Ionics **135**, 719 (2000)
6. D. Waldbillig, A. Wood, D.G. Ivey, Solid State Ionics **176**, 847 (2005)
7. M. Greenberg, E. Wachtel, I. Lubomirsky, J. Fleig, J. Maier, Adv. Funct. Mater. **16**, 48 (2006)
8. A. Kossoy, Y. Feldman, R. Korobko, E. Wachtel, I. Lubomirsky, J. Maier, Adv. Funct. Mater. **19**, 634 (2009)
9. A. Kossoy, Y. Feldman, R. Korobko, E. Wachtel, I. Lubomirsky, J. Maier, Adv. Mater. **22**, 1659 (2010)
10. T. Kawada, N. Sakai, H. Yokokawa, M. Dokiya, I. Anzai, Solid State Ionics **50**, 189 (1992)
11. J.A.M. Vanroosmalen, E.H.P. Cordfunke, Solid State Ionics **52**, 303 (1992)
12. J.A. Labrincha, J.R. Frade, F.M.B. Marques, J. Mater. Sci. **28**, 3809 (1993)
13. K.Q. Huang, M. Feng, J.B. Goodenough, C. Milliken, J. Electrochem. Soc. **144**, 3620 (1997)
14. A. Tsoga, A. Gupta, A. Naoumidis, P. Nikolopoulos, Acta Mater. **48**, 4709 (2000)
15. R.N. Basu, F. Tietz, E. Wessel, D. Stover, J. Mater. Proc. Technol. **147**, 85 (2004)
16. Y. Takeda, R. Kanno, M. Roda, Y. Tomida, O. Yamamoto, J. Electrochem. Soc. **134**, 2656 (1987)
17. F.M.B. Marques, L.M. Navarro, Solid State Ionics **100**, 29 (1997)
18. S.H. Chan, X.J. Chen, K.A. Khor, Solid State Ionics **158**, 29 (2003)
19. H.T. Lim, A.V. Virkar, J. Power Sources **192**, 267 (2009)
20. Y. Tao, H. Nishino, S. Ashidate, H. Kokubo, M. Watanabe, H. Uchida, Electrochim. Acta **54**, 3309 (2009)
21. S. Uhlenbruch, T. Moskalewicz, N. Jordan, H.J. Penkalla, H.P. Buchkremer, Solid State Ionics **180**, 418 (2009)
22. Z. Lu, X.D. Zhou, D. Fisher, J. Templeton, J. Stevenson, N. Wu, A. Ignatiev, Electrochem. Comm. **12**, 179 (2010)
23. S. Taniguchi, M. Kadowaki, H. Kawamura, T. Yasuo, Y. Akiyama, Y. Miyake, T. Saitoh, J. Power Sources **55**, 73 (1995)
24. H. Yokokawa, T. Horita, N. Sakai, K. Yamaji, M.E. Brito, Y.P. Xiong, H. Kishimoto, Solid State Ionics **177**, 3193 (2006)
25. T. Horita, Y.P. Xiong, H. Kishimoto, K. Yamaji, M.E. Brito, H. Yokokawa, J. Electrochem. Soc. **157**, B614 (2010)
26. S. Linderoth, N. Bonanos, K.V. Jensen, J.B. Bilde-Sorensen, J. Am. Ceram. Soc. **84**, 2652 (2001)
27. W.G. Coors, J.R. ÓBrien, J.T. White, Solid State Ionics **180**, 246 (2009)
28. P.M. Delaforce, J.A. Yeomans, N.C. Filkin, G.J. Wright, J. Am. Ceram. Soc. **90**, 918 (2007)

29. X.D. Zhou, S.P Simner, J.W. Templeton, Z. Nie, J.W. Stevenson, P.P. Gorman, J. Electrochem. Soc. **157**, B643 (2010)
30. M. Chen, B. Hallstedt, L.J. Gauckler, Solid State Ionics **176**, 1457 (2005)
31. T.S. Zhang, S.H. Chan, W. Wang, K. Hbaieb, L.B. Kong, J. Ma, Solid State Ionics **180**, 82 (2009)
32. R. Knibbe, J. Drennan, J.G. Love, Solid State Ionics **180**, 984 (2009)
33. D. Pomykalska, M.M. Bucko, M. Rekas, Solid State Ionics **181**, 48 (2010)
34. T.S. Zhang, Z.H. Du, S. Li, L.B. Kong, J. Lu, J. Ma, Solid State Ionics **180**, 1311 (2009)
35. K. Fujita, K. Ogasawara, Y. Matsuzaki, T. Sakurai, J. Power Sources **131**, 261 (2004)
36. S. Taniguchi, M. Kadowaki, T. Yasuo, Y. Akiyama, Y. Itoh, Y. Miyake, K. Nishio, Denki Kagaku **64**, 568 (1996)
37. V. Vedasri, J.L. Young, V.I. Birss, J. Power Sources **195**, 5534 (2010)
38. Y. Wang, M.E. Walter, K. Sabolsky, M.M. Seabaugh, Solid State Ionics **177**, 1517 (2006)
39. S.D. Kim, H. Moon, S.H. Hyun, J. Moon, J. Kim, H.W. Lee, Solid State Ionics **178**, 1304 (2007)
40. A.C. Muller, D. Herbstritt, E. Ivers-Tiffee, Solid State Ionics **152**, 537 (2002)
41. J.R. Kong, K.N. Sun, D.R. Zhou, N.Q. Zhang, J. Mu, J.S. Qiao, J. Power Sources **166**, 337 (2007)
42. Q. Jeangros, A. Faes, J.B. Wagner, T.W. Hansen, U. Aschauer, J. Van Herle, A. Hessler-Wyser, R.E. Dunin-Borkowski, Acta Mater. **58**, 4578 (2010)
43. A. Faes, H.L. Frandsen, M. Pihlatie, A. Kaiser, D.R. Goldstein, J. Fuel Cell Sci. Tech. **7**, 051011 (2010)
44. T. Hatae, Y. Matsuzaki, S. Yamashita, Y. Yamazaki, J. Electrochem. Soc. **157**, B650 (2010)
45. D. Waldbillig, A. Wood, D.G. Ivey, J. Power Sources **145**, 206 (2005)
46. D. Sarantaridis, A. Atkinson, Fuel Cells **7**, 246 (2007)
47. Y. Zhang, B. Liu, B.F. Tu, Solid State Ionics **176**, 2193 (2005)
48. A. Atkinson, S. Barnett, R.J. Gorte, J.T.S. Irvine, A.J. Mcevoy, M. Mogensen, S.C. Singhal, J. Vohs, Nat. Mater. **3**, 17 (2004)
49. O.A. Marina, N.L. Canfield, J.W. Stevenson, Solid State Ionics **149**, 21 (2002)
50. D.P. Fagg, V.V. Kharton, A.V. Kovalevsky, A.P. Viskup, E.N. Naumovich, J.R. Frade, J. Eur. Ceram. Soc. **21**, 1831 (2001)
51. S.Q. Hui, A. Petri, J. Eur. Ceram. Soc. **22**, 1673 (2002)
52. J.P. Ouweltjes, M. van Tuel, M. Sillessen, G. Rietveld, Fuel Cells **9**, 873 (2009)
53. V.V. Kharton, E.V. Tsipis, I.P. Marozau, A.P. Viskup, J.R. Frade, J.T.S. Irvine, Solid State Ionics **178**, 101 (2007)
54. M.R. Pillai, I. Kim, D.M. Bierschenk, S.A. Barnett, J. Power Sources **185**, 1086 (2008)
55. C.D. Savaniu, J.T.S. Irvine, J. Mater. Chem. **19**, 8119 (2009)
56. T. Kolodiazhnyi, A. Petric, J. Electroceram. **15**, 5 (2005)
57. R. Moos, K.H. Hardtl, J. Am. Ceram. Soc. **80**, 2549 (1997)
58. S.Q. Hui, A. Petric, J. Electrochem. Soc. **149**, J1 (2002)
59. Q. Ma, F. Tietz, D. Sebold, D. Stover, J. Power Sources **195**, 1920 (2010)
60. A. Atkinson, Solid State Ionics **95**, 249 (1997)
61. K. Sato, K. Yasiro, T. Kawada, H. Yugami, T. Hashida, J. Mizusaki, J. Power Sources **195**, 5481 (2009)
62. R. Krishnamurthy, B.W. Sheldon, Acta Mater. **52**, 1807 (2004)
63. K. Sato, K. Yashiro, T. Kawada, H. Yugami, T. Hashida, J. Mizusaki, J. Power Sources **195**, 5481 (2010)
64. K. Sato, H. Omura, T. Hashida, K. Yashiro, H. Yugami, T. Kawada, J. Mizusaki, J. Test. Eval. **3**, 246 (2006)
65. J.C. Grenier, N. Ea, M. Pouchard, P. Hagenmuller, J. Solid State Chem. **58**, 243 (1985)
66. F. Prado, N. Grunbaum, A. Caneiro, A. Manthiram, Solid State Ionics **167**, 147 (2004)
67. C. de la Calle, A. Aguadero, J.A. Alonso, M.T. Fernandez-Diaz, Solid State Sci. **10**, 1924 (2008)
68. Z.Q. Deng, W.S. Yang, W. Liu, C.S. Chen, J. Solid State Chem. **179**, 362 (2006)

69. S. McIntosh, J.F. Vente, W.G. Haije, D.H.A. Blank, H.J.M. Bouwmeester, Solid State Ionics **177**, 833 (2006)
70. S. Adler, S. Russek, J. Reimer, M. Fendorf, A. Stacy, Q.Z. Uang, A. Santoro, J. Lynn, J. Baltisberger, U. Werner, Solid State Ionics **68**, 193 (1994)
71. S. Streule, A. Podlesnyak, D. Sheptyakov, E. Pomjakushina, M. Stingaciu, K. Conder, M. Medarde, M.V. Patrakeev, I.A. Leonidov, V.L. Kozhevnikov, J. Mesot, Phys. Rev. B **73**, 094203 (2006)
72. J. Canales-Vazquez, F.M. Figueiredo, J.C. Waerenborgh, W.Z. Zhou, J.R. Frade, J.T.S. Irvine, J. Solid State Chem. **177**, 3105 (2004)
73. J. Canales-Vazquez, M.J. Smith, J.T.S. Irvine, W.Z. Zhou, Adv. Func. Mat. **15**, 1000 (2005)
74. A. Fossdal, M. Menan, I. Waernhus, K. Wiik, M.A. Einarsrud, T. Grande, J. Am. Ceram. Soc. **87**, 1952 (2004)
75. H.L. Lein, K. Wiik, T. Grande, Solid State Ionics **177**, 1795 (2006)
76. S. McIntosh, J.F. Vente, W.G. Haije, D.H.A. Blank, H.J.M. Bouwmeester, Chem. Mat. **18**, 2187–2193 (2006)
77. R. Kriegel, R. Kircheisen, J. Topfer, Solid State Ionics **181**, 64 (2010)
78. V.V. Kharton, A.V. Kovalevsky, M. Avdeev, E.V. Tsipis, M.V. Patrakeev, A.A. Yaremchenko, E.N. Naumovich, J.R. Frade, Chem. Mat. **19**, 2027–2033 (2007)
79. T. Nakamura, K. Yashiro, K. Sato, J. Mizusaki, Solid State Ionics **181**, 402 (2010)
80. V. Kharton, A.A. Yaremchenko, M.V. Patrakeev, E.N. Naumovich, F.M.B. Marques, J. Eur. Ceram. Soc. **23**, 1417–1426 (2003)
81. X. Chen, J. Yu, S.B. Adler, Chem. Mater. **17**, 4537–4546 (2005)
82. A.A. Yaremchenko, V.V. Kharton, E.N. Naumovich, D.I. Shestakov, V.F. Chukharev, A.V. Kovalevsky, A.L. Shaula, M.V. Patrakeev, J.R. Frade, F.M.B. Marques, Solid State Ionics **177**, 549–558 (2006)
83. V.V. Kharton, E.V. Tsipis, I.P. Marozau, A.P. Viskup, J.R. Frade, J.T.S. Irvine, Solid State Ionics **178**, 101 (2007)
84. A.A. Yaremchenko, A.V. Kovalevsky, V.V. Kharton, Solid State Ionics **179**, 2181 (2008)
85. S.R. Bishop, K.L. Duncan, E.D. Wachsman, Electrochim. Acta **54**, 1436 (2009)
86. C.Y. Park, A.J. Jacobson, Solid State Ionics **176**, 2671 (2005)
87. V.V. Kharton, A.A. Yaremchenko, A.L. Shaula, A.P. Viskup, F.M.B. Marques, J.R. Frade, E.N. Naumovich, J.R. Casanova, I.P. Marozau, Def. Dif. Forum **226–228**, 141 (2004)
88. M. Mogensen, L. Lindegaard, U.R. Hansen, G. Mogensen, J. Electrochem. Soc. **141**, 2122 (1994)
89. A. Atkinsom, T.M.G.M. Ramos, Solid State Ionics **129**, 259 (2000)
90. S. Wang, E. Orikawa, T. Hashimoto, J. Electrochem. Soc. **15**, E46 (2004)
91. S.R. Bishop, K.L. Duncan, E.D. Wachsman, Acta Mater. **57**, 3596 (2009)
92. D. Perez-Coll, D. Marrero-Lopez, J.C. Ruiz-Morales, P. Nunez, J.C.C. Abrantes, J.R. Frade, J. Power Sources,**173**, 291 (2007)
93. T.R. Armstrong, J.W. Stevenson, L.R. Pederson, P.E. Raney, J. Electrochem. Soc. **143**, 2919–2925 (1996)
94. S. Miyoshi, J.O. Homg, K. Yashiro, A. Karmal, Y. Nigara, K. Kawamura, T. Kawada, J. Mizusaki, Solid State Ionics **161**, 209 (2003)
95. R.D. Shannon, Acta Cryst. **A32**, 751 (1976)
96. X Dong, Z. Xu, X. Chang, C. Zhang, W. Jin, J. Am. Ceram. Soc. **90**, 3923–3929 (2007)
97. E.V. Tsipis, M.V. Patrakeev, V. Kharton, A.A. Yaremchenko, G.C. Mather, A.L. Shaula, I.A. Leonidov, V.L. Kozhevnikov, J.R. Frade, Solid State Sci. **7**, 355 (2005)
98. J.C. Waerenborgh, F.M. Figueiredo, J.R. Frade, M.T. Colomer, J.R. Jurado, J. Phys. Condens. Matter **13**, 8171 (2001)
99. S.M. Plint, P.A. Connor, S.W. Tao, J.T.S. Irvine, Solid State Ionics **177**, 2005–2008 (2006)
100. V.A. Kolotygin, E.V. Tsipis, A.L. Shaula, E.N. Naumovich, J.R. Frade, S. Bredikhin, V.V. Kharton, J. Sol. State Electrochem. **15**, 313–327 (2011)
101. J.R. Frade, V.V. Kharton, A. Yaremchenko, E. Naumovich, J. Power Sources **130**, 77 (2004)

102. A. Atkinson, A. Selçuk, Solid State Ionics **134**, 59 (2000)
103. A. Atkinson, B. Sun, Matter. Sci. Tech. **23**, 1135 (2007)
104. A. Nakajo, Z. Wuillemin, J. Van Herle, D. Favrat, J. Power Sources **193**, 203 (2009)
105. T. Nagai, W. Ito, T. Sakon, J. Am. Ceram. Soc. **91**, 303 (2008)
106. H. Yacabe, Y. Baba, T. Sakurai, M. Satoh, I. Hirosawa, Y. Yoda, J. Power Sources **131**, 278 (2004)

Current State of Models for the Prediction of Mechanical Failures in Solid Oxide Fuel Cells

Arata Nakajo, Jan Van herle and Daniel Favrat

1 Introduction

The solid oxide fuel cell (SOFC) technology has to face many challenges before its large-scale commercialisation. Cost reduction, along with enhanced reliability, durability, fuel flexibility, load following capabilities and compactness are needed. Yet, despite all the research, the exact underlying mechanisms of the electrochemical reactions have not yet been unambiguously identified. The high-temperature environment promotes physicochemical modifications of the materials that alter the electrochemical and mechanical properties after prolonged use. The driving forces of these degradation processes that arise from chemical interactions between the SOFC materials themselves, on the one hand, and the volatile contaminants transported by the fed gases, on the other hand, have not yet been fully clarified.

The structural reliability of SOFCs is a salient issue. The end of operation of a stack is ultimately caused by the loss of structural integrity of one or several of the cells. This is the result of (i) the accumulation during operation of physicochemical alterations, of plastic and creep deformations, and (ii) the modification of the temperature profile, due to the degradation of the electrochemical performance of the cells. Mechanical issues do not, however, exclusively occur after prolonged use. Inappropriate control during load following, harsh conditions and thermal cycles can induce discrete failures. The in-series

A. Nakajo (✉) · J. Van herle · D. Favrat
Laboratoire d'Energétique Industrielle (LENI),
Ecole Polytechnique Fédérale de Lausanne (EPFL),
Bat. ME A2, Station 9, 1015 Lausanne, Switzerland
e-mail: arata.nakajo@epfl.ch

J. Van herle
e-mail: Jan.Vanherle@epfl.ch

D. Favrat
e-mail: daniel.favrat@epfl.ch

J. T. S. Irvine and P. Connor (eds.), *Solid Oxide Fuels Cells: Facts and Figures*,
Green Energy and Technology, DOI: 10.1007/978-1-4471-4456-4_6,

assembly of the standard repeating units (SRU) in a stack is a striking weakness, the mitigation of which is hindered by the complexity of the failure modes in the ceramic materials. Mechanical failure in a single cell induces a succession of detrimental effects that act in a coupled manner. For instance, delamination or cracking of the electrode layers breaks the ionic and/or electronic conduction paths to the electrochemical reactions sites. The ensuing local loss of performance induces a harmful redistribution of the current density. Once cracked, the electrolyte or sealant does not any longer ensure the separation of the fuel and air compartments to the fullest extent. The local and unsteady combustion provokes a local increase of the temperature, as well as diverse chemical alterations of the cell layers. All these undesirable phenomena promote in turn additional stresses, which reach critical values in an accelerated manner. The SOFC field is, therefore, characterised by strong interactions between the phenomena.

Despite the evidence of mechanical issues in SOFCs, which are experienced even during laboratory button cell tests, this topic is still receiving limited attention. Efforts are seen as stand-alone tasks, owing to the different experimental and modelling techniques needed to gather the essential information, whereas mechanical failures in SOFCs are likely intricately related to physicochemical and electrochemical aspects.

This survey of the current state of models for the prediction of mechanical failures in SOFC stacks comprises five parts. For conciseness, the emphasis is on planar stack design with anode-supported cells. The description can be easily transposed to other situations, differing in terms of type of cell, geometry or operating temperature. In Sect. 2, the diverse origins of stress in the components are briefly presented, along with the most common failure mechanisms. In Sect. 3, a brief overview of the mechanical behaviour of the cell, sealing and gas diffusion layer (GDL) materials is provided, in the view of their implementation in numerical tools based on the finite-element method (FEM). The need for the complete temperature and ageing dependences and refined constitutive laws for rate-independent plasticity and creep is highlighted. Section 4 presents the modelling approaches applied at the SRU scale and discusses the particularities of SOFC modelling: (i) initialisation steps, (ii) boundary conditions representative of stacking of the repeating units and (iii) choice of suitable elements among those available in commercial FEM tools. The discussion is enlarged with (i) simple models, such as derived from the beam theory, and (ii) the recent representative elementary volume (REV) analyses that may provide, in the future, insights into the failure modes. Section 5 illustrates the current modelling capabilities at the SRU scale with studies available in the literature. To conclude, the most stringent model improvement needs are discussed in Sect. 6.

2 Origins of Stresses and Mechanical Failures in SOFC Stacks

Multilayer systems are extensively used in engineering to fulfil combined requirements on different aspects, be it mechanical, electrical or thermal. A planar SOFC stack is a typical example, at different scales. To build a functional unit, the impervious electrolyte, interconnects and sealants achieve the separation of the air and fuel compartments, whereas the in-series electrical stacking of the SRUs further requires the use of GDL. The mechanical interactions between these components depend on the selected technological solutions. Most of these parts are themselves multilayer systems.

In broad terms, the nature and long-term evolution of the stress is governed by the properties of the materials. Similar to other energy conversion devices, SOFC stacks will incur repeated full or partial thermal cycles and variations of the electrical load, which reduce the lifetime. An essential mitigation approach is to ensure reduced or suitable differences between the coefficients of thermal expansion (CTE) of the different materials. This is usually completed by the wise use of compliant linking elements to accommodate the mismatch strains that cannot be avoided. Because of the temperature-dependence of the CTE and the several manufacturing steps of a stack, such as cell sintering, metallic interconnect (MIC) coating, sealing and reduction procedure of the nickel-based anode, this task is not straightforward. In a next step, the understanding and control of irreversible deformation alleviates the history-dependent degradation of the behaviour. Here, a comprehensive analysis must include together electrochemical and mechanical aspects (see Sect. 6).

The following description of the origins of stresses and failure mechanisms per component is not exhaustive, because of their dependence on the choice of the technological solutions and geometry of the components. A classification of the mechanical failures in SOFC stacks developed in Japan is available [46]. Figure 1 depicts some of the common failures observed during experiments.

2.1 Cell

The cell or membrane electrode assembly (MEA) is the central component. It comprises the electrolyte, the electrodes as well as the contacting and required compatibility layers. As any multilayer system made of brittle materials, the MEA is prone to failures related to residual stresses ensuing the manufacturing process, and to further thermal cycling and prolonged exposition to aggressive environments [23]. The electrochemical performance is the first requirement that dictates the choice of the materials and the microstructure. The selection has to be performed in the view of long-term performance to alleviate undesirable chemical reactions and morphological alterations of the microstructure promoted by the high-temperature aggressive environment [47, 68, 80, 109, 118]. This is therefore a trade-off, as high electrochemical activity and chemical stability are usually contradictory aims.

Stresses in a cell embedded in a stack originate from different phenomena. Residual stresses build up during the cool down, after the sintering, because of the mismatches between the CTEs of the materials of the cell. The dissymmetry of the system results in curved specimens. Stresses in the cell then arise from mechanical load due to the stacking and the joining with the other components of the SRU. The reduction of nickel oxide in the anode takes place when fuel is fed for the first time and results in a change in porosity, hence in mechanical properties, and shrinkage [99, 102]. During operation, the uneven distribution of temperature and oxygen vacancy concentration in the layers prone to isothermal expansion, cause additional stresses in the cell [105]. In the long-term, the progressive physico-chemical alterations, creep mainly governed by the temperature and stress profile, and the possible shrinkage of the ceramic materials during thermal cycles [82, 84], can induce history dependences that reduce the resistance against load following and thermal cycling.

The cracking modes in the MEA layers and at the interfaces depend on the design, which consists in the choice of a supporting layer and of materials of suitable CTE. They may evolve during combined cycling and ageing because of the relaxation of the initial shielding compressive stress in the weakest layers [91]. Lanthanum strontium manganite (LSM) has been used for a long time and has proved its stability during operation at high temperature. The use of a composite LSM–yttria-stabilised zirconia (YSZ) electrode enables a considerable improvement of the performance, hence reduction of the operating temperature. Lanthanum strontium cobaltite ferrite (LSCF) cathodes combined to yttria (YDC) or gadolinia-doped ceria (GDC) compatibility layers are the result of concurrent research on more active materials. In an LSM-based cathode deposited on an Ni-YSZ/YSZ anode-support, the stress can change from tensile to compressive, depending on the temperature, whereas an LSCF cathode withstands tensile stress. In the former case, buckling-driven delamination is observed at the interface with the electrolyte [98], whereas cracking of the LSCF cathode may occur [70]. The same issues affect the contacting layers, as shown in Fig. 1. The cracking modes in the Ni-YSZ anode on an electrolyte support exhibit similarities with the latter case. Once at the interface, perpendicular cracks can either further propagate in the support, continue along the interface, or kink [23].

2.2 Interconnect

Structural issues related to the interconnect depend on the choice of the material. High-temperature SOFCs, such as the tubular design of Siemens-Westinghouse [111], make an exclusive use of ceramic materials, apart from the nickel felt used for current collection. The most common material, lanthanum chromite, is brittle and prone to non-uniform isothermal expansion when simultaneously subjected to reducing and oxidising atmospheres on different faces [131].

The decrease of the operating temperature pursued by the anode-supported cell technology enables the use of metallic components, with beneficial implications on costs and design possibilities. Similar to the MEA, an MIC is a multilayer of finite durability [66] wherein, in state-of-the-art solutions, a ceramic coating slows the growth of the oxide scale of lower electrical conductivity and the evaporation of volatile chromium species that contaminates the cathode. At the microscale, the

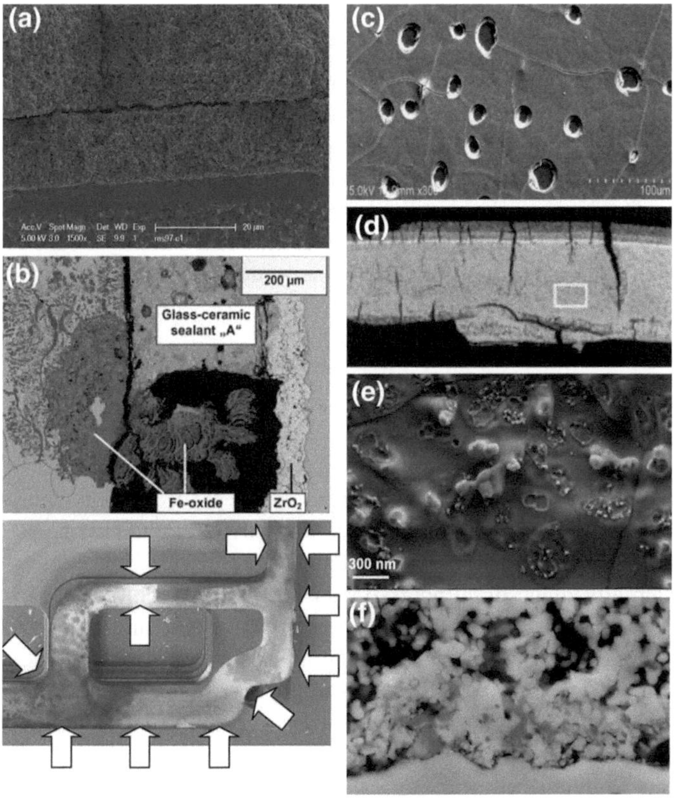

Fig. 1 Examples of mechanical failures and microstructural alterations observed in SOFC stacks. **a** Delamination of cathode current collection layer [21], reproduced here with kind permission from © ASME 2008. **b** cracking of the sealant [10], reproduced here with kind permission from © Elsevier 2006, and loss of gas-tighntess [72], reproduced here with kind permission from © John Wiley and Sons 2009. **c** MIC coating delamination [18], reproduced here with kind permission from © Elsevier 2008. **d** anode support reoxidation cycling [25], reproduced here with kind permission from © Elsevier 2009 **e** formation of zirconates in LSM–YSZ cathode [68], reproduced here with kind permission from © Elsevier 2009. **f** chromium deposit in LSM–YSZ cathode [78], reproduced here with kind permission from © The Electrochemical Society 2007

evolution of the thickness of the layer, coupled to creep and/or rate-independent plasticity, has a strong impact on the nature of the mechanical failures, which exhibit similarities with those acting in thermal-barrier coatings (TBC) of gas turbine blades [24]. Spallation and delamination is experimentally reported [18] (see Fig. 1).

While ductility is an advantage for the manufacturing of complex shapes, structural issues can arise from irreversible deformation generated during operation, as thermal stresses can exceed the elastic limit at distinct locations in the SRU [60, 88, 89, 93]. Cracking of the MIC is seldom observed. In contrast, the creep and increments of plastic strain generated during discrete events and prolonged use, respectively, can progressively lead to cell failure, through buckling, loss of electrical contact or gas-tightness [88, 89, 93]. The creep strain rates in these materials under the stress and temperature of operation is higher than those of the cell [86]. The expected temperature differences of 100 K over the SRU [92] induces very different amounts of irreversible strains in the zones subjected to comparatively high or low temperatures. The need to reduce the thickness of the components to decrease the costs and the thermal inertia of the stack increases the difficulty to prevent such failure modes. As classical stability issues are progressively solved, the contribution of the oxide scale and coatings to the mechanical behaviour of the component will increase.

2.3 Sealants

The sealing solutions span, with increasing level of mechanical interactions between the joined components, from compressive gaskets [13] to glass-ceramic materials [27], through thin metallic elements [127].

The main advantage of compressive gaskets is their compliance, which at first sight, should ease the requirements during the development of an SOFC stack to ensure the integrity of the cell. A loading system is needed to apply a high assembly load. It induces additional complexity, hence weakness, and weight which is not suitable for mobile applications. The ensuing compressive stresses further place tighter requirements on the dimensional tolerances of adjacent SRU components to avoid cell cracking caused by flatness imperfections. Despite the lack of mechanical characterisation (see Sect. 3), the properties of the compressive gasket materials certainly change after the first heat-up and differ depending on the atmosphere. An improper assembly process, the lack of stability of the matting surfaces and of the compressive load can cause the cracking of the gasket. However, the most deleterious structural effects arise from the slightly permeable nature of this sealing solution, which can indirectly induce cell failure, through parasitic combustion and, in the worst case, partial and unsteady reoxidation of the Ni in the anode and reduction of the cathode material [128].

Rigid sealants are alternatives to compressive gaskets. They can be divided into four main categories [28]: glass [14, 35], glass-ceramic [79, 110], metal brazes [54, 123] and bonded-compliant seals (BCS) [125]. All of them are more hermetic than compressive gaskets, at the cost of increased development efforts. They yield stronger mechanical interactions between the different components of the SRU, hence importance of CTE mismatches. A compliant component, such as the metallic foil in the BCS concept, can alleviate the ensuing stress in the cell. Cost and increased flexibility in shape are additional advantages over compressive gaskets, in particular for glass-based sealants, which can be dispensed either by a syringe as a paste or placed on the sealing surfaces as tape cast sheets. Their dense nature and adhesion to the other SRU materials spares the development, weight and price of a compressive loading system. This advantage may be decisive for mobile applications, where dynamic issues related to the application of the mechanical load can arise. The mechanical properties of a glass-ceramic sealant in a stack are expected to exhibit disparities. They depend on the sealing procedure performed on the whole stack, during which completely uniform conditions are impossible to ensure (see Sect. 3).

The composite and evolving structure of the ceramic-glass sealants as devitrification proceeds induces microvoids, which can be possibly healed at operating temperature [95]. The most common failure mode of a glass-ceramic sealant is cracking in the bulk of the sealant or at the sealing interfaces, during thermal cycles. The strength of the interfaces with the sealing areas may suffer from direct or indirect deleterious chemical interactions [17].

2.4 Gas-Diffusion Layer

The variety of mechanical failures related to GDLs is considerable, because solutions are specific to stack manufacturers. These are, most of the time, proprietary, which accounts for the scarce available data. In the case of metallic foams, large cracks in the GDL are unlikely thanks to the usually compressive stress state. The impact of corrosion on the global mechanical behaviour of metallic foams is expected to be severe in comparison with the MIC case because of the thickness of the struts relative to the oxide scales. GDLs can cause cell fracture indirectly. Partially sintered pastes are extensively used to enhance the electrical contact, mitigate the release of volatile poisoning species and slow the growth of oxide scales. They result in weak adhesion between the GDLs and neighbouring elements [20]. The initially designed CTE mismatch, adjusted by consolidation steps to control the stress state and the distribution of the contact pressure on the different components, may be altered by non-uniform creep deformation generated during prolonged operation. The possible ensuing change in stress state during thermal cycles can induce localised losses of electrical contact. Slight differences in relative density, combined to the limited number of struts in the thickness can significantly alter the creep strain rate at high temperature [97]. They affect the uniformity of the gas supply in a stack and, at a smaller scale, the gas flow distribution within the stacked SRUs. This results in accelerated degradation hence increased risks of mechanical failures [90].

3 Materials

The limited knowledge on the essential mechanical properties of the SOFC stack materials and, further, on their evolution during prolonged and cycling conditions, induced by the interplay with physicochemical phenomena, hinders the efficient mitigation of mechanical failures. A brief overview of the properties of the most common materials is provided here. The reader is redirected to compilations on the mechanical properties of SOFC materials available in the literature for more extensive information [7, 86, 90, 106, 120]. The data summarised here are detailed in [86, 90].

3.1 Coefficient of Thermal Expansion

The CTEs of SOFC materials are available for a broad variety of compositions in different atmospheres. The thermal expansion contains a chemical contribution, depending on the environment. Considerable stress and significant ensuing deformation in the SRU can arise from chemical (isothermal) expansions.

Most of the CTE data consists in values for a single temperature difference, typically room temperature (RT) to 1273 K. In comparison, few temperature-dependent or differential values are readily available, despite their importance for a comprehensive stress analysis. The thermal expansions between the reference state and those of interest, i.e. sintering and the temperature of operation, anode reduction, heat treatment of the glass sealant and room temperatures at least, have to be known. The implementation of thermal expansion in commercial FE software is straightforward. It requires either the differential or total CTE from a reference state [1], and can be easily controlled by field variables or dedicated subroutines. Expansion or shrinkages can be modelled by using swelling procedures. Figures 2 and 3 depicts the typical temperature dependence of the CTE of the materials.

The most common electrolyte material is YSZ. Another central one is yttria- (YDC) or gadolinia-doped ceria (GDC). Despite its higher ionic conductivity than 8YSZ, its drawback are non-negligible electronic conduction at low oxygen partial pressure and isothermal expansion. It can also serve as a compatibility layer, to prevent undesirable reactions between the YSZ electrolyte and an LSCF cathode. The CTEs of both materials exhibit temperature dependence. SRU typically have to withstand spatial temperature differences of 100 K, e.g. 973–1073 K. The corresponding variation of the CTE between RT and 1073 K in one data set is of approximately $9.8-11.0 \times 10^{-6}\,K^{-1}$ for 8YSZ and $12.1-12.9 \times 10^{-6}\,K^{-1}$ for GDC, which can lead to imprecision in the calculation of the stress field.

Ni-based cermets are preferred for SOFC anodes owing to their good electrochemical activity for H_2 and CO oxidation, and catalytic activity for the reforming of hydrocarbons. The ionic conducting phase is made of YSZ or GDC. The CTEs of Ni-YSZ anodes in oxidised and reduced state increase for higher NiO or Ni contents. The most striking feature is the anomaly due to the antiferromagnetic to paramagnetic transition of NiO (approx. 530 K), and the ferromagnetic to paramagnetic transition of metallic Ni (approx. 630 K), which entails a change in the structure from rhombohedral to cubic structure in the case of NiO. The CTE between RT and 1073 K is comprised within $11.2-13.3 \times 10^{-6}\,K^{-1}$ [44, 84, 100]. Particular attention is required for composites materials subjected to thermal cycling. In such conditions, Mori et al. [84] report a more severe increase of the CTE of Ni-YSZ material with increasing heating/cooling rates and denser samples.

Fig. 2 Dependence on temperature of the CTE (*right*, *r*) and Young's modulus (*left*, *l*) of MEA materials, from the data compiled in [87, 91]. Reproduced here with kind permission from © John Wiley and Sons 2011

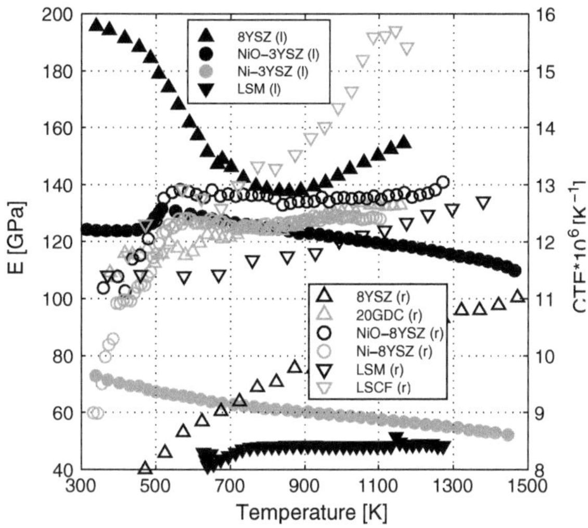

Fig. 3 Dependence on temperature of the CTE (*right*, *r*) and Young's modulus (*left*, *l*) of sealing and MIC materials, from the data compiled in [86]

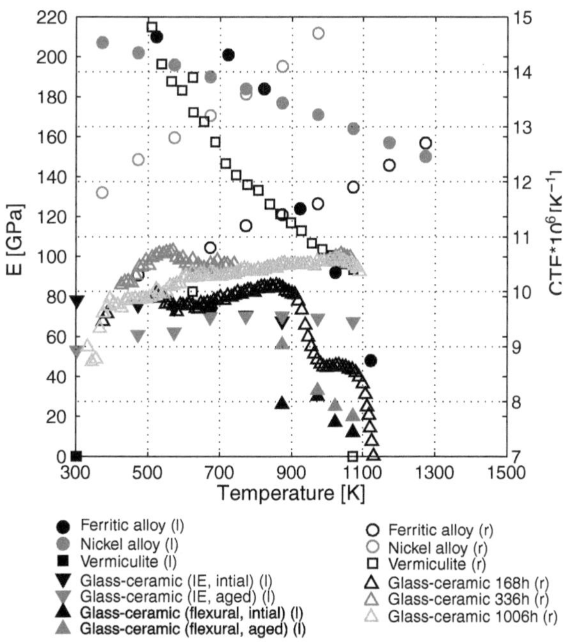

The effect is larger during the first 5–10 cycles, approximately 4.67×10^{-8} and 5.75×10^{-8} per cycle for $2\,\mathrm{K\,min^{-1}}$ and $10\,\mathrm{K\,min^{-1}}$ heating ramps, respectively, between 323 and 1273 K. It then reduces to 4.28×10^{-9} and 5.5×10^{-9} per cycle. The difference between cooling and heating measurements indicates a shrinkage of the anode. Sun et al. [112] report a similar value of 1.2×10^{-8} (1.4×10^{-8} on average for [84]) for cycling between 298 and 1173 K. To our knowledge, the exact origin of this behaviour has not yet been clarified.

The CTE of LSM ($11.6-13.2 \times 10^{-6}\,K^{-1}$, between RT and 1073 K) closely matches that of YSZ or the anode, whereas that of LSCF is higher ($13.0-19.3 \times 10^{-6}\,K^{-1}$, between RT and 1073 K). These CTEs can be tailored, to some extent by modifying the content of Sr on the A-site of $La_{1-x}Sr_xMnO_3$ [83], respectively the Sr and Fe content on the A- and B-site of $La_{1-x}Sr_xCo_{1-y}Fe_yO_3$ [115, 116]. Porous $La_{0.6}Sr_{0.4}MnO_3$ samples undergo severe shrinkage during thermal cycling but not ageing in air or oxygen between 873 and 1373 K [83]. The strain per cycle is as high as 5×10^{-4}.

Most of the mechanical data on rigid sealants concerns the so-called G18, a barium-calcium-aluminosilicate (BCAS) glass-ceramic material ($35BaO - 35SiO_2 - 15CaO - 10B_2O_3 - 5Al_2O_3$, mol%) developed at the Pacific Northwest National Laboratory [76, 110]. Data on reactive air braze can be found in [39, 48, 54, 123, 124, 126]. The addition of boron oxide is a means to decrease the glass transition temperature, control the viscosity and improve the wetting of BCAS ceramic-glasses [28], while that of alumina modifies the rate of crystallisation. The sealing procedure for G18 yields a content of crystallite phase of about 50 vol.%, which increases up to 70 vol.% after 120 h at 1023 K and then stabilises [76]. The CTE between RT and 1073 K is in the range of $9-13.0 \times 10^{-6}\,K^{-1}$ depending on the composition [79].

Compressive gaskets for high temperature applications are usually made of either vermiculite, phlogopite mica or muscovite mica. Exfoliation results in highly compliant structures made of crystal platelets or fine particles. The CTE of such materials is likely very complex and its measurement challenging. Depending on the manufacturing process, an orthotropic behaviour is expected. The scarce available data indicates that it lies within $10-14 \times 10^{-6}\,K^{-1}$ between RT and 1073 K [34].

Preliminary selections retained two main classes of metallic alloys, ferritic and nickel-based chromia-forming alloys [133]. The ferritic alloys have a significant edge on the costs. Most of the initial developments focused on the improvement of the long-term electrical conductivity, mitigation of the volatilization of chromium and CTE adjustments. The typical CTE of chromia-forming ferritic alloys ($11.3-11.9 \times 10^{-6}\,K^{-1}$, between RT and 1073 K) matches that of the anode, whereas that of nickel-based alloys is higher ($15.2 \times 10^{-6}\,K^{-1}$). In both cases, element addition can reduce the mismatch with the anode, electrolyte or cathode materials, depending on the supporting layer and desired stress state in the components.

3.2 Elastic Properties

Different procedures exist for the measurement of the elastic properties. An evaluation of four different techniques is provided in [103], the most common being the impulse excitation technique (IE).

The Young and shear moduli strongly depend on the porosity. A word of caution should be warranted on the use of Young's modulus as an indicator of the presence of microcracks. This approach considers that the decrease of the effective area that can sustain mechanical loads, due to the presence of cracks, results in a decrease of the apparent Young's modulus. While its justification seems straightforward, care is required when using such an approach. Indeed, depending on the fracture toughness of the material, the simplest case, i.e. a unique slender crack in a dense sample, highlights that the crack length which significantly affects the apparent Young's modulus can be far larger than that would have induced failure. The implementation in commercial software of a linear elastic behaviour is straightforward. Figures 2 and 3 depict the typical temperature dependence of the Young's modulus of the materials.

The reported stress–strain curves of YSZ samples are linear over the investigated temperature range [2, 49]. The Young modulus exhibits a strong temperature dependence, whereas, in addition, that of GDC decreases by 80 % at an oxygen partial pressure of 10^{-20}, compared with air [40, 121]. In air, the values are comparable for both materials and range from 140 to 220 GPa for dense samples.

The reduction procedure entails an increase in porosity of the Ni-based anode material, which is well predicted by simple relations [102, 106]. Data on the temperature dependence of the Young's modulus of the common compositions of Ni-YSZ anodes remain scarce, but reported trends are fairly similar [11, 73, 100]. The Young's modulus in oxidised state exhibits a peak related to the antiferromagnetic to paramagnetic transition of NiO, already reflected in the CTE, followed by a decline, though a shift in temperature appears when comparing data between [100] and [11]. Because of the possible differences in porosity, typical values lie in the range of 80–170 GPa (oxidised state) and 30–100 GPa (reduced state).

The data on the elastic properties of cathode SOFC materials is very scarce. Moreover, most of the measurements are performed on dense samples, which differs from the porous microstructure of a functional cathode. The Young's modulus of LSM samples measured by Giraud et al. [33] exhibits singularities. IE measurements could not be performed for temperatures lower than 623 and 873 K for porous and dense samples, respectively, presumably because of their ability to absorb vibrations. To our knowledge, no explanation has been provided for this behaviour. For a porosity of 4–29 %, the Young's modulus ranges from 48 to 120 GPa.

The Young modulus of ferritic alloys for SOFC application drops from 220 to 50 GPa, by increasing the temperature from RT to 1073 K. That of Ni-based alloys degrades significantly less, i.e. around 150 GPa at 1073 K.

At first sight, the mechanical behaviour of metallic foams appears as the simplest among competing GDL solutions, as their structure offers homogenised properties. This is only partially achievable, however, due to their non-periodic structure and the effect of the number of struts [8]. Non-negligible anisotropy, non-uniform deformation patterns and considerable scatter in the measurements of identical samples are reported [8, 96]. Nevertheless, simple relations based on analytical considerations on the possible deformation modes of simplified unit cells provide correct scaling relations for the qualitative prediction of some properties of cellular solids deduced from those of the dense material, e.g. [32], depending on the relative density.

Stress–strain curves of G18 exhibit large differences among samples from similar laboratory batches, which underscore the sensitivity of the mechanical properties to processing. The measurements of the Young's modulus from bend test and IE technique differ drastically and nonlinear behaviour occurs above 973 K, because of significant irreversible deformation. Liu et al. [65] propose a relation for the prediction of the time- and temperature-dependent Young's modulus of BCAS glass-ceramic material. The values gathered by IE are in the range of 50–80 GPa [110].

The mechanical behaviour of compressive gaskets is very complex and likely depends on the environment [13, 34].

3.3 Rate-Independent Plasticity and Creep

Unified equations that couple rate-independent plasticity and creep [114] are not readily available for SOFC materials. The data in the literature allows a simple description that arbitrarily separates the two contributions. In the case of isotropic hardening FEM tools for structural analysis conveniently accept data in the form of tabular data that describes the plastic strain–stress relation for uniaxial loading. This approach suffers limitations, in terms of maximum allowed strain, typically 10 %, predictions in the behaviour during cycling and validity for stress states characterised by large rotations of the principal axes.

Creep occurs when the material is loaded at a temperature near one half of its melting point and can be classified into three main categories, depending on the mechanism: diffusion, viscous and dislocation creep, as shown in deformation mechanism maps [4]. Difficulties arise from the temperature dependence and possible overlapping or changes in the dominant mechanism. In the case of SOFC materials, most of the existing measurements are at temperatures significantly higher than those encountered during normal operation of SOFC stacks. The needed extrapolation can lead to seriously misleading predictions. In general, creep in polycrystalline metals loaded at SOFC operating temperature occurs by grain boundary sliding and diffusion at low stress, while higher stresses are in the regime of thermally activated climb or glide dislocation motion [4].

The evolution of the strain rate during a tensile creep experiment under constant load usually exhibits three regimes: a fast initial deformation, followed by a constant strain rate and a final increase of the strain rate before failure, referred to as primary, secondary and tertiary creep. Most of the data focuses on the second regime. For preliminary investigation, it can be recast in the form of the Norton law. Figure 4 depicts a comparison of experimental creep strain rate and predictions from calibrated Norton law, on the typical range of stress and temperature in an SOFC stack during operation.

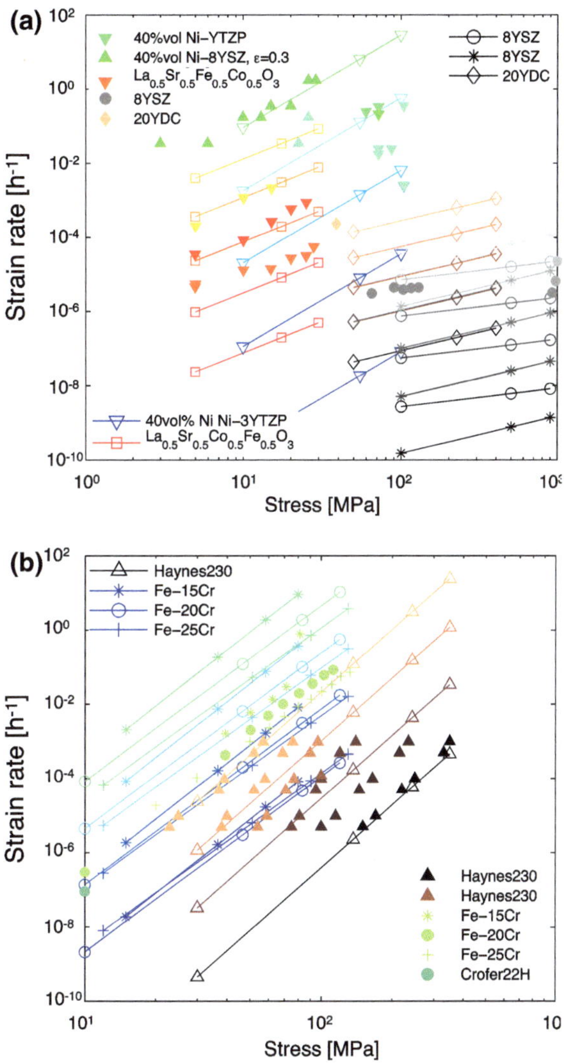

Fig. 4 Comparison between computed (*lines*) and measured (*symbols*) creep strain rates of **a** MEA and **b** MIC materials [86, 90]. Reproduced here with kind permission from © Pergamon 2012

The separation of the contributions of rate-independent plasticity and creep can lead to inconsistent interactions. ABAQUS [1], for instance, properly handles isotropic creep and plasticity together. In the case of either anisotropic plasticity or creep, plasticity is not accounted for in creep simulations.

The investigation on the creep behaviour of the ceramic materials is limited. Data on bending creep of 8YSZ and 20YDC in a temperature range of 1223–1563 K and 1323–1398 K, respectively, is available in [42, 55, 69]. The creep strain rate in the anode is the highest among the cell layers, in their respective expected stress range. It is, therefore, expected to dominate that of anode-supported cells, for geometrical and material reasons, but studies on the topic are scarce [37, 81]. The data found on the bending creep behaviour of LSM materials is not relevant for simulations of SOFC operation, since it is limited to the high-temperature domain, around 1523 K [104]. For $La_{0.5}Sr_{0.5}Fe_{0.5}Co_{0.5}O_3$ measurements at 1173–1323 K in air are available [59].

The maximum elongation before rupture of metallic alloys is high, typically around 20–60 % [41, 119]. The engineering stress–strain curves can be used to estimate those of the metallic GDLs.

For the glass-ceramic sealant, Nguyen et al. [95] propose a refined continuous damage model for dealing, in an homogenised manner, with viscoelastic behaviour and plasticity of the glass phase, microcracking, void formation and decohesion at the interfaces between glass and crystallite phases. A simpler approach is to implement the nonlinear curves in FEM software, with a hypoelastic constitutive relation, and complete these with a simple expression for the creep [110].

The behaviour of high-temperature compressive gaskets is complex. The results from Bram et al. [13] show a fairly reproducible response after the third loading cycle for vermiculite and mica-based compressive gaskets. Their data, however, is not sufficient for structural analysis at the SRU scale which requires the knowledge of the behaviour of the gasket for different loading and unloading paths [12].

3.4 Strength of Brittle Ceramic Materials

The strength of SOFC ceramic materials is not an intrinsic property, but strongly depends on the flaws and defects, that result from the manufacturing process and act as stress concentrators. Their failure stress is statistically distributed as a function of the flaw size distribution and depends on the size of the component. The Weibull analysis [122] is based on the weakest-link model, with an infinite number of links, the weakest one controlling the strength. The classical two-parameter version is valid for one defect population. The preferred test method is the ring-on-ring arrangement which generates a biaxial stress field and tests a larger volume [53]. Table 1 summarises the range of reported characteristic strength, Weibull modulus and reference volume [86]. The Weibull modulus is usually lower than 10, which is the threshold for technological ceramics. For the anode, the amount of data on the strength in the most relevant conditions, i.e. at high temperature and in reduced state, is limited because of the required testing apparatus and efforts.

.Data on the effects of cycling and ageing on the strength of ceramic SOFC materials is limited and contradictory. The decrease of ionic conductivity of YSZ during ageing is well known, though the mechanisms have not yet been unambiguously clarified [3, 38, 50–52]. Among the possible causes of this phenomenon, the cubic to tetragonal phase transformation is expected to have implications on the mechanical behaviour. This is confirmed by the measurements of the tensile strength of 8YSZ samples aged in air at 1273 K for 1000 h by Kondoh et al. [49]. In contrast, studies do not show any significant improvement in similar conditions [69]

Table 1 Example of Weibull parameters of SOFC materials

		Characteristic strength (MPa)	Weibull modulus	Reference volume[a] (mm^3)	
Cathode	RT	52	6.7	1.22	[7]
(LSM)	1073 K	75	3.7	2.84	[7]
Electrolyte	RT	232	5.7	0.54	[7]
(YSZ)	1073 K	154	8.6	0.30	[7]
Anode[b]	RT	290 (277,304)	6.8 (5.4,8.6)	10.11	[86]
(NiO-YSZ)	1073 K	253 (242,266)	6.6 (5.2,8.4)	10.69	[86]
Anode[c]	RT	97.6 (86.4,110.2)	4.3 (2.9,6.4)	9.55	[102]
(NiO-YSZ)	1073 K	–	–	–	
Anode[d]	RT	79 (73,85)	7 (5,10)	4.81	[102]
(Ni-YSZ)	1073 K	–	–	–	
Compatibility layer	RT	134	3.8	1.03	[7]
(GDC)	1073 K	183	5.7	0.58	[7]

[a] Computed afterwards in compliance with ASTM C1499 standards [30]
[b] Porosity of 25 % after reduction
[c] Porosity of 18 %
[d] Porosity of 36 %

or only decreased Weibull modulus [53]. Similarly, the strength of anode materials can decrease during thermal cycling. The literature, however, exhibits opposite observations [73]. Extensive research to find the exact underlying phenomena and causes of these discrepancies has not yet been performed.

Data on the strength of cathode materials mostly consists in flexural bending strength, predominantly for dense materials. Values of the Weibull parameters are seldom found. The data on LSM from Atkinson et al. [7] is one of the few available. Surprisingly low Weibull moduli are reported, without explanation and similar to GDC, the characteristic strength increases with temperature. The low-temperature plastic deformation of LSM samples observed by Cutler et al. [77] is accompanied by a low Young's modulus and bending strength. These results contrast with the higher strength and brittle behaviour measured by De Souza et al. [22] for $La_{0.875}Sr_{0.125}MnO_3$.

3.5 Additional Information

Additional information is needed for structural analysis, depending on the stack technology, the modelling approach and the problems to solve, because of the several manufacturing steps. Section 4 provides the details of the model implementation.

The reduction procedure induces a change of the mechanical properties and, in most cases, a slight shrinkage of the anode. Most of the data on the reduction strain is available in studies of the redox behaviour of anode supports. Pihlatie et al. [99] have proposed a model to predict the cumulative redox strain (CRS). The underlying phenomena of the dimensional change include chemical expansion, rate-independent plasticity, creep and possibly microcracking. The dependence on the reduction conditions, such as temperature and humidity, anode microstructure and cell geometry is recognised [36, 101], but considerable discrepancies exist among the measurements. The largest strain value of -2.34×10^{-2} exceeds the contribution of the typical pursued CTE mismatch between SOFC materials which lies between 10^{-3} and 10^{-2}.

The use of a glass-ceramic material implies additional manufacturing steps which have to be performed on the whole SOFC stack before operation, as the sealants are assembled in the form of paste or tape cast sheets. A typical sequence starts with the burn-off of the binder. The amorphous glass must then flow sufficiently to accommodate initial dimensional irregularities and effectively wet the sealing surfaces above the glass transition temperature (around 850–1010 K), approximately equivalent to the softening point (around 900–960 K) [28]. At the same time, excessive overflow must be avoided for structural integrity. The glass then partially devitrifies during the heating ramp and subsequent heat treatments. A proper control of the final amount of crystallite phases results in sealants of suitable strength and CTE, to alleviate the residual stresses caused by the stronger joining of the SRU components. The key properties of the material, i.e. glass transition and/or softening temperature, crystallization kinetics, viscosity, CTE and strength are interrelated and adjustable by modifications of the composition. The reported ideal sealing temperature of G18 is 1123 K, and effective joining of the materials requires some compressive load [76]. This must be included in the initialisation procedure before the structural analysis (see Sect. 4).

Very few studies have focused on the contact behaviour between the GDL, the cell, the compressive gaskets and the interconnect. The nature of the contact and the friction coefficient, if relevant, are unknown. The values of the former used in stress analyses are guesses, between 0.16 and 0.2 [15, 92]. Some data exists on the strength of the interface between the cathode and the contacting paste [20].

4 Modelling Approaches

The high operating temperature of SOFCs increases the costs of the experiments, the issues related to instrumentation implementation and hinder to draw unambiguous conclusions from post-test analyses. Another characteristic of the SOFC field is the difficulty to uncouple the problems in experiments while keeping the multiphysics nature of the problems. Modelling is of particular interest to overcome these difficulties.

4.1 Information Flow

Structural issues at the SRU scale are commonly investigated by importing the temperature and/or the vacancy profile generated by a thermo-electrochemical model in structural analysis tool. The former uses either the finite-volume, difference method or FEM, whereas the latter are based on the FEM. The effect of stress or partial mechanical failures on the electrochemical performance is not implemented at present. Depending on the issues to solve, the thermo-electrochemical model is dynamic and includes degradation processes. Dynamic mechanical effects are not included, because studies focus on stationary application. The full geometry must be included. Indeed, the consideration of a representative channel somewhat limits the relevance of the analysis, since critical mechanical failures can occur as well in introduction, exhaust and sealing zones. The locations of interest for gas flow distribution, electrochemical behaviour and mechanical stress differ, hence so do the meshes. Here, the simplest and quickest strategy, if applicable, for the importation of the profiles generated by the thermo-electrochemical models is to enable a layer-by-layer importation that spares the need for three-dimensional interpolation.

4.2 Choice of Elements and Contact Algorithm

The different components are meshed and their interaction modelled by the contact algorithm or constrain method of the software [1]. The contact pair tracking methods depend on the software. Except during the assembly of the cell, which can be simplified, the relative motion of the SRU components is small, since it is mostly driven by the differences in thermal and isothermal expansion. This enables the use of a simpler small-sliding tracking method, to reduce the computation time. Similarly, the correction for nonlinear geometric effects is not required in common situations, since large deformations are not expected.

The computation demand in contact simulation is high. Therefore, first order solid or continuum shell elements are favoured [1]. Most of the studies at the SRU scale do not include the effect of the curvature of the cell on the interaction with the other components. Here, at the cell scale, an approach is to use a combination of second-order solid elements for the supporting layer and shell elements for the electrodes or interlayers. In most cases, the electrodes only partially cover the electrolyte that exhibits the strongest and most durable adhesion with the sealing materials. Therefore, a dense mesh is required, which induces memory limitations with standard computing resources. Preliminary tests allow identifying and further restricting the details in the contact simulations to the layers that have the strongest influence on the curvature of the cell. The stress in the omitted layers can be investigated afterwards by using the submodelling capabilities of commercial software [1].

The simplest implementation of rate-independent plasticity and creep enforces, among others, the preservation of the volume. This might not describe the behaviour of GDL materials, such as metallic foams, and result in unrealistic predictions of loss of local contact, when coupled to a high friction coefficient. Depending on the technological solutions, special purpose elements for the modelling of gaskets that uncouple the transverse and through-the-thickness behaviour can serve for preliminary investigations [1].

4.3 Assessment of the Failure of Ceramic Materials

The knowledge of the stress field is a first step towards the analysis of the risks of failure of ceramic materials subjected to multiaxial stress, such as provided by the Weibull theory [122]. A postprocessing procedure based on the the principle of independent action (PIA) is straightforward to implement. In the case future studies highlight the limitations of this simplification, more refined theories exist, such as the Batdorf theory [57]. A complete analysis tool, CARES [94] is available.

4.4 Stacking

A specificity of the modelling of stresses in SOFCs is the stacking of the SRUs. The reduction of the thickness of the components, to decrease the thermal inertia, among others, promotes stacking issues: uneven distribution of the assembly load, increased effect of growing oxide scales, difficulties to ensure the reproducibility of the mechanical properties of some components, such as the GDL and sealants, or structural instabilities can cause contacting issues and direct or indirect cell failures. Few studies have been carried out on the overall deformation of a SRU embedded in a stack, because the number of SRUs that can be meshed is limited by the computation time in contact simulations [60, 63]. A complementary approach is to impose modified periodic boundary conditions on one SRU [88]. This is achieved by implementing linear multipoint constraints [88], which can be derived from those used for the modelling a unit cell in a cellular solid [31], depending on the design. In the classical case of a co- or counter-flow configuration, for instance, rotation around the y-axis (see Fig. 5) may not be prevented, which can be approximated by using help nodes. The simplest version of this approach is to enforce the flatness of the MIC [92].

4.5 Initialisation

The correct handling of the many manufacturing steps of an SOFC stack, i.e. (i) cell sintering, (ii) components assembly and stacking, (iii) sealing procedure and (iv) anode reduction, has a strong influence on the stress field in operation. The use of field variables to control the material properties already provides enough flexibility to handle most of the situations. Figure 6 depicts the different steps of the initialisation procedure.

The cell sintering phase can be simplified by setting artificial CTEs in the electrolyte, anode-compensating layer and cathode, to comply with the use of the small-sliding tracking method. The CTE of the anode equals to zero during this step. Thus, the sintering phase corresponds to an increase of 1 K of the cell [92]. The simpler possibility is to determine the values of the artificial CTEs corresponding to the mismatch strain based on the zero-stress temperature [6, 129, 130]. A better one is to

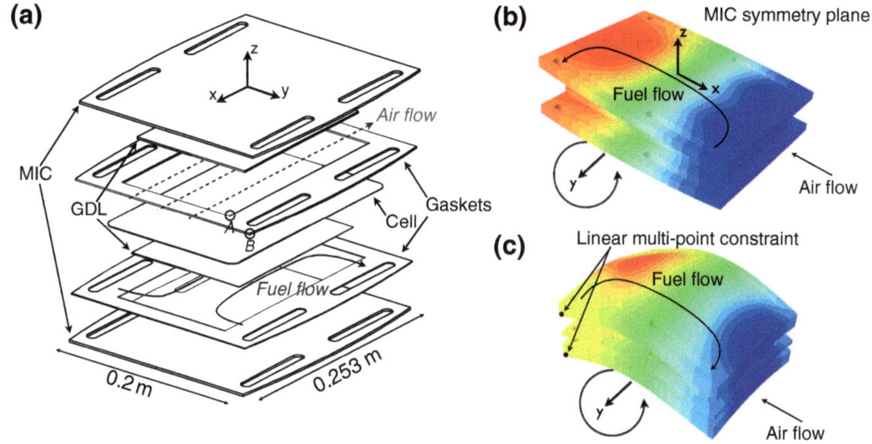

Fig. 5 Schematic view of **a** the Flame SOFC SRU design developed at LENI-EPFL [88, 92], reproduced here with kind permission from © Elsevier 2009 and different boundary conditions to model stacking conditions: **b** enforced flatness of the MIC, **c** modified periodic boundary conditions

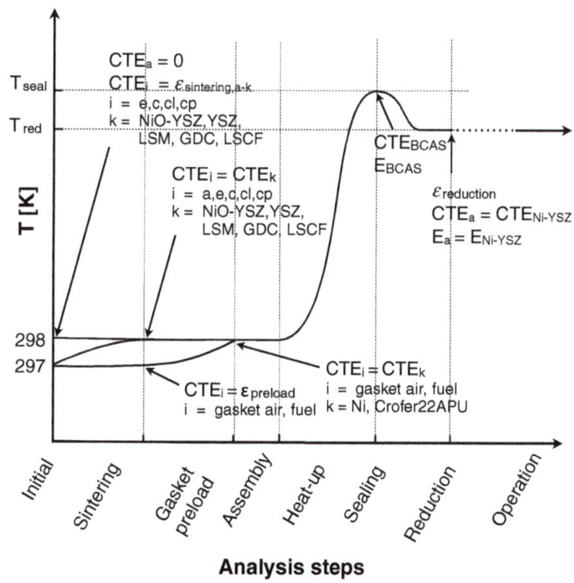

Fig. 6 Description of the initialisation sequence for contact simulations, *i* layer of component, *a* anode, *e* electrolyte, *c* cathode, *cp* compensating layer, *cl* compatibility layer and *k* material [88]. Reproduced here with kind permission from © Elsevier 2012

compute the irreversible strain generated during the cooldown after the sintering, depending on the cell materials and configuration and include it in the calculation of the artificial CTE [91]. The actual CTE are then reinserted in the model for the subsequent steps. The thickness of the gaskets before the assembly can differ from that of the GDL. This is a design variable, which can be used to control the distribution of the assembly pressure on the SRU. An additional initialisation step can be added for this purpose, where an anisotropic and artificial thermal strain is induced prior to the assembly, in a similar manner to the sintering step.

In the case of a glass-ceramic sealant, the mechanical properties change after the sealing procedure. A possibility to approximate the progressive nature of the process is to first discretely modify the mechanical properties, then enable creep for the desired time period. In the case the emphasis is on the sealing system, the initialisation procedure can be simplified to an uniform value, which sets, in a first approximation, the reference state for the mechanical interaction with the other components [125]. If structural failure recurrently occurs during the sealing procedure, a more detailed analysis of the sealing procedure is possible with existing data, by combining calculations of the temperature profile to crystallisation kinetics [9], temperature-dependent viscosity [76], and temperature and composition dependent CTE [79] and Young's modulus [65].

The modification of the CTE and Young's modulus ensuing the anode reduction is straightforward to handle. In the simplified view allowed by the existing data, the shrinkage is uniform. It can be implemented either manually or using the swelling option available in most modelling tools. The temperature profiles generated by the thermo-electrochemical model can then be imported.

4.6 Interaction with Modelling at Smaller Scales

SOFC modelling, be it electrochemical or mechanical, is multiscale. The knowledge gained at the microscale must be implemented at the stack macroscale, through homogenisation techniques or combined homogenisation/localisation procedures, for instance. This is currently seldom achieved in thermo-electrochemical models [47] and almost never applied to mechanical aspects. Therefore, a quick overview of the modelling approaches at the smaller scale is worthwhile.

At the electrode scale, REV modelling of composite materials has been introduced recently in the SOFC field. It echoes that on electrochemical aspects, using the lattice Boltzmann method (LBM) [44, 108, 113]. The microstructure is either artificially generated [45, 67] or gathered from experiments [117], such as focused-ion beam-scanning electron microscope (FIB-SEM). Two different approaches exist, based on either the FEM method [45, 117] or discrete element method (DEM) [67]. The outcome is restricted at present to the prediction of the essential properties, such as CTE [45], the Young's modulus and strength [67], but the techniques have the capability to provide insights into the failure mechanisms.

At the component level, fracture mechanics is applied to study cracking in the glass-ceramic or at the sealing interfaces, commonly observed during thermal cycling. Finite fracture mechanics has been used to identify the critical locations in a glass-ceramic sealing system, in the light of a combined stress and energy criterion for the nucleation and subsequent propagation of cracks [85]. A set of discrete and continuum damage models have been applied to study cracking in the bulk of a ceramic-glass sealant, and at the interface between the glass-ceramic and a YSZ electrolyte [95].

5 Example of Applications and Results

The effective mitigation of mechanical failures requires combined actions at different scales:

- Improvement of the properties of the materials.
- Design of the cell.
- SRU and stack design.
- Stack operating strategy and system implementation.

The present chapter is not exhaustive. The emphasis is on the third and fourth point. The outcome of simple models, which can be applied for the second point, is included, since it can provide input values for contact analyses. In contrast, REV and fracture mechanics simulations, suited to address the first and second point, provide at present material properties or failure criteria. This situation is expected to change in a near future, since the implementation of the knowledge of the failure mechanisms at the stack scale, gained by refined experiments and dedicated models, will require integrated modelling approaches.

5.1 Models Based on the Euler–Bernoulli Theory

Simple models based on the Euler–Bernoulli theory, or on two-dimensional implementation in FEM tools, have many applications. They can provide an assessment of the sensitivity of stresses on the mechanical properties and configuration of the cell layers, and can efficiently estimate parameters from X-ray diffraction (XRD) stress measurements. The implementation of creep, shrinkage or expansion is straightforward, which enables prospective investigations of the reduction step, thermal and/or redox cycling and ageing conditions. Simple models can simplify the initialisation procedure in contact simulation (see Sect. 4.5), by determining the irreversible deformations generated during the cooldown after the sintering step, as a function of the cell materials and configurations, heating ramp and flattening process.

Zhang et al. [134] have computed the stress states, and ensuing probabilities of failure of the different layers in anode-supported cells, which enables an estimation of the required minimum thickness of the anode support, which withstands tensile stress. In their case, the benefit of a reduction of the tensile stress exceeds the detrimental increase in volume.

Liu et al. [64] have proposed failure criteria for different cracking scenario, such as cracks perpendicular to the interface in the cathode and edge cracks at the interfaces either between the anode and the electrolyte or between the electrolyte and the cathode. They illustrate how their approach can determine the maximum allowable curvature and warpage to withstand stack assembly, i.e. avoid cracking in the cathode and edge cracking at the electrolyte interfaces in the anode-supported

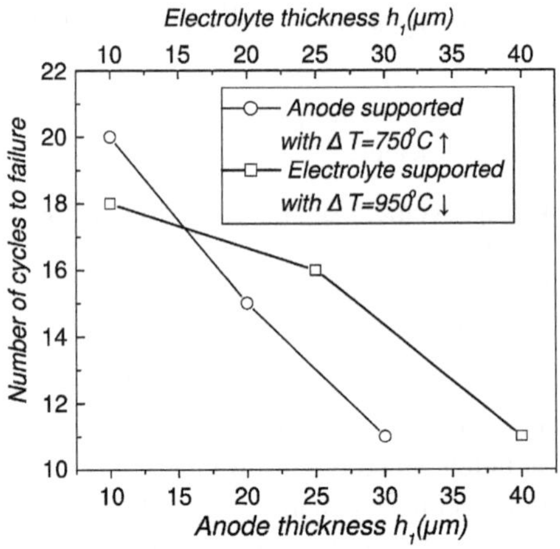

Fig. 7 Anode or electrolyte thickness effects on the number of thermal cycles until failure [62]. Reproduced here with kind permission from © Elsevier 2010

cell configuration. This is a criterion for a simple quality control procedure. In a next step [62], an approach has been developed to provide lifetime predictions during thermal cycling, based on the Paris law for crack nucleation and calculation of the peeling and shear stresses at the interface between the support and electrode, which are more concentrated near the cell edges. In their calculations, which focus on the anode/electrolyte interface in either anode-supported or electrolyte-supported cells, the porosity has overall a detrimental effect on the resistance against cycling, because it influences the initial damage distribution. The increase of the thickness of the thinner layer, that is electrolyte and anode, in anode-supported, respectively electrolyte-supported cells, reduces the number of cycles to failure, as shown in Fig. 7.

Subtle changes in the temperature dependence of the properties, can completely modify the failure modes. In anode-supported cells, models predict that the stress state in a LSM-based cathode can change from tensile to compressive, depending on the temperature [91]. Delamination of the cathode is experimentally observed and may arise from the compressive stress, through buckling, but the approaches developed in the field of TBC do not capture realistically the failure mode. In contrast, an LSCF cathode is always subjected to tensile stress and may crack [70]. Shielding compressive stress develops in the thin YSZ electrolyte, due to the lower CTE than the anode support. The same applies to the GDC or YDC compatibility layer, initially. The situation drastically changes during thermal cycles, after ageing at SOFC operating temperature, during 500–800 h due to the relief of the compressive stress and higher CTE of GDC or YDC, compared with that of the reduced anode, as shown in Fig. 8. Of course, the material properties govern the change in stress state and time scale of the phenomena.

The aforementioned results hold for the bulk of the materials. The stress field differs at the edges. Laurencin et al. [57] (see Fig. 9) have investigated the failure due to the singular stress field at the interfacial edges between the layers in anode and electrolyte-supported cells, under different conditions. The k-dominance radius

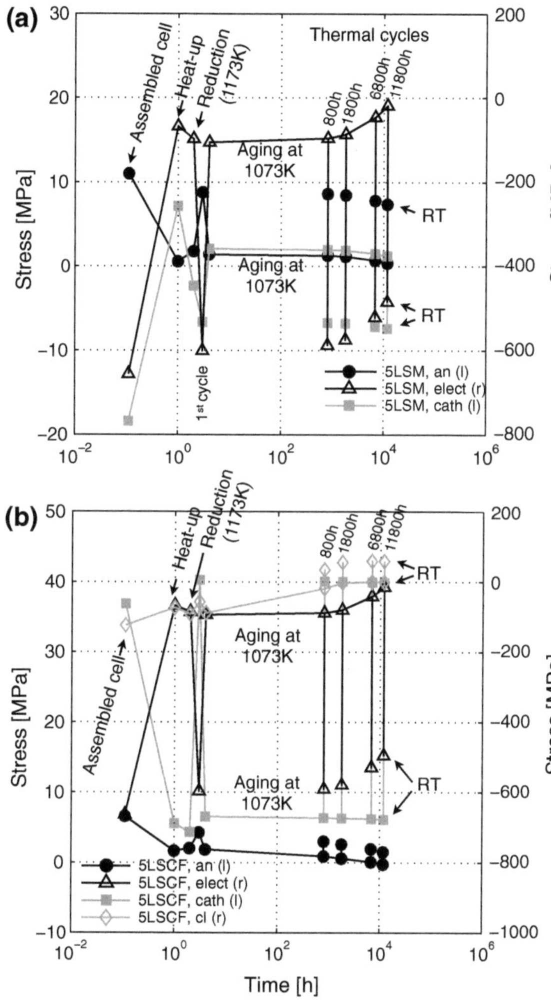

Fig. 8 Evolution of the stress in the different layers of an anode-supported cell with LSM cathode (*top*, 5LSM) and LSCF cathode (*bottom* 5LSCF), from the assembly, followed by the reduction step, and combined ageing at 1073 K and full thermal cycles (*l* left hand side axis, *r* right, *cl* compatibility layer, 5LSM 500 μm Ni-YSZ, 7μm YSZ, 60 μm LSM, 5LSCF 500 μm Ni-YSZ, 7 μm YSZ, 7 μm GDC, 60 μm LSCF) [91]. Reproduced here with kind permission from © John Wiley and Sons 2011

R_k is of 2 μm (anode-supported cell) and 6 μm (electrolyte-supported cell) at the electrolyte–cathode interface. This sets the limit of validity for a Weibull analysis applied on the whole cell. In anode-supported cell, the singularity at the anode/electrolyte interfacial edge does not govern the probability of failure of the anode and electrolyte. The singularity at the interface between the cathode and electrolyte is potentially detrimental for the latter, during redox cycling. In the case of residual stress at room temperature, the radius of the exclusion zone is of 0.5 μm, which practically scales with the size of the particles. Because an electrolyte-supported cell is almost symmetric, the effect of the singularity is identical at the interfaces between the electrolyte and anode or cathode. It does not affect the electrodes, but may initiate the cracking of the electrolyte from defects located in a zone of approximately 6 μm.

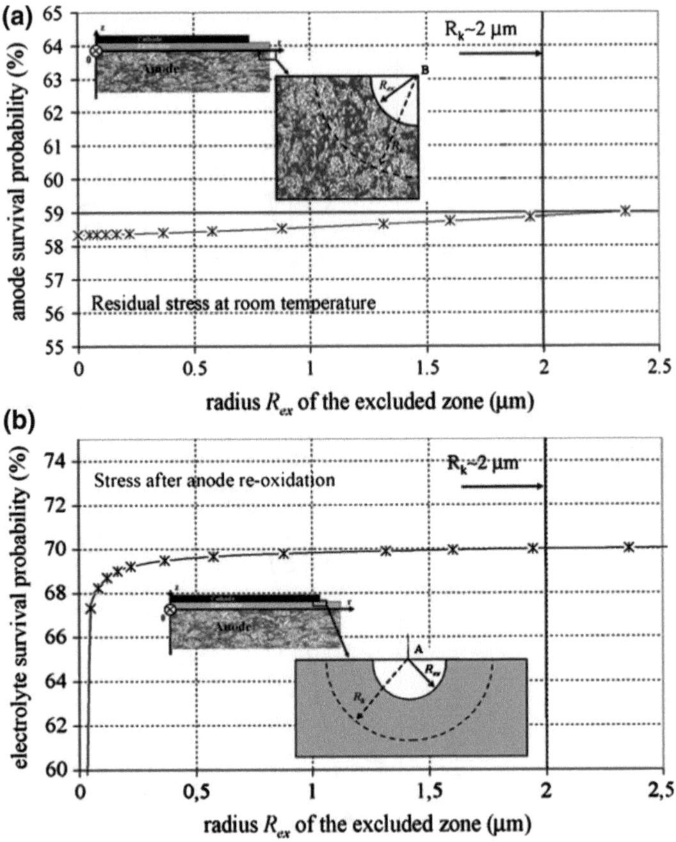

Fig. 9 **a** Survival probability of the anode substrate calculated by removing from the domain of integration a region surrounding the singularity B (case of the residual stress at room temperature). **b** Survival probability of the electrolyte substrate calculated by removing from the domain of integration a region surrounding the singularity A (case of the stress after anode re-oxidation with a bulk expansion of $\epsilon_{ox} = 0.14\,\%$ and an uncracked cathode) [58]. Reproduced here with kind permission from © Elsevier 2007

The study by Sato et al. [105] has shown that isothermal expansion likely governs the failure of electrolyte-supported cells, based on $Ce_{0.8}Sm_0.2O_{2-\delta}$ above 973 K. This is in accordance with other studies on the topic [5].

Malzbender et al. [74] have studied the influence of the cell curvature on the residual stresses before and after the reduction of the anode, yielding an estimation of the reduction strain in the range of 0.1 %. Their model has also been used to examine the oxidation–reoxidation of the anode support [75]. In the case of a thin anode support of 0.27 mm, a model with the anode functional layer of lower porosity provided a better prevision of the evolution of the curvature during reoxidation. Sun et al. [112] have derived values from the analysis of the curvature of the anode support. The behaviour differs, depending on the anode support arrangement, i.e.

presence or not of a compatibility layer. Sun et al. [112] hence suggest that a significant amount of plastic deformation of the Ni phase occurs during reduction.

XRD measurements exhibit a nonlinear relation between the stress and temperature in the electrolyte in anode-supported cells [29, 71, 129, 130]. The variation of the stress that ensues the reduction of the support is smaller than expected, which shows that the shrinkage of the anode support compensates the decrease in Young's modulus. Lara Curzio et al. [56] have monitored the evolution of the stress in the electrolyte during ageing and thermal cycling. As expected, the magnitude of the compressive stress in the electrolyte first decreased at the beginning of both ageing and cycling, but then unexpectedly increased again. The former trend has been attributed to creep, while further work is required to relate the latter to the observed slight coarsening of small Ni grains. Different zero-stress temperatures lead to very strong variations in the computed stress in the electrolyte. The aforementioned trends could be qualitatively reproduced by implementing creep in a simple mode based on the Euler–Bernoulli beam theory, rather than adjusting the zero-stress temperature [91]. The compressive stress in the electrolyte ranges from 700 MPa to less than 50 MPa, depending on the cell geometry, temperature and state of the anode. The irreversible strain is positive (elongation) in the anode and negative (shrinkage) in the electrolyte and compensating layer, and small in the cathode. In the compatibility layer, it depends on the presence or not of a compatibility layer. More refined approaches have been used to improve the sintering process of the support [19].

Liu et al. [66] estimated the lifetime of the MIC before spallation of the oxide scales, and showed that a protective coating can delay the occurrence of spallation by a factor of 3, by slowing the growth of the oxide scale.

5.2 Focus on Single Stack Components

The SRU components can cause the loss of integrity of the cell, directly through mechanical interaction, or indirectly, because of the sequence of deleterious effects activated by its failure. A local loss of gas tightness detrimentally exposes the materials of the cathode (anode) compartment to reducing (oxidising) gases. Unsteady parasitic combustions provoke hot spots [25, 128]. The difficulty to ensure the reproducibility of the mechanical properties of the sealants and GDL materials [8, 110] is a cause of uneven gas supply among the stacked SRUs, of altered electrical contact and of partial loss of contact pressure on compressive gaskets.

At the SRU scale, the sealing solution has received most of the attention. The studies by Weil at al. [124, 125], Jiang et al. [43] and Lin et al. [61] compare the BCS, glass-ceramics and compressive gaskets. The former induce a less stringent joining than glass-ceramic during thermal cycles and partially alleviate the mismatch between the thermal expansions of the parts, while providing a better sealing performance than compressive gaskets. The cell to foil material, typically made of

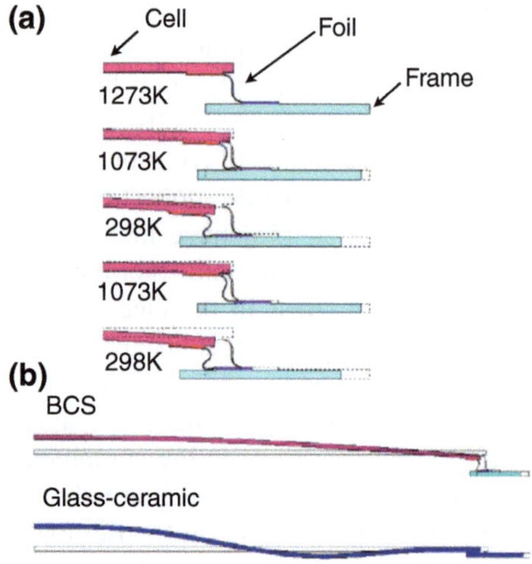

Fig. 10 **a** A series of cross-sectional images based from FE analysis depicting how the BCS components change in size and shape at the corner of the seal as a function of cooling from the stress-free state at 1273 K to room temperature and upon re-heating to 1073 K and cooling back to room temperature. The original size/shape of the components at the stress-free state are denoted by the dotted lines. **b** Schematics of cell deflection for the BCS and glass-ceramic seal designs: cross-sectional view along the diagonal of the cell [125]. Reproduced here with kind permission from © Elsevier 2008

silver braze, undergoes plastic deformation that relieves the stress in the cell [124]. The resulting history-dependence is moderate, as shown in Fig. 10 . The BCS results in a reduction of the stress and deflection compared with the glass-ceramic solution, which lets anticipate less contacting issues with the GDL. Therefore, an increase in foil thickness increases the stress in the cell in the range of 5–10 %. This must be balanced with oxidation issues for the long-term use [132]. The study by Jiang et al. [43] shows that the contribution of the temperature profile, neglecting the residual stress due to the multilayer nature of the cell, modifies the stress in the cell in the range of 30–40 %, approximately. The stress in the cell scales with the voltage, because the higher polarisation losses induce an increase of the temperature, hence decrease of the temperature difference with the reference state. The results of Govindaraju et al. [35] suggest that extreme care is required during the transient operation of a stack with glass-ceramic sealants. The relaxation of the stress in the cell occurs predominantly during less than 1 h after a discrete and significant change in applied temperature profile.

Yakabe et al. [131] have computed large stresses reaching 20–100 MPa in $LaCrO_3$ interconnects. The non-uniform isothermal expansion contributes to 8–80 % of the stress depending on the material and operating conditions. Therefore, the distribution of the vacancy profile and the dependence of the isothermal strain are crucial in the selection of the operating conditions and materials to avoid interconnect cracking, and indirectly, through mechanical interaction, cell failure. MIC are not prone to such issues but the uneven temperature profile can cause its buckling.

The shape of the temperature profile dictates the minimum stable thickness. Typical values range from 1–1.8 mm for a MIC of 400 cm^2 in typical operating conditions, and cannot be reliably predicted from simple indicators, such as the temperature difference over the SRU or maximum thermal gradient [93]. The occurrence of thermal buckling can be acceptable if it does not provoke large displacements or plastic deformation. The predicted remaining deflection of a free buckled MIC exhibiting a purely elastic behaviour can reach 3.5 mm, which is a likely source of loss of electrical contact and gas tightness, and modification of the stress in the cell. The results show that this phenomenon precludes any unconsidered decrease of the thickness of the MIC.

Selimovic et al. [107] have investigated the stress in a free single planar cell, subjected to steady-state and transient temperature profiles. The use of either ceramic or metallic interconnects affects the magnitude of the stress, due to the different thermal conductivities.

5.3 *Stress Analyses of Stacks*

Thermo-electrochemical optimisation is commonly performed at the system level. At the SRU scale, reliability and durability, rather than best initial performance are the main issues to solve. The modelling tools reliably predict the performance, but do not include the multiple possible origins of failures, which is needed to correctly discriminate the candidate designs at the early stage of the development of a stack. Currently, structural analyses of stack and SRU are prospective since they seek to identify the possible issues that may arise during operation. Their reliability has not yet been proven and to our knowledge, calibration is lacking. Studies focusing on button-cell tests with the framework described in Sect. 4 have been published, with the probable intention to further validate the models [15, 16, 58]. The simulations by Chiang et al. [16] show that stress in the cell scales inversely with operating voltage, because of the higher temperature difference, and suggest to keep thermal gradients lower than 10.6 K mm^{-1}. In contrast, the applied assembly load, from 0.5 to 1.0 kg cm^{-2} has a minor effect.

Lin et al. [60, 61] have meshed a planar, intermediate-temperature three-cell stack based on anode supported-cells and a glass-ceramic sealant, with linear continuum shell elements (see Fig. 11). The maximum principal stress in the PEN reaches 170 MPa at room temperature and decreases to 70 MPa in operation, because of the smaller temperature difference with the reference state. The MIC undergoes significant plastic deformation. In operation, the elastic modulus of the glass-ceramic sealant has been reduced by two thirds for a first assessment of the effects of the viscous behaviour. It results in a decrease of the stress of 10 % in the cell. Few studies have been carried out on the overall deformation of a SRU embedded in a functional stack. In the conditions of their simulations, a limited influence of the stack support conditions and position in the stack on the stress profile has been highlighted (see Fig. 11).

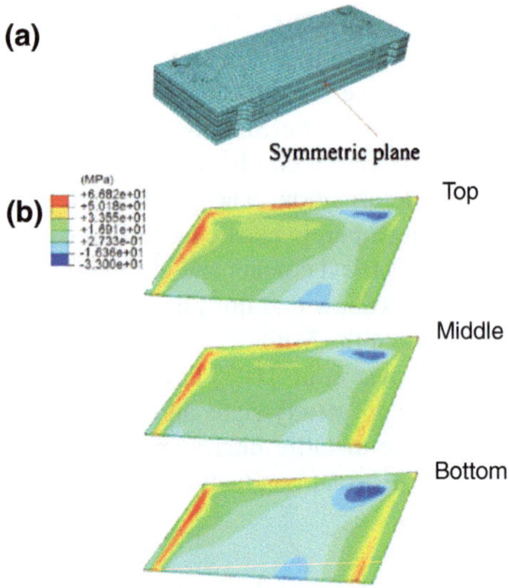

Fig. 11 **a** Schematic of the finite element model for one-half of a three-cell SOFC stack [61]. **b** Distribution of maximum in-plane principal stress in the MEA at steady operation for the *top*, *middle*, and *bottom* cells [60]. Reproduced here with kind permission from © Elsevier 2007, 2009

In Refs [88, 89, 92, 93], a thermo-electrochemical model, which includes degradation phenomena has been coupled with a structural analysis tool to study the mechanical reliability and durability of intermediate temperature, anode-supported SOFC stacks. This has allowed to provide insights into compromises between the requirements from thermo-electrochemical and mechanical aspects. The latter were analysed in the light of cell failure, provided by Weibull analysis, possible loss of electrical contact and gas-tightness, tentatively assessed from the distribution of the contact pressure and maximum tensile stress in the glass-ceramic sealant. Depending on the mechanical properties of the materials, the anode, solely, or together with the cathode, contribute to the cell probability of failure. The specificities of the temperature distribution and the level of mechanical interaction between the SRU components govern the risks of failure, which is illustrated in Figures 12 and 13. The properties of the GDLs, the thickness of the MIC and the deflection of the SRU, due to the cell residual stress and temperature profile, have in comparison a small impact on the cell probability of failure. Figure 14 shows that the anode and the cathode exhibit opposite dependences on operating conditions. The shape of the temperature profile affects the stress in the anode, whereas the stress state in the LSM–YSZ cathode is either tensile or compressive, depending on the thermal mismatch strain with the other layers at a given temperature. Therefore, stress increases in the anode, respectively decreases in the cathode, during current-voltage characterisation as current is drawn from the stack.

Cell mechanical reliability places different constraints on the operating conditions than thermo-electrochemical degradation, as shown in Fig. 13 by the probability of failure approximately one order of magnitude higher at 0.2 W cm^{-2}, than at 0.3 W cm^{-2}. The analysis performed in [92] reveals that transient operation

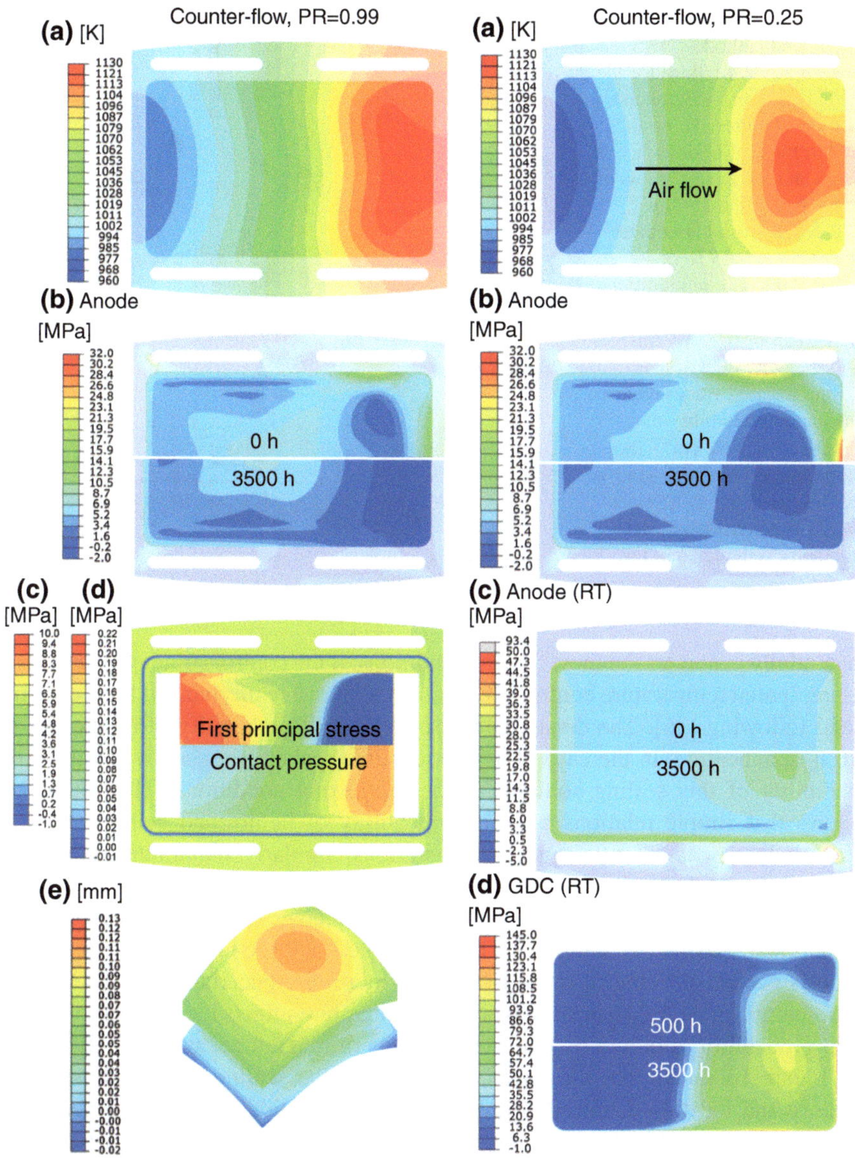

Fig. 12 *Left* Effect of the flow configuration and methane conversion fraction (PR) on the stress. Case of an anode-supported cell with LSM-YSZ cathode and compressive gaskets. **a** Temperature profile and **b** First principal stress in the anode. The MIC is displayed in transparency. **c** First principal stress in the cathode (insert above the symmetry line). **d** Contact pressure on the cathode GDL and compressive gasket and **e** vertical displacement along the z-axis, with an amplification factor of 2,000. *Right column* effect of creep in a cell based on a LSCF cathode and **a** temperature distribution, on **b** the evolution of the first principal stress in the anode support in operation and **c** during thermal cycling to RT and **d** evolution of the first principal stress in the GDC compatibility layer after thermal cycling. The profiles above and below the symmetry axis refer to different operation time [88, 89]. Reproduced here with kind permission from © Elsevier 2012

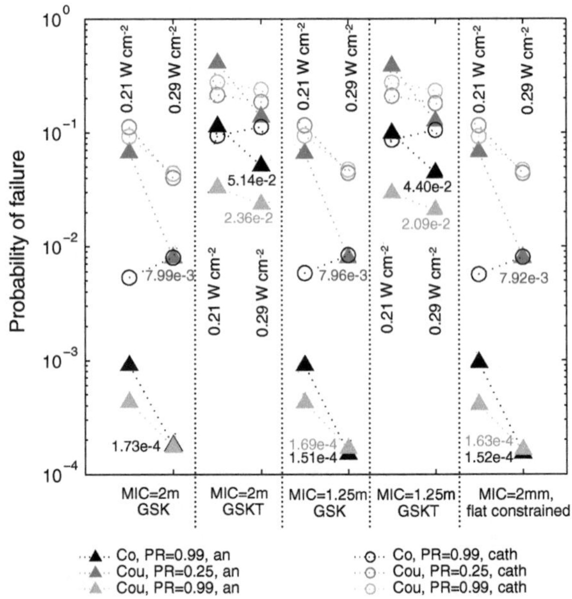

Fig. 13 Comparison of the anode and cathode probabilities of failure for the GSK (compressive gaskets), GSKT (compressive gaskets tied at the sealing interfaces) cases and enforced flatness of the MICs in the co-flow (PR = 0.99 %) and counter-flow (PR = 0.25 % or PR = 0.99 %) configuration. The system specific power is 0.21, 0.25 or 0.29 W cm^{-2} and the MIC thickness is 2 mm. Inserted values refer to the anode probability of failure at 0.29 W cm^{-2} [86]. Reproduced here with kind permission from © Elsevier 2012. The considered SRU geometry is depicted in Fig. 5

potentially generates the most critical conditions, if the control strategy is defective. Here, spatial temperature control is of interest to alleviate mechanical failure during load following [26]. The drastic decrease of the cell probability of failure during 10 h of annealing, in the case of a glass-ceramic sealant, suggests a possible vulnerability of this sealing solution to load following conditions. This altogether shows that simple relations between probability of failure and cell potential or current density do not reflect practical conditions. Similarly, maximum temperature difference and thermal gradients are imperfect indicators for cell cracking probability. For a given value, the probability of failure covers more than two orders of magnitude, as depicted in Fig. 15. This invalidates their use for design purposes and care is needed for control purposes. Furthermore, the uncertainty in the Weibull parameters for SOFC materials available in literature affects the calculation of the cell probability of failure by one to two orders of magnitude, at least [92].

Figure 16 shows that the contact pressure on the sealing gaskets depends on history and location. Controlled preload and/or consolidation steps are needed to ensure an acceptable level of gas-tightness in all zones during operation. The current knowledge on the mechanical properties of the GDLs and on the nature of the interactions at the interfaces does not allow a reliable and comprehensive assessment of possible losses of electrical contact. The temperature profile governs the distribution of the contact pressure on the GDL, which can however be controlled by adjusting the compliance of adjacent components and levels of mechanical interaction at the sealing interfaces. Therefore, in the case of a glass-ceramic sealant, the thermal mismatch strains, rather than the assembly load, govern the magnitude of the compressive stress on the GDL. Significantly higher

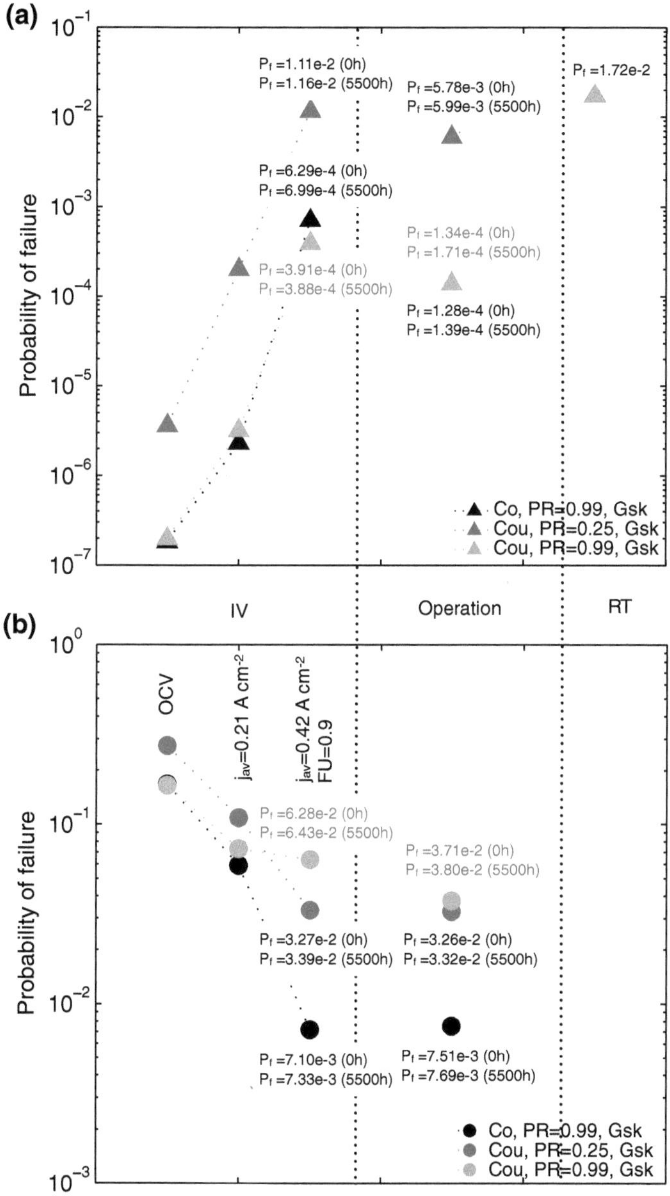

Fig. 14 Evolution of **a** the anode and **b** cathode probability of failure during IV characterisation, operation at the nominal point with creep of the GDL and MIC and thermal cycle to room temperature for the co- (CO) and counter-flow (COU) configuration [89]. Anode-supported cell and compressive gaskets and SRU geometry depicted in Fig. 5. *PR* methane conversion fraction in the reformer, *j* current density, *FU* fuel utilisation, P_f probability of failure. Reproduced here with kind permission from © Elsevier 2012

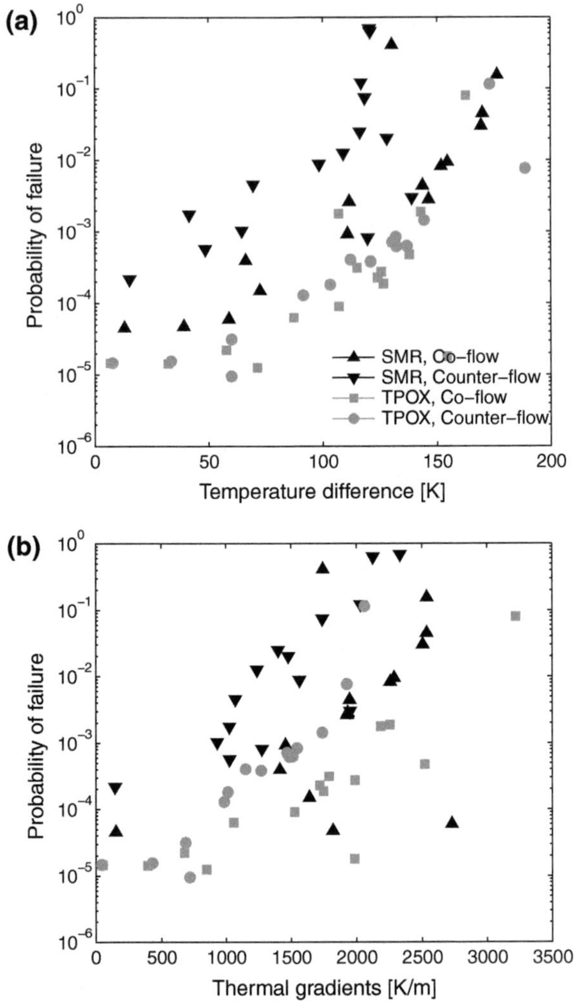

Fig. 15 Correlation between probability of failure and temperature difference (**a**) and thermal gradients (**b**) over the SRU [92]. TPOX thermal partial oxidation, SMR internal steam-methane reforming. Reproduced here with kind permission from © Elsevier 2009

values than anticipated by the assembly load, approximately twice higher in the situation treated in [88], can arise and further generate an enhanced sensitivity to cycling conditions because of irreversible deformation in the GDL.

The modification of the temperature profile due to degradation, and creep in the MIC and GDLs, have a limited impact on the probability of failure of the cell. The latter typically increases by 10–50 % during IV characterisation, after 4,500 h of operation (see Fig. 14), which remains small compared to the variations induced by the operating conditions or choice of a sealing solution (i.e. in the range of one to two orders of magnitude). In contrast, creep in the components governs the evolution of the contact pressure on the GDL. During constant operating conditions, it flattens the contact pressure distribution (see Fig. 17). The generated

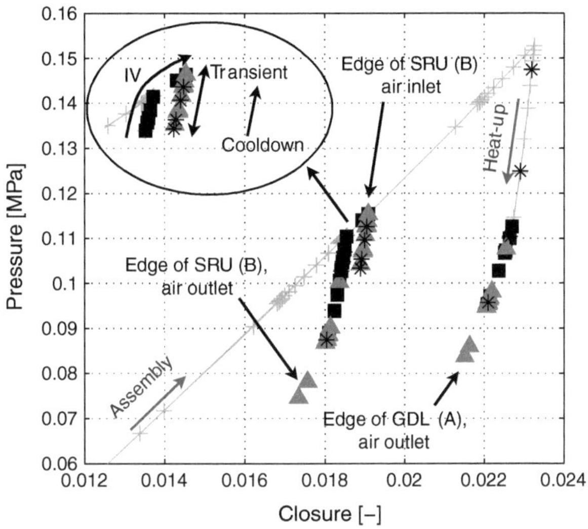

Fig. 16 Details of the evolution of the pressure versus closure in a compressive gasket. Anode-supported cell without anode compensating layer, hermal partial oxidation reformate and counter-flow configuration. Locations A and B refer to Fig. 5. Assembly and heat-up (*grey crosses*), current-voltage characterization (*black squares*), variation of the electrical load (*grey triangles*) and load shutdown and cooldown (*black stars*) [93]. Reproduced here with kind permission from © Elsevier 2009

non-uniform irreversible deformation, combined to the increase of the SRU deflection, then induces a complete release of the compressive stress during IV characterisation and thermal cycling, in the zone subjected to the highest temperature during constant operation, as shown in Fig. 17. Practically, the contact pressure is lost over approximately half of the active area during thermal cycling. Further experiments and model improvements are needed to relate these results with a possible alteration of the electrical contact.

In a cell with LSCF cathode, tensile stress develops in the GDC compatibility layer during thermal cycling. The creep strain rate in the MEA layers is not evenly distributed, due to the temperature profile and thermal stress in operation (see Fig. 12). This typically doubles the tensile stress at room temperature in the zone of highest temperature during operation. The stress, in the range of 100 MPa, likely induces cracking of the GDC layer.

The feedback of the structural model on the thermo-electrochemical model is not considered in the studies, but will receive growing interest. The first investigation of the effect of disparities in the creep behaviour of anode GDL materials is available in [90]. The correction of the anode gas flux proceeds by comparing the nominal situation with a worst case, respectively characterised by a nominal creep strain rate and a twice higher one. For a comprehensive analysis, the overall bending of a SRU in a stack and the creep behaviour of components adjacent to the GDL, as well as thermal coupling between SRUs unevenly supplied in fuel should be accounted for. Figure 18 shows that the degradation of a SRU embedded in a stack, due to a progressive undersupply in fuel, strongly depends on the operating conditions.

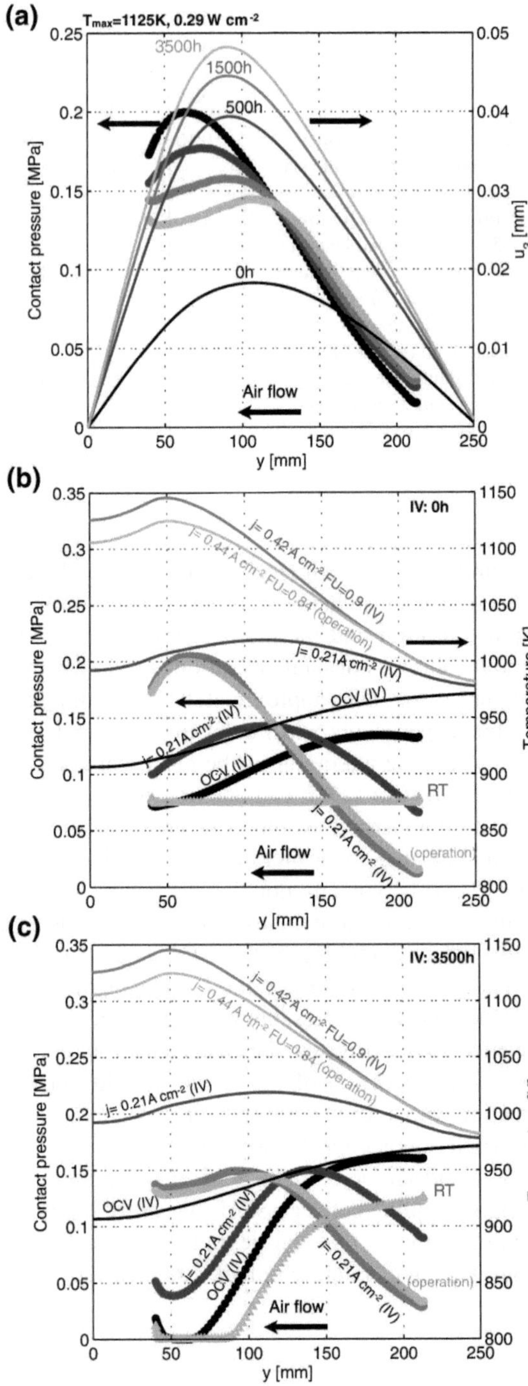

Fig. 17 **a** Evolution of MIC deflection and contact pressure on the cathode GDL, along the x direction (see Fig. 5) for the counter-flow configuration, with compressive gaskets and a methane conversion percentage in the reformer of 25 % and a maximum allowed solid temperature of 1,125 K in the SRU. **b**, **c** Evolution of the contact pressure on the cathode GDL and temperature profile, along the x direction (see Fig. 5), during an IV characterisation and thermal cycle to room temperature. (**b**) Initial time (**c**) 3,500 h at the nominal point, depicted in *light gray*. Counter-flow configuration, compressive gaskets and methane conversion percentage in the reformer of 25 % [89]. *j* current density, *FU* fuel utilisation, *OCV* open-circuit voltage. Reproduced here with kind permission from © Elsevier 2012

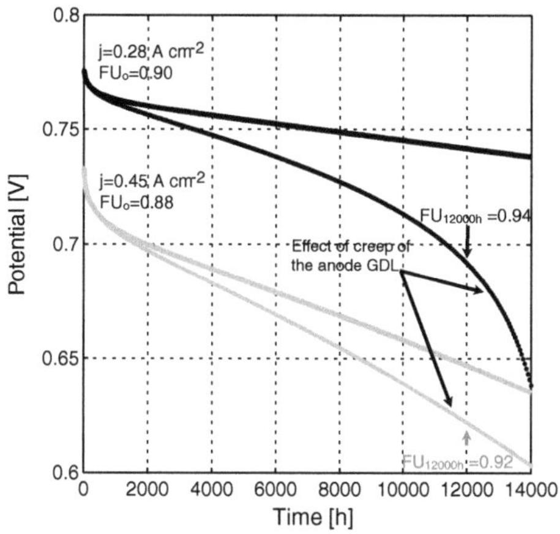

Fig. 18 Effect of fuel undersupply of a SRU in a stack due to creep in the anode GDL (*lines* no creep, *dots* with creep), during operation at a constant current density, corresponding to initial specific system power of 0.2 W cm^{-2} (*black*) and 0.3 W cm^{-2} (*grey*). Counter-flow configuration, 25 % of pre-reformed methane and maximum allowable temperature of 1150 K. *j* current density, *FU* fuel utilisation [90]. Reproduced here with kind permission from © Elsevier 2012

6 Future Work

Current SRU models can address several of the observed mechanical failures and highlight their strong sensitivity to subtle changes in the properties of the materials, of the interfaces, operating conditions and history. This qualitatively complies with the difficulties to mitigate in practice mechanical failures in SOFC stacks. The reliability of the prediction and exploitation of the modelling tools however suffer from the lack of knowledge on (i) the mechanical properties of all the materials, e.g. elastic, rate-independent plastic, creep behaviour and strength, (ii) the type of interactions between the components, (iii) effect of ageing or cycling, whereas (iv) experimental data dedicated for calibration is not available.

The potential for model improvements allowed by todays modelling capabilities is considerable, which lets anticipate significant improvements in the accuracy and variety of the simulated predictions. Once the failure mechanisms are understood, simplifications will be possible for the efficient implementation of both electrochemical and mechanical aspects in design models, in the view of both reliability and durability.

The particularity of mechanical issues in SOFC is their intricate link with the physico-chemical and electrochemical properties of the materials. Indeed, unlike for TBC or ceramic combustors, the main requirement for the ceramic materials is not their temperature and mechanical resistance, but electrochemical properties. Therefore, to go further towards reliable lifetime predictions, hence effective mitigation of mechanical failure, the multiphysics nature of the problem must be fully acknowledged and not restricted to thermal, or isothermal stress. Dedicated experiments combined to REV simulations can provide insights into the complex failure modes that affect the composite ceramic materials, in view of the morphological and compositional alterations that result from prolonged use. Multi-scale modelling approaches must be used to integrate the outcome at the stack level. The growing interest in such techniques is encouraging. Similarly, the procedures to tackle efficiently stacking issues, in regard of dimensional tolerances, with a feedback on the thermo-electrochemical calculations, must be developed.

The literature does not comprise many experimental data to test first the relevance of the assumptions, and then calibrate the models. Despite the importance of confidentiality issues in electrode and stack engineering, massive efforts on such tasks are needed to propose general modelling frameworks.

Thermo-electrochemical performance or degradation considerations can be included in optimisation problems fairly easily. They entail trade-offs, which must be further completed by mechanical aspects. This knowledge will provide comprehensive guidance in design and synthesis of control.

References

1. ABAQUS Inc., v6.8, Hibbitt, Karlsson and Sorensen, Rhode Island
2. J. Adams, Young's modulus, flexural strength, and fracture of yttria-stabilized zirconia versus temperature. J. Am. Ceram. Soc. **80**(4), 903–908 (2000)
3. C. Appel, N. Bonanos, A. Horsewell, S. Linderoth, Ageing behaviour of zirconia stabilised by yttria and manganese oxide. J. Mater. Sci. **36**(18), 4493–4501 (2001)
4. M. Ashby, H. Frost, *Deformation Mechanism Maps: The Plasticity and Creep of Metals and Ceramics* (Pergamon Press, Oxford, 1982)
5. A. Atkinson, Chemically-induced stresses in gadolinium-doped ceria solid oxide fuel cell electrolytes. Solid State Ionics **95**(3–4), 249–258 (1997)
6. A. Atkinson, A. Selçuk, Residual stress and fracture of laminated ceramic membranes. Acta Materialia **47**(3), 867–874 (1999)
7. A. Atkinson, A. Selçuk, Mechanical behaviour of ceramic oxygen ion-conducting membranes. Solid State Ionics **134**(1–2), 59–66 (2000)
8. X. Badiche, S. Forest, T. Guibert, Y. Bienvenu, J.D. Bartout, P. Ienny, M. Croset, H. Bernet, Mechanical properties and non-homogeneous deformation of open-cell nickel foams: application of the mechanics of cellular solids and of porous materials. Mater. Sci. Eng. A **289**(1–2), 276–288 (2000)
9. N.P. Bansal, E.A. Gamble, Crystallization kinetics of a solid oxide fuel cell seal glass by differential thermal analysis. J. Power Sources **147**(1–2), 107–115 (2005)
10. P. Batfalsky, V. Haanappel, J. Malzbender, N. Menzler, V. Shemet, I. Vinke, R. Steinbrech, Chemical interaction between glass-ceramic sealants and interconnect steels in SOFC stacks. J. Power Sources **155**(2), 128–137 (2006)
11. S. Biswas, T. Nithyanantham, N. Saraswathi, S. Bandopadhyay, Evaluation of elastic properties of reduced NiO-8YSZ anode-supported bi-layer SOFC structures at elevated temperatures in ambient air and reducing environments. J. Mater. Sci. **44**(3), 778–785 (2009)
12. A. Bouzid, H. Champliaud, Contact stress evaluation of nonlinear gaskets using dual kriging interpolation. J. Press. Vessel Technol. **126**(4), 445–450 (2004)
13. M. Bram, S. Reckers, P. Drinovac, J. Mönch, R.W. Steinbrech, H.P. Buchkremer, D. Stöver, Deformation behavior and leakage tests of alternate sealing materials for SOFC stacks. J. Power Sources **138**(1–2), 111–119 (2004)

14. H.-T. Chang, C.-K. Lin, C.-K. Liu, High-temperature mechanical properties of a glass sealant for solid oxide fuel cell. J. Power Sources **189**(2), 1093–1099 (2009)
15. L.-K. Chiang, H.-C. Liu, Y.-H. Shiu, C.-H. Lee, R.-Y. Lee, Thermo-electrochemical and thermal stress analysis for an anode-supported sofc cell. Renew. Energy **33**(12), 2580–2588 (2008)
16. L.-K. Chiang, H.-C. Liu, Y.-H. Shiu, C.-H. Lee, R.-Y. Lee, Thermal stress and thermo-electrochemical analysis of a planar anode-supported solid oxide fuel cell: effects of anode porosity. J. Power Sources **195**(7), 1895–1904 (2010)
17. Y.-S. Chou, J.W. Stevenson, P. Singh, Effect of aluminizing of Cr-containing ferritic alloys on the seal strength of a novel high-temperature solid oxide fuel cell sealing glass. J. Power Sources **185**(2), 1001–1008 (2008)
18. C.-L. Chu, J.-Y. Wang, S. Lee, Effects of $La_{0.67}Sr_{0.33}MnO_3$ protective coating on SOFC interconnect by plasma-sputtering. Int. J. Hydrogen Energy **33**(10), 2536–2546 (2008)
19. M. Cologna, V.M Sglavo, Vertical sintering to measure the uniaxial viscosity of thin ceramic layers. Acta Materialia **58**(17), 5558–5564 (2010)
20. G. Delette, J. Laurencin, M. Dupeux, J. Doyer, Measurement of the fracture energy at the interface between porous cathode layer and electrolyte in planar solid oxide fuel cells. Scripta Materialia **59**(1), 31–34 (2008)
21. S. Diethelm, J. Van herle, Z. Wuillemin, A. Nakajo, N. Autissier, M. Molinelli, Impact of materials and design on solid oxide fuel cell stack operation. J. Fuel Cell Sci. Technol. **5**(3), 3–6 (2008)
22. C. D'Souza, N. Sammes, Mechanical properties of strontium-doped lanthanum manganite. J. Am. Ceram. Soc. **83**(1), 47–52 (2000)
23. A. Evans, J. Hutchinson, The thermomechanical integrity of thin films and multilayers. Acta Metallurgica et Materialia **43**(7), 2507–2530 (1995)
24. A.G. Evans, D.R. Mumm, J.W. Hutchinson, G.H. Meier, F.S Pettit, Mechanisms controlling the durability of thermal barrier coatings. Prog. Mater. Sci. **46**(5), 505–553 (2001)
25. A. Faes, A. Nakajo, A. Hessler-Wyser, D. Dubois, A. Brisse, S. Modena, J. Van herle, Redox study of anode-supported solid oxide fuel cell. J. Power Sources **193**(1), 55–64 (2009)
26. M. Fardadi, F. Mueller, F. Jabbari, Feedback control of solid oxide fuel cell spatial temperature variation. J. Power Sources **195**(13), 4222–4233 (2010)
27. J.W Fergus, Metallic interconnects for solid oxide fuel cells. Mater. Sci. Eng. A **397**(1–2), 271–283 (2005)
28. J.W Fergus, Sealants for solid oxide fuel cells. J. Power Sources **147**(1–2), 46–57 (2005)
29. W. Fischer, J. Malzbender, G. Blass, R. Steinbrech, Residual stresses in planar solid oxide fuel cells. J. Power Sources **150**, 73–77 (2005)
30. A.S for Testing and Materials. Standard Test Method for Monotonic Equibiaxial Flexural Strength of Advanced Ceramics at Ambient Temperature. ASTM standard C1499-04
31. M. Galli, J. Botsis, J. Janczak-Rusch, An elastoplastic three-dimensional homogenization model for particle reinforced composites. Comput. Mater. Sci. **41**, 312–321 (2008)
32. L. Gibson, M. Ashby, *Cellular Solids: Structure and Properties* (Cambridge University Press, Cambridge, 1999)
33. S. Giraud, J. Canel, Young's modulus of some SOFCs materials as a function of temperature. J. Eur. Ceram. Soc. **28**(1), 77–83 (2008)
34. R. Goodall, C. Williams, J. Fernie, T. Clyne, et al., Thermal expansion and stiffness characteristics of a highly porous, fire-resistant composite material, in *International SAMPE Symposium and Exhibition (Proceedings)*, vol. 47, pp. 1001–1010
35. N. Govindaraju, W. Liu, X. Sun, P. Singh, R. Singh, A modeling study on the thermomechanical behavior of glass-ceramic and self-healing glass seals at elevated temperatures. J. Power Sources **190**(2), 476–484 (2009)
36. L. Grahl-Madsen, P. Larsen, N. Bonanos, J. Engell, S. Linderoth, Mechanical strength and electrical conductivity of Ni-YSZ cermets fabricated by viscous processing. J. Mater. Sci. **41**(4), 1097–1107 (2006)

37. F. Gutierrez-Mora, J.M. Ralph, J.L Routbort, High-temperature mechanical properties of anode-supported bilayers. Solid State Ionics **149**(3–4), 177–184 (2002)
38. C. Haering, A. Roosen, H. Schichl, Degradation of the electrical conductivity in stabilised zirconia systems: Part I: yttria-stabilised zirconia. Solid State Ionics **176**(3–4), 253–259 (2005)
39. J.S. Hardy, J.Y. Kim, E.C. Thomsen, K.S Weil, Improved wetting of mixed ionic/electronic conductors used in electrochemical devices with ternary air braze filler metals. J. Electrochem. Soc. **154**(3), 32–39 (2007)
40. T. Hashida, K. Sato, Y. Takeyama, T. Kawada, J. Mizusaki, Deformation and fracture characteristics of zirconia and ceria-based electrolytes for SOFCs under reducing atmospheres. ECS Trans. **25**(2), 1565–1572 (2009)
41. Haynes International. Material data sheet no. h-3000 h
42. F. Iguchi, Y. Endo, T. Ishida, T. Yokobori, H. Yugami, T. Otake, T. Kawada, J. Mizusaki, Oxygen partial pressure dependence of creep on yttria-doped ceria ceramics. Solid State Ionics **176**(5–6), 641–644 (2005)
43. T.L. Jiang, M.-H Chen, Thermal-stress analyses of an operating planar solid oxide fuel cell with the bonded compliant seal design. Int. J. Hydrogen Energy **34**(19), 8223–8234 (2009)
44. J. John R. Izzo, A.S. Joshi, K.N. Grew, W.K.S. Chiu, A. Tkachuk, S.H. Wang, W. Yun. Nondestructive reconstruction and analysis of SOFC Anodes using X-ray computed tomography at Sub-50 nm resolution. J. Electrochem. Soc. **155**(5), B504–B508 (2008)
45. J. Johnson, J. Qu, Effective modulus and coefficient of thermal expansion of Ni–YSZ porous cermets. J. Power Sources **181**(1), 85–92 (2008)
46. T. Kawada, S. Watanabe, S.ichi Hashimoto, T. Sakamoto, A. Unemoto, M. Kurumatani, K. Sato, F. Iguchi, K. Yashiro, K. Amezawa, K. Terada, M. Kubo, H. Yugami, T. Hashida, J. Mizusaki, Classification of mechanical failure in SOFC and strategy for evaluation of operational margin. ECS Trans. **25**(2), 467–472 (2009)
47. J. Kim, W. Liu, C. Lee, Multi-scale solid oxide fuel cell materials modeling. Comput. Mech. **44**(5), 683–703 (2009)
48. J.Y. Kim, J.S. Hardy, K.S Weil, Novel metal-ceramic joining for planar SOFCs. J. Electrochem. Soc. **152**(6), 52–58 (2005)
49. J. Kondoh, Aging strengthening of 8 mol% yttria-fully-stabilized zirconia. J. Alloys Compd. **370**(1–2), 285–290 (2004)
50. J. Kondoh, T. Kawashima, S. Kikuchi, Y. Tomii, Y. Ito, Effect of aging on yttria-stabilized zirconia. 1: a study of its electrochemical properties. J. Electrochem. Soc. **145**(5), 1527–1536 (1998)
51. J. Kondoh, S. Kikuchi, Y. Tomii, Y. Ito, Effect of aging on yttria-stabilized zirconia. J. Electrochem. Soc. **145**(5), 1550–1560 (1998)
52. J. Kondoh, S. Kikuchi, Y. Tomii, Y. Ito, Effect of aging on yttria-stabilized zirconia. 2: a study of the effect of the microstructures on conductivity. J. Electrochem. Soc. **145**(5), 1536–1550 (1998)
53. J. Kübler, R. Primas, B. Gut, Mechanical Strength of Thermally Aged and Cycled Thin Zirconia Sheets, Advances in Science and Technology, Ceramics: Charting the Future, ed. P. Vincenzini, Techna, Florence, Italy, ISBN 88-86538-02-2, 923–928 (1995)
54. B. Kuhn, F. Wetzel, J. Malzbender, R. Steinbrech, L. Singheiser, Mechanical performance of reactive-air-brazed (RAB) ceramic/metal joints for solid oxide fuel cells at ambient temperature. J. Power Sources **193**(1), 199–202 (2009)
55. A. Lakki, R. Herzog, M. Weller, H. Schubert, C.Reetz, O. Görke, M. Kilo, G. Borchardt, Mechanical loss, creep, diffusion and ionic conductivity of ZrO_2-8 mol% Y_2O_3 polycrystals. J. Eur. Ceram. Soc. **20**(3), 285–296 (2000)
56. E. Lara-Curzio, M. Radovic, M. Trejo, C. Cofer, T. Watkins, K. More, Effect of thermal cycling and thermal aging on the mechanical properties of, and residual stresses in, Ni-YSZ/YSZ bi-layers. IV.A.11 Reliability and durability of materials and components foe solid oxide fuel cells. Adv. Solid Oxide Fuel Cells II **27**(4), 383–391 (2007)

57. J. Laurencin, G. Delette, M. Dupeux, An estimation of ceramic fracture at singularities by a statistical approach. J. Eur. Ceram. Soc. **28**(1), 1–13 (2008)
58. J. Laurencin, G. Delette, F. Lefebvre-Joud, M. Dupeux, A numerical tool to estimate sofc mechanical degradation: case of the planar cell configuration. J. Eur. Ceram. Soc. **28**(9), 1857–1869 (2008)
59. H. Lein, K. Wiik, M. Einarsrud, T. Grande, E. Lara-curzio, High-temperature creep behavior of mixed conducting $La_{0.5}Sr_{0.5}FeCo_xO_{3-\delta}$ $(0.5 \leq x \leq 1)$ materials. J. Am. Ceram. Soc. **89**(9):2895–2898 (2006)
60. C.-K. Lin, T.-T. Chen, Y.-P. Chyou, L.-K Chiang, Thermal stress analysis of a planar SOFC stack. J. Power Sources **164**(1), 238–251 (2007)
61. C.-K. Lin, L.-H. Huang, L.-K. Chiang, Y.-P Chyou, Thermal stress analysis of planar solid oxide fuel cell stacks: effects of sealing design. J. Power Sources **192**(2), 515–524 (2009)
62. L. Liu, G.-Y. Kim, A. Chandra, Modeling of thermal stresses and lifetime prediction of planar solid oxide fuel cell under thermal cycling conditions. J. Power Sources **195**(8), 2310–2318 (2010)
63. W. Liu, X. Sun, M. Khaleel, Effect of creep of ferritic interconnect on long-term performance of solid oxide fuel cell stacks. Fuel Cells **10**(4), 703–717 (2010)
64. W. Liu, X. Sun, M. Khaleel, J. Qu, Global failure criteria for positive/electrolyte/negative structure of planar solid oxide fuel cell. J. Power Sources **192**(2), 486–493 (2009)
65. W. Liu, X. Sun, M.A Khaleel, Predicting Young's modulus of glass/ceramic sealant for solid oxide fuel cell considering the combined effects of aging, micro-voids and self-healing. J. Power Sources **185**(2), 1193–1200 (2008)
66. W. Liu, X. Sun, E. Stephens, M. Khaleel, Life prediction of coated and uncoated metallic interconnect for solid oxide fuel cell applications. J. Power Sources **189**(2), 1044–1050 (2009)
67. X. Liu, C.L. Martin, G. Delette, D. Bouvard, Elasticity and strength of partially sintered ceramics. J. Mech. Phys. Solids **58**(6), 829–842 (2010)
68. Y. Liu, A. Hagen, R. Barfod, M. Chen, H. Wang, F. Poulsen, P. Hendriksen, Microstructural studies on degradation of interface between LSM-YSZ cathode and YSZ electrolyte in SOFCs. Solid State Ionics **180**(23–25), 1298–1304 (2009)
69. F. Lowrie, R. Rawlings, Room and high temperature failure mechanisms in solid oxide fuel cell electrolytes. J. Eur. Ceram. Soc. **20**(6), 751–760 (2000)
70. A. Mai, V.A. Haanappel, S. Uhlenbruck, F. Tietz, D. Stöver, Ferrite-based perovskites as cathode materials for anode-supported solid oxide fuel cells: Part I. Variation of composition. Solid State Ionics **176**(15–16), 1341–1350 (2005)
71. J. Malzbender, W. Fischer, R. Steinbrech, Studies of residual stresses in planar solid oxide fuel cells. J. Power Sources **182**(2), 594–598 (2008)
72. J. Malzbender, L. Singheiser, R. Steinbrech, A review of advanced techniques for characterising SOFC behaviour. Fuel Cells **9**(6), 785–793 (2009)
73. J. Malzbender, R. Steinbrech, L. Singheiser. Failure probability of solid oxide fuel cells, in *Ceramic Engineering and Science Proceedings*, vol. 26, pp. 293–298. American Ceramic Society, 2005
74. J. Malzbender, T. Wakui, R. Steinbrech, Curvature of planar solid oxide fuel cells during sealing and cooling of stacks. Fuel Cells **6**(2), 123–129 (2006)
75. J. Malzbender, E. Wessel, R. Steinbrech, Reduction and re-oxidation of anodes for solid oxide fuel cells. Solid State Ionics **176**(29–30), 2201–2203 (2005)
76. K. Meinhardt, D.-S. Kim, Y.-S. Chou, K. Weil, Synthesis and properties of a barium aluminosilicate solid oxide fuel cell glass-ceramic sealant. J. Power Sources **182**(1), 188–196 (2008)
77. D. Meixner, R. Cutler, Low-temperature plastic deformation of a perovskite ceramic material. Solid State Ionics **146**(3–4), 285–300 (2002)
78. N.H. Menzler, de L.G. Haart, D. Sebold, Characterization of cathode chromium incorporation during mid-term stack operation under various operational conditions. ECS Trans. **7**(1), 245–254 (2007)

79. J. Milhans, S. Ahzi, H. Garmestani, M. Khaleel, X. Sun, B. Koeppel, Modeling of the effective elastic and thermal properties of glass-ceramic solid oxide fuel cell seal materials. Mater. Des. **30**(5), 1667–1673 (2009)
80. A. Mitterdorfer, L.J. Gauckler, $La_2Zr_2O_7$ formation and oxygen reduction kinetics of the $La_{0.85}Sr_{0.15}Mn_yO_3$, O_2(g)YSZ system. Solid State Ionics **111**(3–4), 185–218 (1998)
81. A. Morales-Rodriguez, A. Bravo-Leon, A. Dominguez-Rodriguez, S. Lopez-Esteban, J. Moya, M. Jimenez-Melendo, High-temperature mechanical properties of zirconia/nickel composites. J. Eur. Ceram. Soc. **23**(15), 2849–2856 (2003)
82. M. Mori, Effect of B-site doing on thermal cycle shrinkage for $La_{0.8}Sr_{0.2}Mn_{1-x}M_xO_{3+\delta}$ perovskites (M= Mg, Al, Ti, Mn, Fe, Co, Ni; $0 \leq x \leq 0.1$). Solid State Ionics **174**(1–4), 1–8 (2004)
83. M. Mori, Y. Hiei, N. Sammes, G. Tompsett, Thermal-expansion behaviors and mechanisms for Ca-or Sr-doped lanthanum manganite perovskites under oxidizing atmospheres. J. Electrochem. Soc. **147**, 1295 (2000)
84. M. Mori, T. Yamamoto, H. Itoh, H. Inaba, H. Tagawa, Thermal expansion of nickel–zirconia anodes in solid oxide fuel cells during fabrication and operation. J. Electrochem. Soc. **145**, 1374 (1998)
85. A. Müller, W. Becker, D. Stolten, J. Hohe, A hybrid method to assess interface debonding by finite fracture mechanics. Eng. Fract. Mech. **73**(8), 994–1008 (2006)
86. A. Nakajo, Thermomechanical and electrochemical degradation in anode-supported solid oxide fuel cell stacks. Ph.D Thesis, 4930, Ecole Polytechnique Fédérale de Lausanne, 2011
87. A. Nakajo, J. Kuebler, A. Faes, U.F. Vogt, H.J. Schindler, L.-K. Chiang, S. Modena, J. Van herle, T. Hocker, Compilation of mechanical properties for the structural analysis of solid oxide fuel cell stacks. constitutive materials of anode-supported cells. Ceram. Int. **38**(5), 3907–3927 (2012)
88. A. Nakajo, F. Mueller, J. Brouwer, J. Van herle, D. Favrat, Mechanical reliability and durability of SOFC stacks. Part I: modelling of the effect of operating conditions and design alternatives on the reliability. Int. J. Hydrogen Energy **37**(11), 9249–9268 (2012)
89. A. Nakajo, F. Mueller, J. Brouwer, J. Van herle, D. Favrat, Mechanical reliability and durability of SOFC stacks. Part II: modelling of mechanical failures during ageing and cycling. Int. J. Hydrogen Energy **37**(11), 9269–9286 (2012)
90. A. Nakajo, F. Mueller, J. Brouwer, J. Van herle, D. Favrat, Progressive activation of degradation processes in solid oxide fuel cells stacks: Part I: lifetime extension by optimisation of the operating conditions. J. Power Sources **216**(0), 449–463 (2012)
91. A. Nakajo, J. Van herle, D. Favrat, Sensitivity of stresses and failure mechanisms in SOFCs to the mechanical properties and geometry of the constitutive layers. Fuel Cells **11**(4), 537–552 (2011)
92. A. Nakajo, Z. Wuillemin, J. Van herle, D. Favrat, Simulation of thermal stresses in anode-supported solid oxide fuel cell stacks. Part I: probability of failure of the cells. J. Power Sources **193**(1), 203–215 (2009)
93. A. Nakajo, Z. Wuillemin, J. Van herle, D. Favrat, Simulation of thermal stresses in anode-supported solid oxide fuel cell stacks. Part II: loss of gas-tightness, electrical contact and thermal buckling. J. Power Sources **193**(1), 216–226 (2009)
94. N. Nemeth, J. Manderscheid, J. Gyekenyesi, Ceramics Analysis and Reliability Evaluation of Structures (CARES). Users and programmers manual (1990)
95. B. Nguyen, B. Koeppel, S. Ahzi, M. Khaleel, S. Prabhakar, Crack growth in solid oxide fuel cell materials: from discrete to continuum damage modeling. J. Am. Ceram. Soc. **89**(4), 1358–1368 (2006)
96. O.B. Olurin, D.S. Wilkinson, G.C. Weatherly, V. Paserin, J. Shu, Strength and ductility of as-plated and sintered CVD nickel foams. Compos. Sci. Technol. **63**(16), 2317–2329 (2003)
97. S.M. Oppenheimer, D.C Dunand, Finite element modeling of creep deformation in cellular metals. Acta Materialia **55**(11), 3825–3834 (2007)

98. K. Park, S. Yu, J. Bae, H. Kim, Y. Ko, Fast performance degradation of SOFC caused by cathode delamination in long-term testing. Int. J. Hydrogen Energy **35**(16), 8670–8677 (2010)
99. M. Pihlatie, H. Frandsen, A. Kaiser, M. Mogensen, Continuum mechanics simulations of NiO/Ni-YSZ composites during reduction and re-oxidation. J. Power Sources **195**(9), 2677–2690 (2010)
100. M. Pihlatie, A. Kaiser, M. Mogensen, Mechanical properties of NiO/Ni–YSZ composites depending on temperature, porosity and redox cycling. J. Eur. Ceram. Soc. **29**(9), 1657–1664 (2009)
101. M. Pihlatie, T. Ramos, A. Kaiser, Testing and improving the redox stability of Ni-based solid oxide fuel cells. J. Power Sources **193**(1), 322–330 (2009)
102. M. Radovic, E. Lara-Curzio, Mechanical properties of tape cast nickel-based anode materials for solid oxide fuel cells before and after reduction in hydrogen. Acta Materialia **52**(20), 5747–5756 (2004)
103. M. Radovic, E. Lara-Curzio, L. Riester, Comparison of different experimental techniques for determination of elastic properties of solids. Mater. Sci. Eng. A **368**(1–2), 56–70 (2004)
104. J.L. Routbort, K.C. Goretta, R.E. Cook, J. Wolfenstine, Deformation of perovskite electronic ceramics—a review. Solid State Ionics **129**(1–4), 53–62 (2000)
105. K. Sato, K. Yashiro, T. Kawada, H. Yugami, T. Hashida, J. Mizusaki, Fracture process of nonstoichiometric oxide based solid oxide fuel cell under oxidizing/reducing gradient conditions. J. Power Sources **195**(17), 5481–5486 (2010)
106. A. Selcuk, A. Atkinson, Elastic properties of ceramic oxides used in solid oxide fuel cells (SOFC). J. Eur. Ceram. Soc. **17**(12), 1523–1532 (1997)
107. A. Selimovic, M. Kemm, T. Torisson, M. Assadi, Steady state and transient thermal stress analysis in planar solid oxide fuel cells. J. Power Sources **145**(2), 463–469 (2005)
108. N. Shikazono, D. Kanno, K. Matsuzaki, H. Teshima, S. Sumino, N. Kasagi, Numerical assessment of SOFC anode polarization based on three-dimensional model microstructure reconstructed from FIB-SEM images. J. Electrochem. Soc. **157**(5), B665–B672 (2010)
109. D. Simwonis, F. Tietz, D. Stöver, Nickel coarsening in annealed Ni/8YSZ anode substrates for solid oxide fuel cells. Solid State Ionics **132**(3–4), 241–251 (2000)
110. E. Stephens, J. Vetrano, B. Koeppel, Y. Chou, X. Sun, M. Khaleel, Experimental characterization of glass-ceramic seal properties and their constitutive implementation in solid oxide fuel cell stack models. J. Power Sources **193**(2), 625–631 (2009)
111. C. Stiller, B. Thorud, O. Bolland, R. Kandepu, L. Imsland, Control strategy for a solid oxide fuel cell and gas turbine hybrid system. J. Power Sources **158**(1), 303–315 (2006)
112. B. Sun, R. Rudkin, A. Atkinson, Effect of thermal cycling on residual stress and curvature of anode-supported SOFCs. Fuel Cells **9**(6), 805–813 (2009)
113. Y. Suzue, N. Shikazono, N. Kasagi, Micro modeling of solid oxide fuel cell anode based on stochastic reconstruction. J. Power Sources **184**(1), 52–59 (2008)
114. R. Swindeman, M. Swindeman, A comparison of creep models for nickel base alloys for advanced energy systems. Int. J. Press. Vessels Pip. **85**(1–2), 72–79 (2008)
115. L.W. Tai, M.M. Nasrallah, H.U. Anderson, D.M. Sparlin, S.R. Sehlin, Structure and electrical properties of $La_{1-x}Sr_xCo_{1-y}Fe_yO_3$. Part 1. The system $La_{0.8}Sr_{0.2}Co_{1-y}Fe_yO_3$. Solid State Ionics **76**(3–4), 259–271 (1995)
116. L.W. Tai, M.M. Nasrallah, H.U. Anderson, D.M. Sparlin, S.R. Sehlin, Structure and electrical properties of $La_{1-x}Sr_xCo_{1-y}Fe_yO_3$. Part 2. The system $La_{1-x}Sr_xCo_{0.2}Fe_{0.8}O_3$. Solid State Ionics **76**(3–4), 273–283 (1995)
117. N. Takano, M. Zako, F. Kubo, K. Kimura, Microstructure-based stress analysis and evaluation for porous ceramics by homogenization method with digital image-based modeling. Int. J. Solids Struct. **40**(5), 1225–1242 (2003)
118. P. Tanasini, M. Cannarozzo, P. Costamagna, A. Faes, J. Van herle, A. Hessler-Wyser, C. Comninellis, Experimental and theoretical investigation of degradation mechanisms by particle coarsening in SOFC electrodes. Fuel Cells **9**(5), 740–752 (2009)

119. ThyssenKrupp. Material data sheet no. 4046.
120. F. Tietz, Thermal expansion of SOFC materials. Ionics **5**(1), 129–139 (1999)
121. S. Watanabe, F. Iguchi, J. Mizusaki, K. Sato, K. Yashiro, T. Kawada, Y. Takeyama, T. Hashida, Development of in-situ mechanical testing method for SOFC components. *Proceedings of the ASME 2010 Eigth International Fuel Cell Science*, 2010
122. W. Weibull, A statistical theory of the strength of materials. Proc. Roy. Swed. Inst. Eng. Res. **151**, 1–45 (1939)
123. K. Weil, J. Hardy, J. Rice, J. Kim, Brazing as a means of sealing ceramic membranes for use in advanced coal gasification processes. Fuel **85**(2), 156–162 (2006)
124. K. Weil, B. Koeppel, Comparative finite element analysis of the stress-strain states in three different bonded solid oxide fuel cell seal designs. J. Power Sources **180**(1), 343–353 (2008)
125. K. Weil, B. Koeppel, Thermal stress analysis of the planar SOFC bonded compliant seal design. Int. J. Hydrogen Energy **33**(14), 3976–3990 (2008)
126. K.S. Weil, C.A. Coyle, J.T. Darsell, G.G. Xia, J.S Hardy, Effects of thermal cycling and thermal aging on the hermeticity and strength of silver-copper oxide air-brazed seals. J. Power Sources **152**, 97–104 (2005)
127. K.S. Weil, C.A. Coyle, J.S. Hardy, J.Y. Kim, G.-G Xia, Alternative planar SOFC sealing concepts. Fuel Cells Bulletin **2004**(5), 11–16 (2004)
128. Z. Wuillemin, N. Autissier, A. Nakajo, M.-T. Luong, J. Van herle, D. Favrat, Modeling and study of the influence of sealing on a solid oxide fuel cell. J. Fuel Cell Sci. Technol. **5**(1), 011016–9 (2008)
129. H. Yakabe, Y. Baba, T. Sakurai, M. Satoh, I. Hirosawa, Y. Yoda, Evaluation of residual stresses in a SOFC stack. J. Power Sources **131**(1–2), 278–284 (2004)
130. H. Yakabe, Y. Baba, T. Sakurai, Y. Yoshitaka, Evaluation of the residual stress for anode-supported SOFCs. J. Power Sources **135**(1–2), 9–16 (2004)
131. H. Yakabe, I. Yasuda, Model analysis of the expansion behavior of $LaCrO_3$ interconnector under solid oxide fuel cell operation. J. Electrochem. Soc. **150**(1), A35–A45 (2003)
132. Z. Yang, M.S. Walker, P. Singh, J.W. Stevenson, T. Norby, Oxidation behavior of ferritic stainless steels under SOFC interconnect exposure conditions. J. Electrochem. Soc. **151**(12), B669–B678 (2004)
133. Z. Yang, K.S. Weil, D.M. Paxton, J.W Stevenson, Selection and evaluation of heat-resistant alloys for SOFC interconnect applications. J. Electrochem. Soc. **150**(9), A1188–A1201 (2003)
134. T. Zhang, Q. Zhu, W.L. Huang, Z. Xie, X. Xin, Stress field and failure probability analysis for the single cell of planar solid oxide fuel cells. J. Power Sources **182**(2), 540–545 (2008)

Alternative Materials for SOFCs, Opportunities and Limitations

John T. S. Irvine and Paul Connor

Abstract In this chapter, we highlight some critical aspects of materials for use in solid oxide fuel cells. In relation to oxide ion conducting electrolytes, we address topics including clustering of defects in zirconias and the resultant limitations on ionic conductivity. We also discuss the ionic conduction window for various electrolyte systems. The positive and negative attributes of different anode materials are considered, highlighting the opportunities for alternative materials to be utilised in certain parts of the SOFC system. Some suitable system concepts are presented and a strategy to optimise performance and durability in the same electrode structures is presented.

1 Introduction

Fuel cell systems are highly efficient energy conversion devices, which produce high quality power in a clean and silent manner. They are highly scalable and so can be used to from a decentralised power grid, with careful placement enhancing efficiency do to use for CHP. Solid oxide fuel cells (SOFC) also have fuel flexibility, allowing a number of chemical energy feedstocks.

J. T. S. Irvine (✉) · P. Connor
School of Chemistry, University of St Andrews,
St Andrews, Fife, KY16 9ST, United Kingdom
e-mail: jtsi@st-and.ac.uk

P. Connor
e-mail: pac5@st-andrews.ac.uk

J. T. S. Irvine and P. Connor (eds.), *Solid Oxide Fuels Cells: Facts and Figures*,
Green Energy and Technology, DOI: 10.1007/978-1-4471-4456-4_7,

Table 1 Technology Drivers

Performance	Durability	Cost	Fuel Flexibility
Materials	Materials	Manufacture	Materials
Microstructure and processing	Temperature	Materials	System management
System management	System management		

The focus of fuel cell systems must be the supply of clean energy, which requires a whole cycle analysis. When this is done, the direct use of the more transportable, storable and available hydrocarbon fuels is often the most efficient.

The main technological drivers are listed in the Table 1. To compete with existing energy conversion devices, fuel cell systems must have high performance (both power density and efficiency), durability, low cost and, to fit in most niches, fuel flexibility. Whilst there are significant engineering challenges to achieve all these points, they all require careful selection of the component materials in the cells.

2 Practical Constraints for SOFC

The SOFC operates at relatively high (400–1,000 °C) temperatures, as these are required to get sufficient ionic conduction from the ceramic materials. These high temperatures coupled with the air and fuel input gases produce both highly oxidising and highly reducing atmospheres inside the cells. The materials used to form the device must, therefore, be able to stand up to the temperatures used and the oxidising and reducing nature of the gases. They must still perform their role in the cell be it electrode, electrolyte or interconnect and they must also last for upwards of 10,000 h of use under these extreme operating conditions.

There are many types of SOFC based on different materials, but they must all adhere to a similar set of conditions to make a working cell. The most significant conditions are set out in the Table 2 below.

In terms of durability, all materials must be both stable under operating conditions and not react (significantly) with the other cell parts they are in contact with. Their thermal expansion coefficients (TEC) must also match well enough to ensure heating and cooling cycles do not damage the cell. The electrodes must conduct electrons, ions and gases (i.e. be porous) and have reasonable kinetics for the required electrochemical reactions. The electrolyte must be able to conduct ions, while keeping electrons and gases separated, all the while being stable in both the fuel and oxidising atmospheres.

Higher operating temperatures will enhance the conductivity of the components and hence lower the resistances of the cell and will also typically increase the kinetics of the various reactions, and so will tend to improve the performance of the system. This is generally true but only as long as the materials in the stack are stable. The higher operating, and hence exhaust, temperatures can also be used to improve total efficiency of the system by allowing better use of the waste heat.

Table 2 Desirable characteristics for electrolyte, cathode and anode

Cathode	Electrolyte	Anode
Electronic Conductor	Electronic insulator	Electronic Conductor
Ionic Conductor	Ionic Conductor	Ionic Conductor
Porous	Gas Tight	Porous
Stable in air at High *T*	Stable in air and fuel at High *T*	Stable in fuel at High *T* Ideally also stable in air
No reaction with other cell components	No reaction with other cell components	No reaction with other cell components
Matched TEC	Matched TEC	Matched TEC
Catalyse O_2 reduction		Catalyse H_2 oxidation
		Catalyse fuel processing if needed

A good example of this is the Rolls Royce Fuel Cell Systems concept [1] with the exhaust fed to a gas turbine, which increases the overall electrical efficiency. Another advantage of moving to higher temperatures is that the ease of the reforming of hydrocarbons is improved, with less tendency for cracking [2], allowing for better fuel flexibility.

However, the complete SOFC stack must not significantly degrade over extended times, at the operating temperature. It is the limit of material stability that gives the maximum operating temperature of a fuel cell system. An all-ceramic based cell may be stable to above 900 °C for considerable times, whereas 700–750 °C seems a practical maximum temperature where metals can be incorporated into the conduction path of the stack (See "Challenges Imposed by Thermochemical Expansion of Solid State Electrochemical Materials"). These metals are used as interconnects and/or cell supports, and as metals are easier to manufacture and have typically higher conductivities, similar performance to all ceramic cells can be achieved at lower cost and operating temperature. However, as the temperature comes down so does the speed of many of the reactions in the cell and so the resistance of any ceramic part will increase. To overcome this, all the electrolyte, in particular, must become thinner [3], which, while reducing material costs, does tend to make the processing more expensive and difficult.

The use of lower temperatures will also slow the rate of most chemical degradation processes and so should also increase cell and stack life. The lower temperature will slow diffusional processes such as grain coarsening and diffusion across interfaces, which can lead to solid state reactions. However, some impurities may not be removed at these lower temperatures building up on the electrodes leading to degradation. It should also be noted that lower performance at lower temperatures may necessitate the use of higher electrochemical loads than would be required for optimal durability.

The choice of operating temperature will naturally come out of the system design specifications, including fuel source, as a compromise between cost, performance and fuel source. In some cases, a material or processing step is chosen for commercial or IP reasons, such as use of metal interconnects, or of thicker YSZ electrolytes, which then sets the operating temperature and then applications found for the system. In both cases, there is a close relation between material properties and operating temperatures.

3 Performance

The overriding criterion for a fuel cell/stack is the amount of electric energy it can supply over a practical lifetime. The time frame depends on a lot of factors but mainly end use and initial cost. For a long-term stationary generation system, operating time could be tens of thousands of hours, but may only be hundreds of hours for small portable units. The causes and rates of the various types of degradation are an area of ongoing study.

A reasonable power density expectation for a (planar) fuel cell is about 500 mW cm^{-2} at 0.7 V, which corresponds to about 0.7 A cm^{-2} and so a total cell resistance of about 0.5 ohm cm^{-2}. Spreading this resistance equally among the cell components gives a maximum of about 0.15 ohm cm^{-2} as targets for each of the cathode, anode and electrolyte [3–5]. with a little leftover for interconnects and contact resistances. This specific resistance, however, does give a good upper limit for any one component in a stack and so this value will be used for the rest of this work.

4 Electrolyte Material Selection

The electrolyte, along with the interconnect, has to withstand both highly reducing and highly oxidising atmospheres at the same time in a working fuel cell. As summarised in Table 2, it must be stable in both of these environments and also must not react with the electrodes it is in contact with. It must also be mechanically robust and dense enough to keep the gases separate. It must also be a purely ionic conductor over the whole range of partial oxygen pressures (pO_2) it experiences. This is a very taxing set of conditions, which only a few materials satisfy.

The choice and form of electrolyte materials will typically set the style and operating conditions for the rest of the fuel cell system. Thicker, self-supporting electrolytes are robust and relatively easy to make, but require higher temperatures to achieve suitable levels of ionic conductance (i.e. low enough resistance). Lower temperature operation requires thinner electrolytes, and/or higher conductivity materials, typically supported on another component of the cell. Where not being limited by other materials in the cell, thinner electrolytes can also be used at higher temperatures to improve performance.

The trade-off between material conductivity, temperature and cell thickness is illustrated the plot below. Using the above cut-off resistance of 0.15 Ω cm^{-2} for the electrolyte, it is possible to calculate a maximum thickness for any given

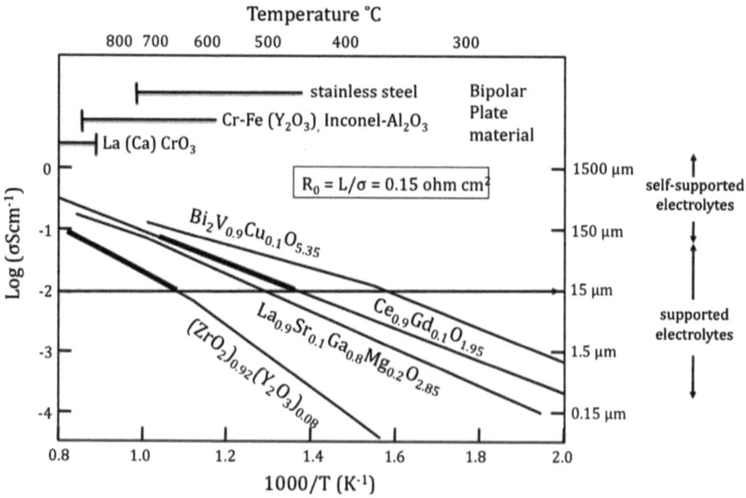

Fig. 1 Conductivity plots of various electrolytes and interconnects, with corresponding thicknesses for a resistance of less than 0.15 ohm cm^2. Based on 6 modified with data from [17]. Reproduced here with kind permission from © Nature Publishing Group 2001

conductivity level, or conversely minimum conductivity for given electrolyte thickness [6].

$$\sigma = \frac{t\,\mu\text{m}}{10{,}000\,\mu\text{m}\,\text{cm}^{-1} \times 0.15\,\Omega\,\text{cm}^{-2}}\,\text{S}\,\text{cm}^{-1}$$

A given preparation thickness of electrolyte can be represented by a horizontal line (the example is 15 μm, $\sigma = 0.01$ S cm^{-1} to give cut-off resistance) with any material with conductivity above that line would be expected to give reasonable performance. Using the 15 μm line on the plot as an example YSZ needs to be over 700 °C to achieve better than 0.15 Ω cm^{-2} whereas GDC needs only be above 450 °C. These two regions are shown as heavier lines. Alternatively, if the electrolyte material and temperature are defined, the conductivity can be worked, and hence approximate required electrolyte thickness determined. As always, higher temperatures give better performance, lower resistance (Fig. 1).

For a good electrolyte, the material must be able to be (cheaply) processed to form a gas tight membrane. Thicker electrolytes, above about 100 μm, can be strong enough to support both themselves and the electrodes; however electrolytes thinner than this need to be supported by another (thicker) layer to achieve the required mechanical strength and still be gas tight.

Other properties of electrolytes can also impose a maximum operating temperature to ensure continued operation. Proton conducting oxide electrolytes, for example decrease in hydration on heating, decreasing proton carrier concentration. Simultaneously, competing conduction mechanisms such as hole electronic and oxide ionic become more significant with temperature. This effectively means that pure protonic behaviour is typically observed only up to 700 °C [7]. Doped cerates, e.g. Gd–CeO_2, and some other materials show significant levels of mixed conduction, with electronic conduction becoming relatively more significant with increasing temperature [8]. This increasing electronic conduction eventually reaches a level that causes unacceptable efficiency, which gives a practical upper limit to operating temperature of say 600 °C. Where the system is operating at low current or on standby for significant periods, losses through non-electrochemical oxidation of fuel are particularly important. Zirconias do not show such high temperature limitations and are still the most widely used materials.

5 Zirconia Electrolytes: YSZ/SSZ

Yttria-Stabilised Zirconia (YSZ) is the state-of-the-art electrolyte for most solid oxide fuel cell systems. This is due to its excellent mechanical and ionic conduction properties, despite its conductivity limitations below 750 °C. Also the development time from new material identification, cell design and stack building and the high cost of any unforeseen problems on the way means that tried and tested YSZ will remain the electrolyte of choice for a long time to come.

Scandia-Stabilised Zirconia (SSZ) shows higher ionic conductivities than YSZ [9], but its high cost and lack of availability, coupled with degradation problems [10] have limited its widespread use. Political changes have improved the availability of scandium, and so increased use in recent years. Scandia zirconias also have much smaller stability ranges for cubic phase formation than their yttria analogues; see the equilibrium phase diagrams shown in Fig. 2.

The most ionically conductive phase in zirconia is the cubic fluorite phase, which is not the stable phase at room temperature for pure zirconia. Doping the zirconia with larger cations and/or introducing oxygen vacancies makes the cubic fluorite structure more stable. The dopant is typically a divalent or trivalent element (Ca^{2+}, Y^{3+} or Sc^{3+}). These vacancies tend to increase the effective ionic size, making the fluorite phase more stable, and also enhance the ionic conductivity by creating more sites for ion motion. Comparing the Y_2O_3 and Sc_2O_3 systems in Fig. 2 the influence of ion size can be seen with a much more extensive cubic region being observed for the Y_2O_3 systems as Y is much larger than Zr, whereas Sc is essentially the same size[13]. The stabilisation of the high temperature cubic phase which gives the materials their name (Scandia-, Yttria- or Calcia-Stabilised Zirconias) reduces the stresses from phase changes during heating and cooling, compared to the unstabilised forms.

The conductivity of stabilised zirconias increases to a maximum with increasing dopant levels see Fig. 3. Increasing the amount of doping induces more and more oxygen vacancy formation increasing conductivity. At higher vacancy levels, however, the vacancies start to interact significantly, lowering their mobility, and so reducing conductivity. In the zirconia system, the maximum is often not achieved within the cubic phase field as transformation to lower

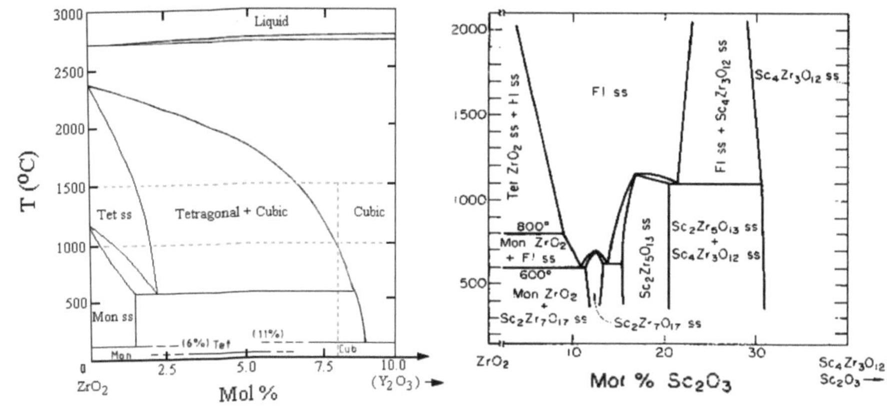

Fig. 2 Binary phase diagrams for the Y_2O_3/ZrO_2 and Sc_2O_3/ZrO_2 systems [11, 12]. Reproduced here with kind permission from © Academic Press 1970

conductivity polymorphs coincides with the conductivity maximum as dopant content is decreased, see for example that the highest conductivity in Fig. 3 is observed for a rapidly cooled composition of 7.7 mol % Y_2O_3 and that on annealing this transforms to a two-phase sample with much decreased conductivity. The best conductivities are typically assumed to occur at about 8 mol % YSZ (8YSZ), see Fig. 3, and about 11 % for SSZ [10]. For pure systems, both these compositions are strictly not stable in the cubic phase at lower temperatures, although chemical interaction between electrodes and electrolyte on processing might assist in stabilising the cubic polymorph to lower temperature.

Increased dopant levels lower the strength, with tetragonal forms tending to be much stronger and despite a penalty in conductivity [14], such compositions are used in some development programmes, especially where strength and small particle size are beneficial.

Samples, which are cubic throughout the measured temperature range, basically show two linear conductivity regions in Arrhenius conductivity plots, Fig. 4. In co-doped systems, a key observation is that the low-temperature activation energy decreases and the high-temperature activation energy increases as yttrium content increases and scandium content decreases [16]. This correlates with the strength of short-range order as indicated by neutron [17] and electron diffraction studies [18]. Although scandia substitution increases conductivity and decreases high-temperature activation energy, it also increases the tendency to short-range ordering at lower temperatures, resulting in a significant increase in activation energy for conduction. This is attributed to the ionic size of the Sc ion, which favours a lower coordination number than that associated with ideal fluorite phases. It should also be realised that Zr, which has similar size to Sc, also prefers a lower coordination number than is ideal for fluorite hence driving the tendency for short-range order in zirconia fluorites.

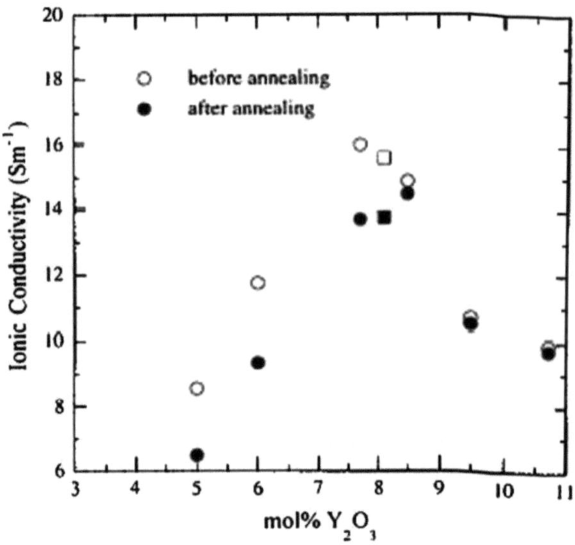

Fig. 3 The change in conductivity as a function of Y_2O_3 content zirconia of unannealed and annealed 5-11 mol % (○, ●) and co-precipitated 8 mol % YSZ(□, ■) at 1000 °C. The maximum conductivity moves from 7.7 to 8.5 mol % after anealing [15]. Reproduced here with kind permission from © Elsevier 1998

6 Doped Ceria Oxides

The doped ceria (CeO_2) based electrolytes, most typically Gadolinium doped ceria (GDC), but also Sm doped (SDC), are probably the next most commonly considered electrolyte. Ceria has the same fluorite structure as doped zirconia, with a similar conduction mechanism. However, the ceria fluorite phase is stable at all temperatures and so the dopants are used purely to increase the vacancy concentration, and hence conductivity. Ceria-based electrolytes exhibit much better ionic conductivity at lower temperatures than YSZ, able to operate down to about 500 °C.

The main problem with these ceria-based electrolytes is the increasing electronic (*n*-type) conduction under reducing conditions, where Ce^{4+} is reduced to Ce^{3+}. While this starts on the anode side of the electrolyte, it can penetrate through to the cathode side giving an electronic conduction path through the electrolyte [8]. This internal current leakage reduces the cell efficiency by both using up fuel, without giving external current, but also the voltage of the cell is reduced, so lowering power output. While the cell is operating, oxide ions are pumped to the anode side, and so raise the local oxygen activity (pO_2) and decrease the extent of the ceria reduction and hence electronic conduction. This helps to alleviate some losses; however, if current is stopped or reduced, fuel is oxidised without yielding useful power.

As Ce^{3+} is significantly larger than Ce^{4+} (Ionic radius 1.01 Å vs 0.87 Å) [13], there is also a considerable lattice expansion on reduction. This causes strain inside the electrolyte, which can lead to mechanical stability problems over time. As the level of reduction changes with cell load, this mechanical stress will vary during cell operation, limiting cell life. Again, this stress will change across the electrolyte due to variation in oxygen activity.

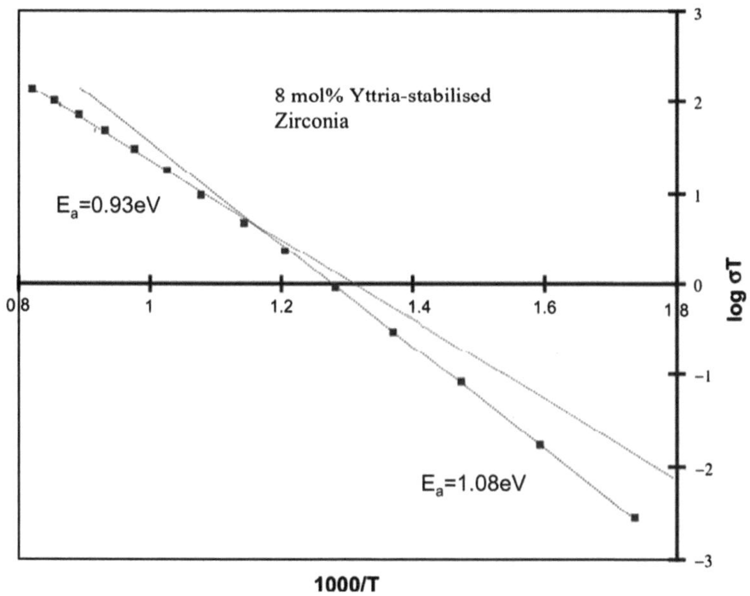

Fig. 4 Arrhenius plot of 8 mol % YSZ, showing two different slopes at high and lower temperature [17]. Reproduced here with kind permission from © Springer New York 1996

The practical maximum temperature for a ceria-based electrolyte is at most 600 °C, at which temperature the reduction causes too much electronic conductivity and the cell efficiency drops too far. Protective layers of zirconia are sometimes applied to prevent such short circuits

7 Anode Selection

Anodes in SOFCs encounter quite a range of different atmospheres, depending upon source of fuel gas, degree of reforming or partial oxidation and cleanup processes. The fuel gas will thus comprise of differing ratios of H_2, CO, CH_4 and steam with smaller amounts of higher hydrocarbons, sulphur species, as well as nitrogen and possibly even oxygen if partial oxidation has been utilised. The carbon rich fuels can add extra complications as coke formation is likely. Irreversible reactions can occur with sulphur species and at high steam contents anode oxidation may occur. The cell will preferably be produced in air, and so the anode should be stable in both air and the highly reducing fuel environment, although a change in nature between these environments might be tolerable.

An anode has at least two functional parts: the "active region" or electrochemical functional layer where the fuel is oxidised, and the current collector region which removes the electrons but still allows the gas access to the inner active region.

Table 3 Strengths and weaknesses of Ni–YSZ cermet anodes

Strengths	Weaknesses
Good current collection	Not redox stable
Can reform methane	Formation of carbon fibres
Steam reforming catalyst	Low sulphur tolerance
	Triple phase boundary easily poisoned

Depending on the fuel used there might also be a region in the anode gas path for a fuel reforming catalyst.

The state-of-the-art anode is the Ni/YSZ cermet (CERamic–METal composite). This mixture of ionically conducting YSZ ceramic, and both electronically conducting and catalytic Ni metal gives the required anode functionality. It is also easily formed from a nickel oxide–YSZ composite ceramic framework, which can be fired in air, followed by an in situ reduction to form the desired Ni–YSZ cermet. This in situ reduction of NiO to Ni metal has the added advantage of increasing the porosity of the electrode at the same time aiding gas transport. As the Ni acts as both current collector and fuel oxidation catalyst, the separation of active region and current collector is sometimes hard to define.

Nickel metal has a large TEC, much larger than that of the YSZ electrolyte, and so needs the YSZ (and pores) to reduce the TEC of the electrode to that of the rest of the cell. There is also, however, a minimum amount of Ni content required in the electrode to ensure an electronic conduction path (percolation) through all the Ni particles. This gives a compromise between enough Ni to percolate, but not too much to give a too high TEC, while still keeping the electrode porous enough for gas transport.

The strengths of the cermet are its good performance with hydrogen as a fuel coupled with its excellent current collection, and hence low electrode resistance. This gives high power density for H_2 and reformate fuels. It has catalytic ability not only to oxidise hydrogen, but to also reform methane as well as being a good steam reforming catalyst [19].

It does, however, have significant weaknesses. A major problem is the fact it is not redox stable, with the active Ni metal particles easily oxidised back to NiO.

$$2Ni + O_2 \rightarrow 2NiO$$

This oxidation can occur on fuel starvation, such as on shut down or simply through lack of fuel, it can also occur under high steam conditions, such as under high load. This oxidation has two main effects. The NiO takes up significantly more volume than the Ni, causing stress in the electrode, which can lead to cracking, eventually affecting the structure of the electrode even the whole cell. The other effect is the highly catalytically active small Ni particles tend to coarsen on repeated oxidation and reduction, which leads to a lowering of the active area of the cell, and hence a drop in performance (Table 3).

Although Ni is a good oxidation catalyst, it is also a good catalyst for the production of carbon fibres

$$C_nH_{2n+2} \rightarrow C$$

This carbon builds up inside the porous electrode blocking the gas transport into the electrode, severely compromising performance. If left unchecked, this can even physically damage the electrode structure, especially as the carbon often grows underneath Ni particles. This limits the usefulness of Ni cermets for the readily available higher hydrocarbon fuels, and requires careful control of conditions, e.g. high steam, even for the lower hydrocarbons.

The other reaction of concern for Ni cermets is

$$Ni + S \rightarrow NiS$$

Sulphur is a common impurity in many carbon-based fuels, both as naturally occurring compounds e.g. in biogas or coal gas and the mercaptans added to natural gas as an odorant [19].

The Ni–YSZ electrodes suffer from various degradation mechanisms, but they mainly affect the triple phase boundary (TPB) in the electrode. This is the region where the electrons, oxide ions and the fuel gases meet and is the site of the fuel reduction. In the cermet, this region is limited to the interfaces between Ni and YSZ particles which are connected to current collector and electrolyte, respectively, and are open to pores. This region is poisoned easily as the electrochemical activity concentrates impurities.

As the TPB is limited to the contact between Ni and YSZ particles, anything that changes the size of the Ni particles will change the amount of contact. As the Ni particle size increases with electrode ageing, and especially redox cycling the TPB decreases lowering the electrode performance. This is always going to be a problem for mixed ionic/electronic conducting phases.

8 Oxide Anodes

A number of good alternative single-phase perovskite materials exist as possible anode materials. These are dimensionally stable (and often even redox stable) and so can easily be used in solid oxide fuel cells. They have good electrode kinetics (catalytic) properties and are often both ionic as well as electronic conductors. Their conductivity can be good for oxides, but not as good as metals (e.g. Ni) and so they are best used as true anodes, i.e. the electrochemical functional layer but not are generally not good enough to act as current collectors.

Target conductivity for an electrode is around 100 S cm^{-1} but this value can be reduced to as low as 1 S cm^{-1} through careful morphological design, and careful use and design of current collection. This allows the use of oxide materials as anodes, removing the metal content and making a more stable cell.

The most common structural type for possible anode materials is the perovskite structure. This structure is very flexible, allowing most transition metals to be included in the structure, but also show good stability and electronic conductivity.

Table 4 Transition metal elements in a perovskite framework, showing ability to reduce oxide coordination from 6 and stability in fuel atmosphere

	Sc	Ti	V	Cr	Mn	Fe	Co	Ni	Cu	Zn
MO_5			V^{V}		✔	✔	✔	✔	Cu^{II}	
MO_4tet			V^{V}		?	✔	?	?		✔
MO_4Sq								?	Cu^{II}	
Fuel	✔	✔	✔	✔	✔	?				

Table 4 shows some elements that can be included in perovskite materials, highlighting those that would be expected to allow a decrease in coordination number in the perovskite lattice, and hence afford catalytic behaviour, with the elements stable in the reducing conditions of an SOFC anode indicated on the bottom line. The use of these species produce a dimensionally stable anode, that does not change significantly between fully oxidised and fully reduced. This reduces many degradation and lifetime issues.

A good example is $La_{0.75}Sr_{0.25}Cr_{0.5}Mn_{0.5}O_3$ (LSCM) which shows excellent redox stability, good conductivity and good electrode kinetics. It has been prepared both by standard high temperature methods [20] as well as by impregnation into a porous framework [21] and shows comparable performance to Ni–YSZ cermet anodes especially when prepared with high surface area and with appropriate catalysts [21].

A particularly important property of LSCM is that it has catalytic activity for direct oxidation of hydrocarbons, which is more significant than its reforming activity [22, 23]. This is shown in Fig. 5 where similar performance is achieved with direct methane (b) as with hydrogen (a). This means that it is a direct oxidation catalyst with less dependence on the conversion of fuels to hydrogen, e.g. via water gas shift or reforming, with the stack or cell.

9 Fuel

Fuel choice is an important aspect. Hydrogen is often associated with fuel cells due to its green credentials and simple chemistry; however, hydrogen has several practical difficulties. It is not widely available, as it tends to only be made in large quantities where it is needed for a chemical process. Also due to its physical properties, it is hard to transport and store in an energetically efficient manner. Also, most H_2 produced at present is made from reforming natural gas, and so there is very little green H_2 available.

It is generally more efficient to move natural gas about and use it directly and efficiently. For transportable energy sources, liquid fuels are much more practical, which also requires the use of hydrocarbon fuels. These can be produced from renewable (e.g. biomass) sources. With increasing demand and limited supplies hydrocarbons fuels will become even more expensive, emphasising the importance of enhanced efficiency from devices such as SOFCs.

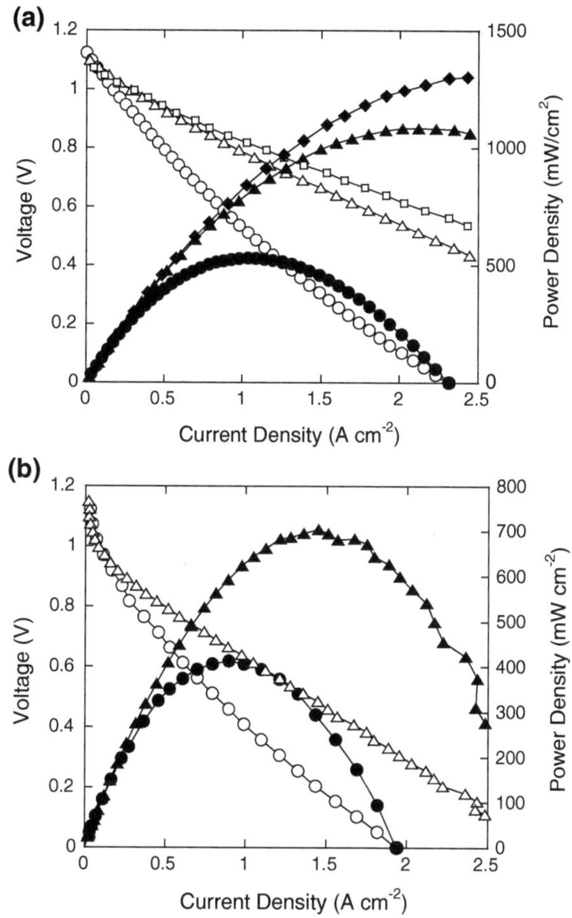

Fig. 5 V-I polarisation curves for a cell with an LSCM anode containing 0.5 wt % Pd and 5 wt % ceria, **a** measured in humidified H_2 (3 % H_2O) at: (○) at 973 K, (Δ) at 1073 K, and (◇) at 1173 K. **b** Measured in humidified (3 % H_2O) CH_4 at: (○) at 973 K and (Δ) at 1073 K. Filled points give the power density [21]. Reproduced here with kind permission from © The Electrochemical Society 2008

Hydrocarbons utilisation in a fuel cell can follow several parallel pathways (See Fig. 6), each with many individual steps. The simplest is direct oxidation, but this requires good oxidation catalyst anodes, and careful control of fuel conditions, but releases the most energy as it has no endothermic processes. The most common method of using hydrocarbon fuels is to reform them by heating them with steam or CO_2. This transforms the fuel to H_2 and CO which can be used directly in a fuel cell or via water gas shift from CO to H_2. This is also an endothermic reaction and so needs a heat input, which preferably can be coupled with the heat given out by the fuel cell oxidations. A similar process is partial oxidation with air, which be used to simplify overall system structure. Cracking is the thermal decomposition of methane and short-chain hydrocarbons to yield carbon and is generally undesirable as this leads to structural disruption of the anode; however, a small degree of cracking to yield tars can be beneficial [24].

A fuel cell produces waste heat, carbon dioxide and steam, enabling reformation directly in stack. This has several merits, simplifying both fuel and heat flow

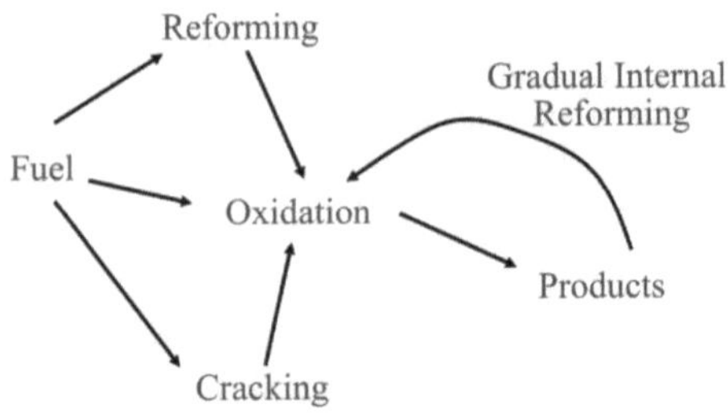

Fig. 6 Alternative pathways for hydrocarbon utilisation in SOFC systems [19]. Reproduced here with kind permission from © Nature Publishing Group 2004

in a finished system. Allied to this is the concept of gradual internal reforming, whereby a relatively dry fuel stream becomes increasingly hydrated, allowing improved reforming as the fuel passes through the stack or across the cell.

When fuels are oxidised at different electrodes, all these pathways may occur; however, one or two processes will tend to dominate at a given electrode material, this is illustrated for certain electrodes in Table 5.

10 Fuel Utilisation

It is fairly common when testing fuel cells to get high efficiency in laboratory tests when running at low fuel utilisation; however, when using a fuel cell stack in a real system, with real, expensive fuels the utilisation needs to be as high as possible, ideally more than 80 %. This high fuel utilisation produces large amounts of water (steam) in the stack, coupled with low levels of fuel. The pO_2 at the outlet of the cell/stack can rise high enough to oxidise a nonredox stable electrode (e.g. Ni-based electrodes). Therefore, all the system then needs to be designed to cope with this high steam content. One solution is anode gas recycling, where the cell is run at a safe degree of utilisation, and part of the outlet stream is recycled back to the fuel input, giving the fuel a second chance to be utilised. This does add water and CO_2 into the fuel input stream, which will reduce voltage and render the fuel atmosphere more oxidising. This injection of water can have a positive effect on reforming and aids capture of waste heat in the system.

Another option is to use redox stable anodes to raise the actual fuel utilisation, without making the system more complex. This is perhaps one of the best reasons for the use of redox stable anodes as they can tolerate these high steam conditions.

A practical method to optimise performance and fuel utilisation is to use a two-stage approach with conventional Ni-based electrodes to get high performance, followed by a stable oxide electrode to maximise the degree of utilisation of the fuel. For a system with gradual internal reforming, a graduated electrode could be used with the bulk of the area of the anode being Ni based, to give good reforming and performance, with a strip of oxide based anode along the exhaust end of the anode, where there will be a high steam content and all the fuel is reformed. This last strip can increase the total fuel utilisation to over 90 % and stay stable.

Table 5 Dominating characteristics for a few selected electrode materials for use in hydrocarbon oxidation

Key characteristics of selected hydrocarbon oxidation electrodes	
Direct Oxidation Catalysts	CeO_2, LSCM
Reforming Catalysts	Ni
Cracking catalysts	Ni
Current Collection	Cu, Ni, $SrTiO_3$, $Nb_2Ti_2O_6$
Mixed Electronic/Ionic Conductors	$(Gd,Ce)O_{2-x}$
Electrochemical enhancement	

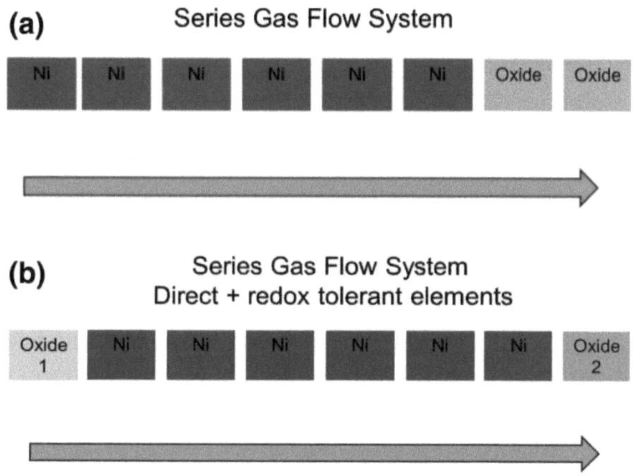

Fig. 7 Schematic of series gas flow through several fuel cells: **a**—to maximise fuel utilisation, **b**—to reduce reforming temperature shock at the inlet in addition to increasing utilisation

The same effect can be achieved using a serial flow set as below where the gas flows over the Ni-based cells as the fuel is used up and then some oxide based cells at the end, Fig. 7a. Another useful scenario is to have an oxide anode in the fuel inlet position, this might utilise a direct oxidation anode at the inlet, to minimise reforming shock and initiate gradual internal reforming, with a redox tolerant anode at the exhaust, Fig. 7b.

Another possibility to improve fuel utilisation using a simple planar stack design is a series–parallel gas flow system that utilises two different stacks, with the gas flowing through a standard Ni–YSZ based cell stack first, with a reasonable, but safe utilisation, followed by an oxide based cell stack to extract the last of the unused fuel from the exhaust.

Anode gas recycle can also be used with these systems to further enhance the fuel utilisation or to supply the steam for any reforming processes.

11 Simultaneous Optimisation of Performance and Durability

The limiting factor for fuel cell performance is often the extent of triple phase boundary (TPB) available to be the site of cathode or anode reactions. Whilst the properties of the materials used, whether cermet, composite oxides or a mixed ion

Fig. 8 (La,Sr)$Mn_{0.5}Cr_{0.5}O_3$ impregnated onto a YSZ skeleton after firing at 1250 °C in air, **a** (*left*), after subsequent reduction at 800 °C in 5 % H_2, **b** (*middle*) and after subsequent reoxidation in air at 800 °C, **c** (*right*) [21]. Reproduced here with kind permission from © The Electrochemical Society 2008

conductor, and the final microstructure achieved are important, this may be dominated by the extent of surface area of the electrode. The principal way to significantly increase the TPB length is to decrease the particle size and the use of nanomaterials is the best way to get the highest surface area.

For most applications, a fuel cell system must also have a long life. In a lifetime that may be over 40,000 h, degradation in performance must be small, on the order of a percent every 1,000 h. In an SOFC, the two largest causes of performance loss (degradation) is the increase in cell resistance due to new phases at critical interfaces, and the loss of TPB reducing the active area of the cells. The new phases are typically formed by the slow reaction of component parts of the cells (and metallic interconnect corrosion) to form blocking or resistive materials. The loss of TPB can be due to impurities building up at interface or due to grain growth of the active particles, to reduce their surface energy. Many of the materials used in an SOFC can react at very high temperatures, but are typically stable at the lower operating temperature of the cells. An example of this is LSM and YSZ which can react at high temperature to form a pyrochlore ($La_2Zr_2O_7$), which is quite resistive compared to YSZ. This forms at the interface of the electrolyte and LSM, which is the location of the TPB [25]. The fact that small particles coalesce and grow when heated is well known (Ostwald ripening). The atoms in small Ni particles are quite mobile and so the particles tend to coarsen slowly under SOFC operating conditions. The best way to slow both of these problems is to use larger particles, which will minimise grain growth, and lowers the reaction area for any side reactions.

So for high performance, one requires small (nano-) particles but to keep performance for an appreciable time one needs large, stable particles. This sets two contradictory demands on microstructure.

An elegant solution to this problem is the in situ formation of nanostructured materials on the electrodes under fuel cell conditions, especially when this is an equilibrium process. Such behaviour has been observed in YSZ scaffolds impregnated with Mn containing perovskites [21]. When (La,Sr)$Mn_{0.5}Cr_{0.5}O_3$ is prepared by solution impregnation onto a YSZ skeleton and sintered in air, it forms a think uniform coating, Fig. 8a, due to a favourable surface interaction between the

Mn-containing perovskite and the YSZ. On reducing at 800 °C, the coating transforms to a nanoparticulate structure, Fig. 8b, and this reverts towards a smooth coating on reoxidation at this temperature, Fig. 8c.

This in situ grown nanostructure was associated with the high performance shown above in Fig. 5. Clearly, nanostructures that are formed under operating temperatures would be intrinsically more stable than those formed ex situ. Thus, this research provides an important future strategy in matching performance and durability.

12 Summary

In conclusion, it is important to note that many subtle interplays exist that determine whether new materials have a role to play in fuel cell systems that might be used commercially. There are several competing factors to consider and it is generally inappropriate to consider that one material is a simple "one size" fit to needs. Instead, it is important to optimise microstructure as well as composition and to consider pertinent application condition, if one is really identify the opportunities that overcome the limitations of a particular system.

References

1. F.J. Gardner, M.J. Day, N.P. Brandon, M.N. Pashley, M. Cassidy, SOFC technology development at Rolls-Royce. J. Power Sources **86**, 122–129 (2000)
2. K. Sasaki, K. Watanabe, Y. Teraoka, Direct-alcohol SOFCs: current-voltage characteristics and fuel gas compositions. J. Electrochem. Soc. **151**, A965–A970 (2004)
3. BCH Steele (1994) Proc 1st European solid oxide fuel cell forum (ed U Bossel) 375–397
4. Brian C.H. Steele, Survey of materials selection for ceramic fuel cells II Cathodes and anodes. Solid State Ionics **86–88**, 1223–1234 (1996)
5. Allan J. Jacobson, Materials for solid oxide fuel cells chem. Mater **22**, 660–674 (2010)
6. C. H. Brian Steele Angelika H, Materials for fuel-cell technologies. Nature **414**:345–352 (2001)
7. Angela Kruth, John T.S. Irvine, Water incorporation studies on doped barium cerate perovskites. Solid State Ionics **162–163**, 83–91 (2003)
8. Z. Xinge, M. Robertson, C. Deĉes-Petit, W. Qu, O. Kesler, R. Maric, D. Ghosh, Internal shorting and fuel loss of a low temperature solid oxide fuel cell with SDC electrolyte. J. Power Sources **164**:668–677 (2007)
9. S.P.S. Badwal, F.T. Ciacchi, D. Milosevic, Scandia-zirconia electrolytes for intermediate temperature oxide fuel cell operation. Solid State Ionics **136–137**, 91–99 (2000)
10. T.I. Politova, J.T.S. Irvine, Investigation of scandia-yttria-zirconia system as an electrolyte material for intermediate temperature fuel cells. Influence of yttria content in system $(Y_2O_3)_x(Sc_2O_3)_{(11-x)}(ZrO_2)_{89}$. Solid State Ionics **168**, 153–165 (2004)
11. H.G. Scott, Phase relationships in the zirconia-yttria system. J. Mater. Sci. **10**, 1527–1535 (1975)
12. F.M. Spiridonov, L.N. Popova, R.Y. Popil'skii, On the phase relations and the electrical conductivity in the system ZrO_2-Sc_2O_3. J. Solid State Chem. **2**, 430–438 (1970)

13. R.D. Shannon, Revised effective ionic radii and systematic studies of interatomic distances in halides and chalcogenides. Acta Cryst. **A32**, 751–767 (1976)
14. W. Weppner, Tetragonal zirconia polycrystals: a high performance solid oxygen ion conductor. Solid State Ionics **52**, 15–21 (1992)
15. I.R. Gibson, G.P. Dransfield, J.T.S. Irvine, Influence of Yttria concentration upon electrical properties and susceptibility to ageing of Yttria-stabilised Zirconia. J. Europ. Ceram. Soc. **18**, 661–667 (1998)
16. John T.S. Irvine, Jeremy W.L. Dobson, Tatiana Politova, Susana García-Martíın, Atef Shenouda, Co-doping of Scandia–Zirconia electrolytes for SOFCs. Faraday Discuss. **134**, 41–49 (2007)
17. I.R. Gibson, J.T.S. Irvine, Study of order/disorder transition in Yttria-stabilised Zirconia by Neutron Diffraction. J. Mater. Chem. **6**, 895–898 (1996)
18. S. Garcia-Martin, D.P. Fagg, J.T.S. Irvine, Characterization of diffuse scattering in Yttria-stabilized zirconia by electron diffraction and high-resolution transmission electron microscopy. Chem. Mater. **20**, 5933–5938 (2008)
19. A. Atkinson, S. Barnett, R.J. Gorte, J.T.S. Irvine, A.J. McEvoy, M. Mogensen, S.C. Singhal, J. Vohs, Advanced anodes for high-temperature fuel cells. Nat. Mater. **3**, 17–27 (2004)
20. S. Tao, J.T.S. Irvine, A Redox-stable, efficient anode for solid-oxide fuel cells. Nat. Mater. **2**, 320–323 (2003)
21. G. Kim, G. Corre, J.T.S. Irvine, J.M. Vohs, R.J. Gorte, Engineering composite oxide SOFC anodes for efficient oxidation of methane. Electrochem. Solid State Lett. **11**, B16–B19 (2008)
22. S. Tao, JTS. Irvine, S.M. Plint (2005) Methane oxidation at redox stable fuel cell electrode $La_{0.75}Sr_{0.25}Cr_{0.5}Mn_{0.5}O_{3-\delta}$. J. Phys. Chem. 110:21771–21776
23. Shanwen Tao, John T.S. Irvine, Catalytic properties of the perovskite $La_{0.75}Sr_{0.25}Cr_{0.5}Mn_{0.5}O_{3-\delta}$ in relation to its potential as a solid oxide fuel cell anode material. Chem. Mater. **16**, 4116–4121 (2004)
24. Steven McIntosh, Hongpeng He, Shung-Ik Lee, Olga Costa-Nunes, Venkatesan V. Krishnan, John M. Vohs, Raymond J. Gorte, An examination of carbonaceous deposits in direct-utilization SOFC anodes. J. Electrochem. Soc. **151**, A604–A608 (2004)
25. A. Mitterdorfer, L.J. Gauckler, $La_2Zr_2O_7$ formation and oxygen reduction kinetics of the $La_{0.85}Sr_{0.15}Mn_yO_3$, $O_2(g)$|YSZ system. Solid State Ionics **111**, 185–218 (1998)

Materials for Next Generation SOFCs

Stephen J. Skinner, Stuart Cook and John A. Kilner

Abstract New materials are constantly identified as potential components of next generation energy technologies. Significantly, new engineering solutions have also been proposed that will enhance the performance of both these new and also existing materials. Here the latest developments in both materials and device engineering are summarised focusing on the potential for enhanced ionic mobility achieved through engineered structures utilising thin film deposition technologies. In the second section of this work the promise of new materials will be discussed, considering the layered and double perovskite structure types as next generation electrodes.

1 Introductory Comments

Solid oxide fuel cells have been the subject of extensive research activities over the past 40 years, with significant advances made in the development of materials for anodes and cathodes and the identification of novel electrolyte materials. Developers have selected a relatively narrow compositional space to explore, focussing on the fluorite, AO_2, and perovskite, ABO_3, structural families. Indeed the materials currently used in SOFCs can be narrowed down to the choice of one of three electrolytes: yttria stabilised zirconia (YSZ), gadolinium substituted ceria (GDC) or substituted lanthanum gallates [1], with most interest in YSZ and GDC. For the electrodes there are currently limited choices for developers, with Ni

S. J. Skinner (✉) · S. Cook · J. A. Kilner
Department of Materials, Imperial College London, Prince Consort Road,
London, SW7 2BP, UK
e-mail: s.skinner@imperial.ac.uk

J. T. S. Irvine and P. Connor (eds.), *Solid Oxide Fuels Cells: Facts and Figures*,
Green Energy and Technology, DOI: 10.1007/978-1-4471-4456-4_8,

cermets utilised for the anode and at the cathode side $La_{1-x}Sr_xMO_{3-d}$ (M = Fe, Co, Mn) typically selected. From these few material types the majority of individual cells are engineered, but it is clear that these materials have significant limitations in their performance [2]. If SOFC technology is to make a significant and timely contribution to future energy production and security, further materials development leading to higher performance and lower temperature of operation are required. To meet these challenges a new approach to materials development is necessary.

There are two main strategies for the development of next generation SOFC materials: engineering solutions and materials solutions. In the following sections both approaches will be discussed and contrasted, with potential future directions for SOFC development outlined.

2 Interfacial Conductivity

2.1 Electrolyte Materials

There is a continuing demand for new materials that show improved oxygen ion transport properties. This is especially true for devices for clean energy systems, including solid oxide fuel cells, where there is a drive towards lower temperatures ($\leq$500 °C) of operation. Two main methods have been used to develop materials to meet this demand. The first, and more conventional, is the optimisation of the lattice mobility for the oxygen ion, i.e. by the further development of existing materials. The second, and more recent, is by the investigation of nanomaterials and nanostructures. This latter approach is unusual because it involves the optimisation of the *interfacial* transport of oxygen, which has not received much attention over the past 20 years. Grain boundaries in oxide materials have been recognised as short circuit paths for diffusing ions in a range of materials, but this is mainly true for materials where the bulk diffusion coefficients are very low. In fast ion conductors grain boundaries and interfaces were mostly viewed as being detrimental to the transport of ions, and were characterised by their blocking effect transverse to the direction of the ion transport. Little attention was given to the possibility of fast oxygen ion transport along interfaces parallel to the transporting ions (e.g. see [3]).

Interest in the possibility of high transport rates along interfaces initiated some years ago in studies of cation conductors, when it was observed that composite materials of an ionic conductor and an insulator could enhance the ionic conductivity when the insulating phase reached the percolation limit. The system that was most thoroughly investigated was the LiI:Al_2O_3 composite system, and theories were advanced that interfacial layers of very high conductivity [4] should give rise to enhancement of the ionic conductivity of the composite. Similar ideas were proposed for effects at grain boundaries and this led to the investigation of

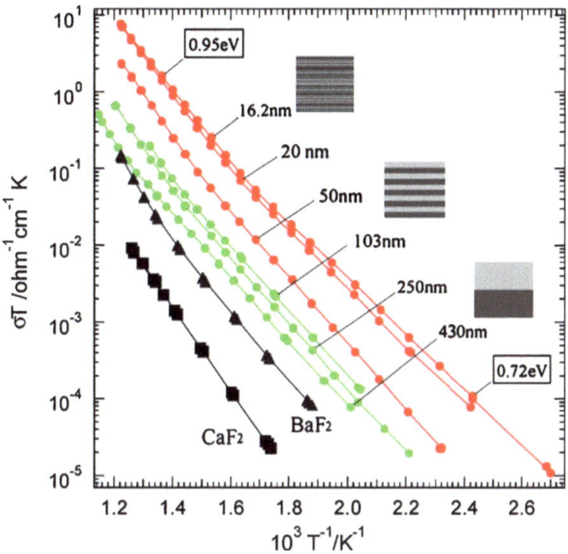

Fig. 1 Enhancement of the conductivity of BaF_2/CaF_2 heterostructures [6]. Reproduced here with kind permission from © Elsevier 2011

materials with nano-sized grains to search for measurable effects but up to the year 2000 the results were contradictory as shown in a review by Tuller [5].

The idea of using space charge to influence the conductivity was developed by Sata and co-workers [6] who investigated the ionic conductivity of CaF_2/BaF_2 multilayer heterostructures and demonstrated that the conductivity was enhanced along the interfaces. Sata et al. interpreted this effect as being caused by the enhancement of the concentration of mobile carriers at the interface due to the presence of the space charge layers. This effect is shown in the data presented in Fig. 1. Here it should be noted that, as the materials are not substituted, the Debye length is considerable and the dimensions of the space charge layer are comparable with the layer thickness. This is less likely to be true in materials such as yttria stabilised zirconia (YSZ) where the high level of substitution leads to a much reduced Debye length.

Following the landmark paper by Sata [6], investigations were started in oxide systems by measuring the conductivity in thin (down to ~15 nm) epitaxial oxide layers. The first studies to show conductivity enhancement were in thin monolayers of yttria stabilised zirconia investigated by Kosacki et al. [7] and Karthikeyan et al. [8]. Kosacki et al.'s findings for highly textured YSZ films on an MgO substrate showed an increase in conductivity as the layer thickness decreased (Fig. 2). They interpreted this result in terms of a two component model, the bulk material and a thin interfacial layer, of different conductivity. The oxygen ion conductivity of this interfacial layer, of ~1 to 2 nm thickness, between the MgO substrate and the YSZ film, was strongly enhanced by up to 3 or 4 orders of magnitude. The activation energy was more than halved from 1.04 eV to 0.45 eV. This change in activation energy is important for achieving high values of conductivity at low temperatures.

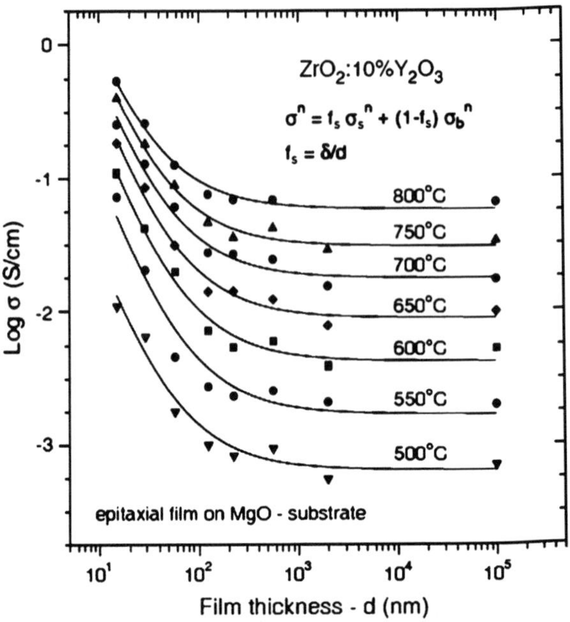

Fig. 2 Thickness dependence of the conductivity of highly textured YSZ thin films at a range of measurement temperatures [7]. Reproduced here with kind permission from © Elsevier 2011

These early findings, although interesting, did not exclude the possible contribution of the free surface of the film to the enhancement. This prompted a number of subsequent investigations into the ionic conductivity of multilayer heterostructures, mostly produced by pulsed laser deposition, of an oxygen ion conductor (mainly YSZ) sandwiched between layers of insulating materials. The temperature dependence of the ionic conductivity for a range of these monolayer and multilayer materials is shown in Fig. 3a,b. Most of the results in this figure fall within a fairly tight group, however, the results obtained by Garcia-Barriocanal et al. [9] stand out, as is shown clearly in Fig. 3b, and appear to show a much larger enhancement. These workers investigated very thin layers of YSZ alternated with strontium titanate (STO) in samples produced by RF sputtering. They measured epitaxial trilayers of 10 nm STO/1 nm YSZ/10 nm STO and found the conductivity was highly enhanced (trace 9a) in Fig. 3b. They claim over 8 orders of magnitude of enhancement of the ionic conductivity at 100 °C. This is clearly of great interest if the conductivity they measure is indeed ionic. The prospect of low temperature devices, albeit in thin film form, is very appealing if this finding were to be verified. At the date of writing there is still a great deal of debate over these findings, they have not been repeated and there is controversy over the nature of the carrier, an electronic carrier would be much easier to reconcile with these exceptionally high conductivities [10–12]. In addition the interfacial strain at the STO/YSZ interface is very large, described as having a 45° rotation in the growth axis, resulting in a lattice mismatch on the order of 7 %, which should lead to a partially coherent or fully incoherent interface. Despite the lack of experimental verification there have been some recent theoretical studies, including simulations

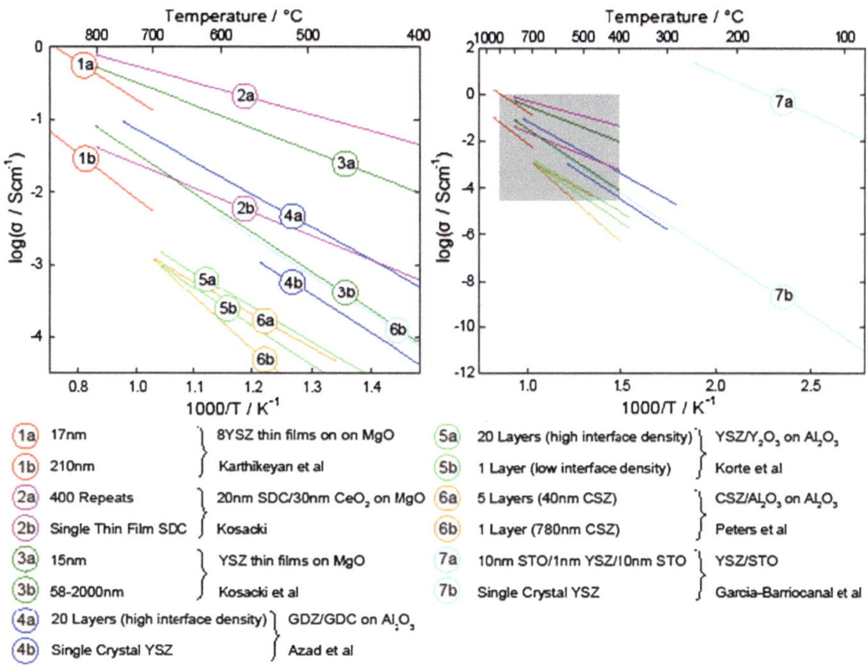

Fig. 3 Compilation of oxygen ion conductivity data for multilayer heterostructures and epitaxial thin films [7–9, 16, 17, 22, 83–85]

by Density Functional Theory (DFT) which would tend to support the idea that the observed effects could indeed be ionic [13], although the simulations by Kushima and Yildiz [14] have suggested that the maximum degree of enhancement might be much less than that observed.

Korte and co-workers [15–17], have investigated a number of different interfaces by fabricating stabilised zirconia-insulator heterostructures using pulsed laser deposition. This is the most systematic of current investigations of heterostructures and recently Korte et al. [18] have reviewed the status of current work on these structures and have classified them according to the degree of strain and hence coherency of the crystal structure across the interfaces. This scheme is shown in Fig. 4. By an analysis of the conductivity results they show that the lattice strain in the zirconia layers is important in determining the magnitude of the change in both conductivity and in activation energy. Tensile strain leads to a lowering of the activation energy and hence an enhancement of the conductivity, by up to an order of magnitude, compressive strain can lower the conductivity with respect to the bulk single crystal value. This is complementary to both early works studying cationic conduction under hydrostatic stress (e.g. [19]) and more recently, the experimental and theoretical work of Araki and colleagues [20–22] who have measured changes in the conductivity of polycrystalline samples of YSZ under uniaxial tensile and compressive stress and observed the enhancement

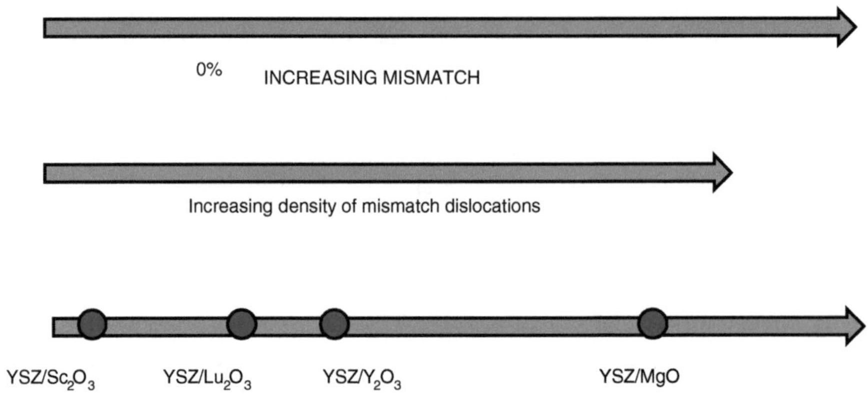

Fig. 4 Classification of interfaces following Korte et al. [18]

(maximum ~18 %) and deterioration of the conductivity respectively. They have confirmed this finding by molecular dynamics studies of oxygen migration under uniaxial stress [22] and report clear effects when the tensile strain is larger than 0.3 %, well below the strains expected at most of the superlattice interfaces although it is important to note that the volume of material strained, and thus the scale of the effect on conductivity this will display, will be different in each case. It seems clear from these results that dilation of the YSZ lattice facilitates the migration of the large oxygen ion, although Araki reports no decrease in the activation energy for conductivity.

Of the data reviewed by Korte [18] the largest *enhancement* seen is in structures formed with calcia stabilised zirconia and Al_2O_3 as the insulating layer. Here the interfacial strain would be very large (>8 %) which results in an incoherent interface and a disordered region with a high density of mismatch dislocations. The enhancement of conductivity seen is around a factor of 60 at 575 °C with a lowering of the activation energy of almost 0.5 eV, a significant decrease leading to substantial low temperature conductivity. Korte et al. [18] speculate on the role of any misfit dislocations in the conductivity enhancement.

More recently Sillasen et al. [23], have extended the studies of epitaxial YSZ films on MgO substrates prepared by magnetron sputtering. They reported extensive enhancement of the conductivity by up to 3 orders of magnitude, again with a lowering of the activation energy. They conclude that there are two contributions to the conductivity, one due to the bulk and one, like the earlier studies of Kosacki [7] due to the film substrate interface which dominates at lower temperatures due to a lower activation energy. The interface between the MgO and the YSZ is very disordered and by careful characterisation of the interface by TEM they conclude that the enhancement of the ionic conductivity along the interface is

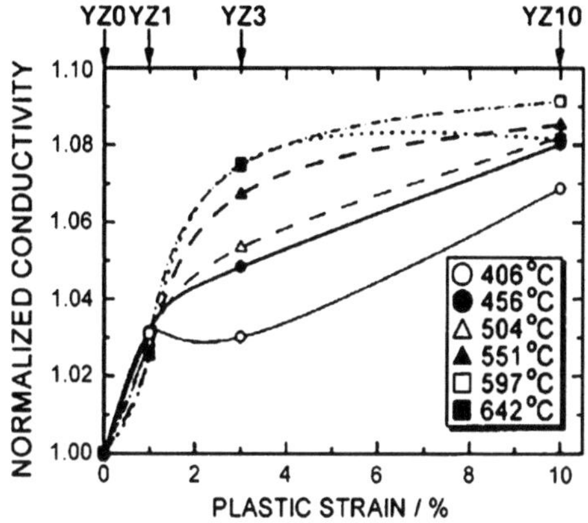

Fig. 5 Plastic strain dependence of the conductivity of single crystal YSZ from [25]. Reproduced here with kind permission from © Elsevier 2011

caused by a combination of elastic strain at the interface *and* the presence of misfit dislocations.

This explanation of the role of misfit dislocations seems to be a likely explanation, as conductivity enhancement has been seen in YSZ single crystals that have been plastically deformed at high temperatures. Otsuka and co-workers [24, 25] have shown that the conductivity of the material, plastically deformed, with dislocation densities of 8×10^{12} m^{-2} or greater show an enhancement of the conductivity. This effect is shown in Fig. 5 where the enhancement seen is small, around 8 %, but they estimate that in order for this effect to be observable, the conductivity in the core region of the dislocation would need to be enhanced by a factor of 10^2–10^4 at a temperature of 597 °C, depending upon the estimated size of the core region (radius 1–10 times the dislocation Burgers vector).

The local density of dislocations at an interface can be high, especially for cases where the misfit is high, but may not be simple. Conchon et al. [26] examined the epitaxial MgO/ZrO_2 interface (semi-coherent) and conclude the presence of two dislocation networks, one a randomly distributed square network of misfit dislocations with a fairly low density. The other, a periodic array of dislocations of spacing 1.2 nm giving a high density of dislocations responsible for the strain relaxation of the layer. The dislocations themselves, particularly edge dislocations, have an associated strain field (with a tensile component) which has been shown to be significant and extensive in another fluorite oxide UO_2 [27]. These fields can interact as has been shown for networks of dislocations at low angle tilt boundaries in YSZ bicrystals [28] when the dislocation spacing is on the order of a few nanometres. It can be speculated that this interaction will lead to significant deviations from the expected shape of the strain field and in periodic dislocation arrays, such as those described above, lead to a continuous interconnected interfacial pathway where the host lattice experiences a tensile strain enhancing the ionic mobility.

2.2 Electrode Materials

Unlike the work on electrolyte materials there has been little work on the effect of interfaces on mixed conducting oxides, perhaps because of the dominance of the electronic carriers which will mask most of the effects on the oxygen transport. There has however been some work on the effect that heterointerfaces might have on the exchange of oxygen with mixed conducting structures. The first observation of an effect was noted by Sase et al. [29] who noticed an enhancement of the oxygen exchange at the $(La,Sr)CoO_3/(La,Sr)_2CoO_4$ interface in polycrystalline composite materials. The value of the exchange coefficient calculated at the interface is estimated to increase by around 3 orders of magnitude. This effect has been further investigated by la'O et al. [30] who made heterojunctions of similar materials and found that the electrochemical activity of these interfaces is also enhanced.

3 Layered Perovskites

The perovskite type of electrode material has been thoroughly reviewed, with extensive studies of the materials chemistry and engineering of these materials available [31–33]. Several advantages of the ABO_3 type electrodes have been proposed, with the relatively simple 3D structure viewed as desirable for ionic conduction, and the size ratio of the A and B sites allowing for the introduction of substituents that promote oxygen vacancy formation, variable valence transition metals to promote electronic conductivity and in essence a flexible host structure that can be modified to optimise materials properties for the operating regime of interest for the fuel cell developer. Whilst this is an attractive proposition, there appear to be some fundamental limits such as with the maximum achievable oxide ion diffusivity as evidenced in the $La_{1-x}Sr_xCoO_{3-d}$ family [34, 35].

An alternative strategy retaining some of the characteristics of perovskite may be to look towards related materials with structural features that would promote both ionic and electronic conduction. There are clear routes to achieving this with the obvious choices being the layered perovskites with the A_2BO_4 structure type or the double and triple perovskite materials. With the former, investigation of interstitial conduction mechanisms is essential whilst with the latter there is debate about the effect of order/disorder of the oxygen lattice sites. Each of these structural types will be considered in subsequent sections.

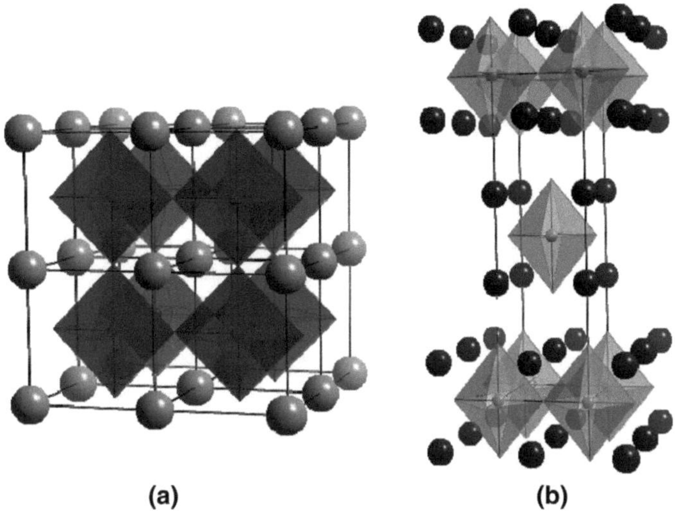

Fig. 6 **a** $LaMO_3$ where $M = (n = \infty)$ and **b** La_2MO_4 $(n = 1)$ crystal structures of the Ruddlesden-Popper family of potential cathode materials where M = Transition metal. Polyhedra show M-O6 coordination. Spheres indicate La positions, oxygen positions omitted for clarity

3.1 Ruddlesden–Popper Electrodes

Materials with the general formula $A_{n+1}M_nO_{3n+1}$, are referred to as Ruddlesden-Popper (RP) phases and were first observed in the $Sr_{n+1}Ti_nO_{3n+1}$ titanates [36]. Many of these materials have since been studied as potential superconductors and as a consequence the structural chemistry of these materials at low temperatures is well known. The crystal structure of the Ruddlesden–Popper phases consists of n consecutive perovskite layers, denoted $(AMO_3)_n$, alternating with rock salt layers (AO), with the alternation of ABO_3 and AO being along the crystallographic *c* direction. The well-known perovskite AMO_3 and A_2MO_4 structures correspond to $n = \infty$ and $n = 1$ members of the series respectively, Fig. 6. Stacking of the ABO_3 and AO layers, coupled with mixed valence cations on the B site, leads to opportunities to incorporate oxygen excess in the form of interstitial species, leading to alternative conduction mechanisms, including either interstitial or interstialcy. Many of these phases can adopt both hypo- and hyper-stoichiometry depending on the chemical potential of the system, and thus these phases may be viewed as flexible and open to optimisation.

Several A_2BO_4-structured materials of the RP type have been proposed as SOFC cathodes, including those with A = La, Sr, Ba, Pr, Nd and M = Ni, Cu, Co, Fe [37–40]. These Ruddlesden-Popper (RP) phases have been proven to have good electrochemical and transport properties, but chemical stability over the temperature range of 600–1000 °C and oxygen chemical potentials necessary for SOFC applications remains an issue [41–43]. Initial investigation of the RP phases

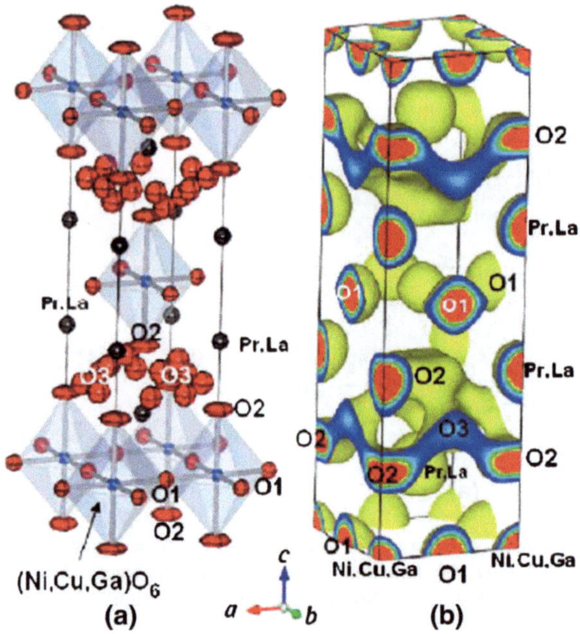

Fig. 7 Representation of the (**a**) crystal structure and (**b**) isosurface of nuclear density at 0.05 fm Å^{-3} for the $(Pr_{0.9}La_{0.1})_2(Ni_{0.74}Cu_{0.21}Ga_{0.05})O_{4+\delta}$ compound determined in situ at 1015.6 °C [47]. Reproduced here with kind permission from © American Chemical Society

concerned the La_2NiO_{4+d} family of compounds. Total conductivity of the nickelates was determined to peak at ~100 S cm^{-1} at 450 °C and remains predominantly *p*-type electronic in the entire $p(O_2)$ range [44, 45], meeting the minimum criteria for SOFC cathodes. In these compounds, ionic transport has been found to occur via anisotropic diffusion of interstitial ions in the rock-salt-type layers and vacancies in the perovskite layers [39]. Whilst hole transport is found to be lower than in the perovskite analogues, it is still of a level adequate for practical applications. Introduction of small amounts of divalent A site dopants leads to higher conductivity and decreased oxygen content, consequently reducing oxygen diffusivity, which is determined by the interstitial anion migration [46], presumably through a loss of interstitial charge carriers. Although the conduction mechanism for oxygen was assumed to be via interstitial position in the lattice, the first direct evidence for the diffusion path of oxide ions in a K_2NiF_4-type mixed conductor was reported by Yashima et al. [47] $(Pr_{0.9}La_{0.1})_2(Ni_{0.74}Cu_{0.21}Ga_{0.05})O_{4+d}$ is a complex substituted nickelate which was proposed as an optimised composition and investigated with the use of high-temperature neutron powder diffraction. From these data the nuclear-density distribution can be refined and modelled to highlight the two-dimensional (2D) network of O2-O3-O2 diffusion paths of oxide ions where O2 represents the oxygen at a (0, 0, *z*) position, and O3 represents the interstitial oxygen located at a 16*n* site (see Fig. 7). This result is

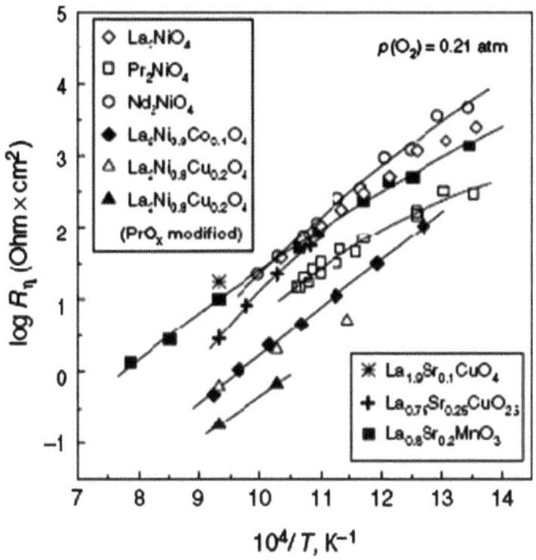

Fig. 8 Selected polarisation data for nickelate cathodes with YSZ and LSGM electrolytes higlighting the competitive performance compared to conventional perovskite type materials [2]. Reproduced here with kind permission from © Elsevier 2011

also consistent with the anisotropic transport of oxide ions in La_2NiO_{4+d} [48, 49] as determined in both single crystals, thin films and polycrystalline samples.

Figure 8 presents electrochemical results for a range of nickelate-based electrodes and it is clear that for the Ln_2NiO_{4+d} (Ln = La, Nd, Pr) series, relatively low polarisation resistances were observed [2]. The high performance of the Pr analogue may be a result of the metastability of the praseodymium nickelate, which decomposes into PrO_x and $Pr_4Ni_3O_{10-d}$, having higher ionic and electronic conductivities [42]. Kim et al. [50] performed AC impedance measurements and suggested that the electrode reaction is limited by the surface exchange reaction for the La_2NiO_{4+d}(LNO). These results are in agreement with the enhancement of the electrochemical activity of $La_2Ni_{0.8}Cu_{0.2}O_{4+d}$ [51] and $LaNi_{0.5}Fe_{0.5}O_{3-d}$ [52] by surface modification with praseodymium oxide.

Much of the work on these materials has focused on their chemical and electrical characterisation. Compatibility of these materials with the electrolytes is also an important issue to be addressed. For this purpose, Munnings et al. [53] reported on the stability and reactivity of La_2NiO_{4+d} with the LSGM electrolyte. They found an effect on the surface stoichiometry and that this stoichiometry change adversely affected the cathode performance. They also observed that the best performance of the La_2NiO_{4+d} cathode was on surfaces with elevated levels of strontium and magnesium. Thermodynamic modelling calculations made by Solak et al. [54] also show that La_2NiO_{4+d} is not chemically compatible with the LSGM electrolyte not only at fabrication conditions but also at operation conditions. On the contrary, Sayers et al. [43] recently found that there is no evidence of secondary phase formation on La_2NiO_{4+d}-LSGM mixtures from the obtained diffraction data over a period of 72 h at temperatures of up to 1000 °C. They also found that there is significant reactivity between La_2NiO_{4+d} and CGO after 24 h at 900 °C, with the formation of a higher order Ruddlesden–Popper ($La_{n+1}Ni_nO_{3n+1}$) phase as one of the reaction products. Further to these observations Aguadero et al. [55] suggest that indeed the LSGM electrolyte is an excellent choice for use with

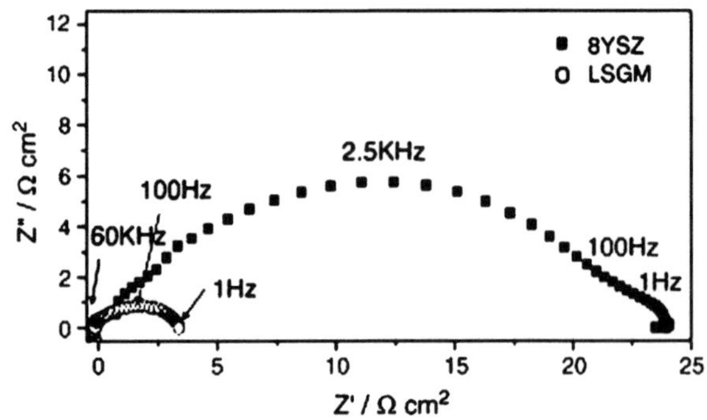

Fig. 9 Comparison of the polarisation resistance of a Cu substituted La_2NiO_4 cathode deposited as a symmetrical cell on both 8YSZ and LSGM electrolytes, highlighting the enhanced performance with LSGM. [55]

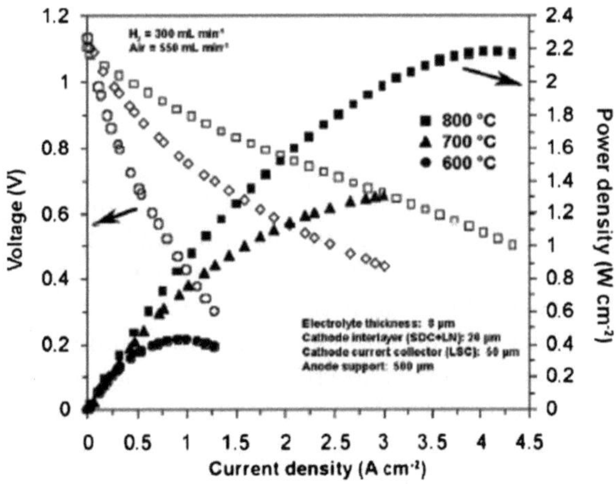

Fig. 10 Cell voltage and power density measured for a La_2NiO_4 bilayer cathode on YSZ electrolyte with a LSC current collector [61]. Reproduced here with kind permission from © Elsevier 2011

LNO electrodes, particularly with Cu substitution for Ni, achieving 1–2 orders of magnitude improvement in polarisation resistance over the same electrodes deposited on YSZ at 750 °C, Fig. 9.

Detailed investigation of the LNO system by Sayers et al. [43, 55, 56] has highlighted the sensitivity of this material to oxygen partial pressure and to the presence of catalytically active metals. Indeed in the presence of $Ce_{1-x}Gd_xO_{2-d}$ complex exsolution behaviour is observed resulting in the reduction of the LNO phase and the segregation of Gd_2O_3, leaving a deficient GDC material. Some degree of reversibility is also reported, which may be associated with catalytic activity of the oxides towards oxygen reduction. It is also clear that in the presence of a Pt catalyst LNO will undergo oxidation to the higher order RP phases at relatively low temperatures [57]. Evidently phase transformation and stability in the nickelates is complex and requires further systematic studies.

Another interesting material with the K_2NiF_4-type structure is $Sr_{2-x}La_xMnO_{4+d}$ which Munnings et al. [58] report presents similar thermal expansion behaviour to the most common electrolyte materials and is chemically stable over a wide oxygen partial pressure range. In an effort to evaluate materials from this family as electrodes Liping et al. studied the reactivity between $Sr_{1.4}La_{0.6}MnO_{4+d}$ and a

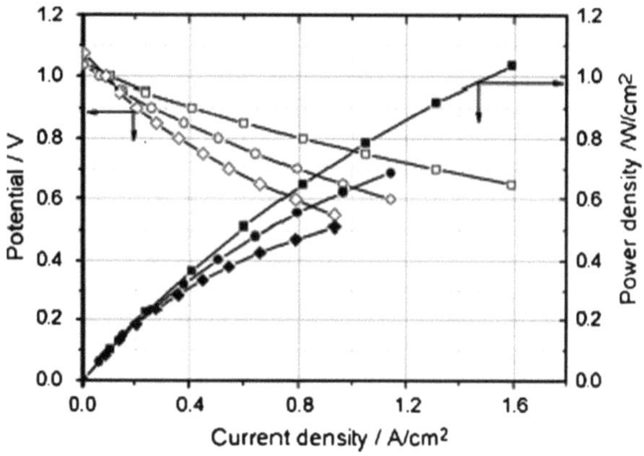

Fig. 11 Current–Voltage-Power density curves for a $Nd_{1.95}NiO_{4+d}$ cell sintered at 1100 °C measured at 700, 750 and 800 °C [62]. Reproduced here with kind permission from © Elsevier 2011

CGO electrolyte with no reaction identified after calcinations at 1000 °C for 12 h [59]. Of course this is a relatively short timescale and further reactivity studies would be required to fully evaluate degradation behaviour. Unfortunately these compositions have relatively low electronic conductivity of the order of 2 S cm^{-1} and are therefore unlikely to be used as cathodes.

Development of electrodes based on the nickelate system has also featured the use of novel electrode structures, with Rieu et al. suggesting that a LNO cathode formed from a compact micron thick layer, followed with an open porous structure of the same chemical composition enhances the electrochemical performance of the cell, reducing the polarisation resistance [60]. Further investigation of nickelate cathodes has shown that excellent performance can be achieved through the use of engineered structures. Laberty et al. [61] demonstrated that when using a La_2NiO_{4+d} cathode on a ceria-samaria electrolyte, Fig. 10, considerable care should be taken to ensure that appropriate barrier layers are included. Indeed Laberty et al. find that a cathode interlayer of La_2NiO_{4+d}/SDC combined with a lanthanum cobaltate current collector produced good results. By contrast Lalanne et al. [62] adopted cation deficiency in a $Nd_{2-x}NiO_{4+d}$ system and report power densities of up to 1 W cm^{-2}, Fig. 11, with this cathode in a cell based on a YSZ electrolyte. Again it was deemed necessary to add a current collection layer to this cell to achieve maximum performance.

Increasing the number of perovskite blocks in the RP phases leads to an increase in the B-site valence, and in the nickelate series of materials this results in a change in the electronic conductivity, with significant increases reported for the $n = 2$ and 3 compounds, leading to the possibility that these materials may produce higher performance cathodes. Recently both the $n = 2$ and 3 nickelate phases have shown some promise as cathodes operating on ceria-based electrolytes at between 500–700 °C, with Takahashi et al. [63] showing the best cathode behaviour with a $n = 3$ material, Fig. 12. However it should be noted that the overall power density achieved in all of the cells reported, including the $n = 1$ phase, was significantly lower than previously reported, as discussed above. It was also noted that to date there have been no reports of the ionic conductivity in these higher order phases and hence there is considerable scope for future development of these Ruddlesden-

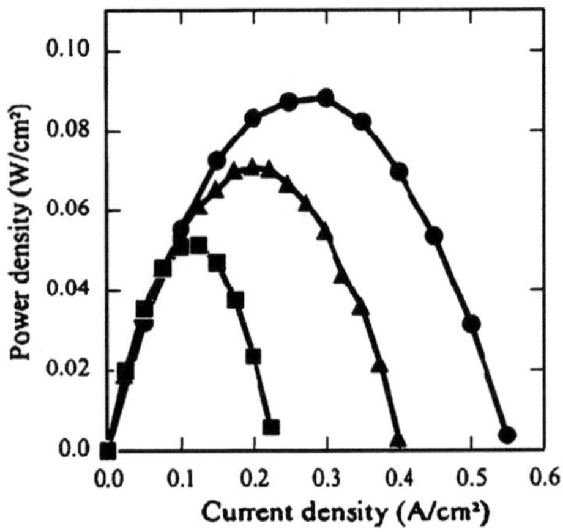

Fig. 12 Power density curves for the La_2NiO_4 (*squares*), $La_3Ni_2O_7$ (*triangles*) and $La_4Ni_3O_{10}$ (*circles*) cathodes measured at 700 °C [63]. Reproduced here with kind permission from © Elsevier 2011

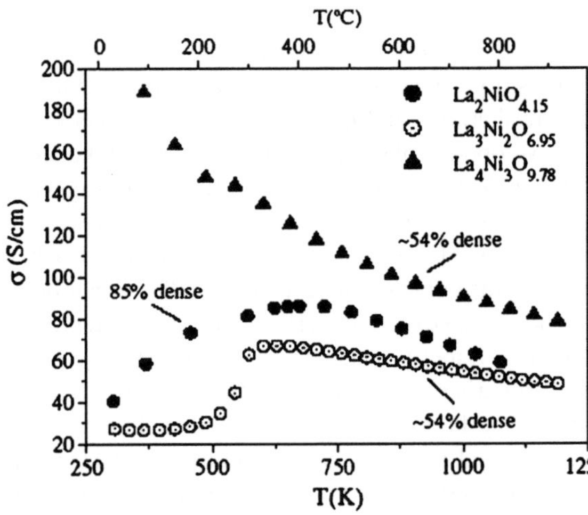

Fig. 13 Total conductivity of the nickelate phases where $n = 1, 2, 3$ in the Ruddlesden–Popper series, highlighting the increased conductivity as n increases. [43]

Popper materials, either as single phase cathodes or incorporated as an electronic conductor to enhance current collection. Amow and Skinner [64] highlighted the increased total conductivity with increasing number of perovskite blocks, Fig. 13, associated with increased Ni valence. Taken together these observations highlight the potential of layered perovskites as next generation fuel cell materials.

3.2 Double Perovskites

A further extension of the concept of developing layered perovskites leads to the consideration of double ($AA'B_2O_{5+d}$) and triple perovskites. In the context of fuel cell electrodes the first double perovskite to be investigated was $GdBaCo_2O_{5+d}$ (GBCO) which adopts orthorhombic symmetry, in which Co ions order in square

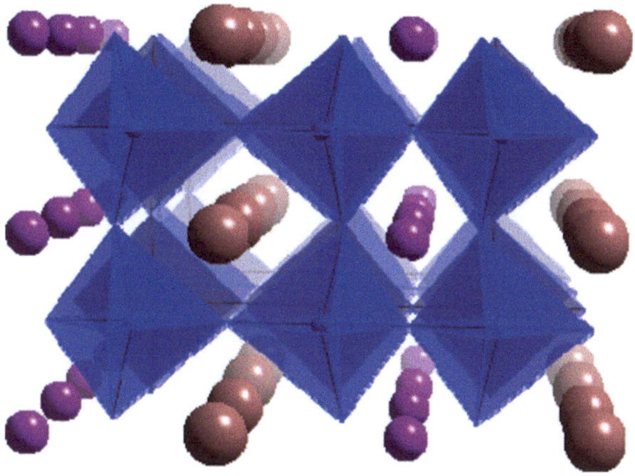

Fig. 14 Schematic representation of the double perovskite structure AA'$B_2O_{5.5}$ where *A* Rare earth, *A'* Alkaline earth, *B* Transition metal

pyramids (CoO_5) and octahedra (CoO_6), alternating along the *b*-axis [65]. The Ba cations order in alternating (001) layers resulting in the oxygen vacancies predominantly ordering along (100) in the $[GdO]_x$ planes (see Fig. 14) [66]. Chang et al. [67] proposed that GBCO could be an effective IT-SOFC cathode suggesting that it possessed fast oxygen transport kinetics as they found through AC impedance spectroscopy that GBCO cathode materials exhibit good performance at low temperatures on CGO electrolytes [68]. Supporting this data Tarancon et al. [69] reported fast oxygen surface exchange kinetics ($k^* = 2.8 \times 10^{-7}$ cm s^{-1}) and oxygen ion diffusivity ($D^* = 4.8 \times 10^{-10}$ cm^2 s^{-1}) for GBCO measured at 575 °C. Having optimised the electrode microstructure these authors report a competitive area specific resistance (ASR) of 0.25 Ω cm^2 at an operating temperature of 625 °C for a GBCO/CGO/GBCO cell. Li et al. [70] investigated the same composition on an $Sm_xCe_{1-x}O_{2-d}$ (SDC) electrolyte finding similar results. Of course ASR is only one factor to consider when evaluating electrode performance: it is essential that the potential cathode materials are compatible with electrolytes. GBCO has been evaluated in contact with LSGM by Peña-Martínez et al. [71] with no reactivity reported and good electrochemical performance, indicating that this is a viable, exciting alternative to conventional perovskite cathodes.

In order to enhance the performance of these double perovskites A site substitutions have been studied, with Sr substitution for Ba in GBCO an obvious choice, based on the extensive experience of fuel cell researchers in successfully optimising La-based perovskites with Sr substitution. Initially the structural chemistry was investigated [72], with $GdBa_{1-x}Sr_xCo_2O_{5+\delta}$ found to exhibit a structural change from orthorhombic *Pmmm* ($x = 0$) to tetragonal *P4/mmm* ($x = 0.2$–0.6) followed by a further change to the orthorhombic *Pnma* ($x = 1$) structure. On testing as a fuel cell cathode it was observed that the compositions with $x = 0.2$ and 0.6 exhibited higher power density when than either the $x = 0$ or $x = 1$ samples. This was attributed to faster oxygen transport within the tetragonal structure. Contrary to the work of Peña-Martínez et al. [71] it has been observed that the parent $GdBaCo_2O_{5+\delta}$ sample suffers from interfacial reaction with LSGM

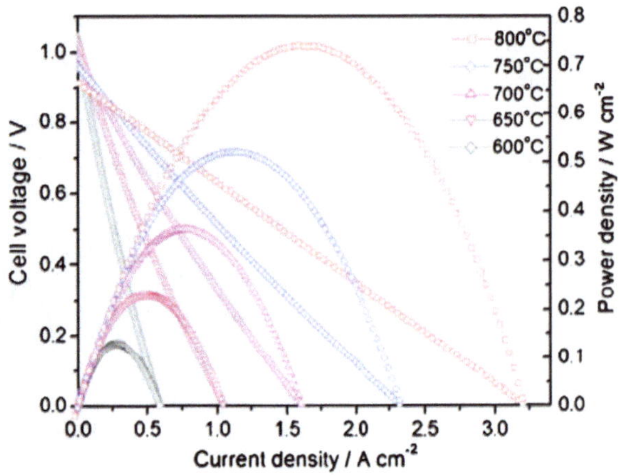

Fig. 15 Performance of a PBCO-SDC composite fuel cell on SDC electrolyte with an LSGM barrier layer [77]. Reproduced here with kind permission from © Elsevier 2011

and CGO electrolytes at temperatures of >1100 °C, whilst Sr substitution has been observed to significantly improve the chemical stability of $GdBa_{1-x}Sr_xCo_2O_{5+\delta}$. These combined results indicate that a Sr-substituted GBCO cathode with $0.2 < Sr < 0.6$ would be an attractive proposition.

Replacing the Gd with Pr producing $PrBaCo_2O_{5+\delta}$ (PBCO) has been suggested as an alternative SOFC cathode for use at low temperatures due to the reported unusually rapid oxygen transport kinetics at low temperatures (300–500 °C) [73, 74]. Kim et al. suggested oxygen diffusion coefficients 1 order of magnitude greater than the already impressive GBCO, however, their data originate from samples with relatively low densities suggesting that these values may be enhanced through diffusion via pores. Kinetic parameters are, as mentioned above, only one aspect of potential cathode performance and Zhu et al. [75] have demonstrated the suitability of PBCO for IT-SOFC applications. In this case they measured a PBCO/SDC/Ni-SDC single cell obtaining a maximum power density of approximately 850 mW cm^{-2} at 650 °C. Investigating the possibilities of rare-earth substitution further, $SmBaCo_2O_{5+\delta}$ has demonstrated good electrochemical properties and was proposed as a cathode material for IT-SOFC [76]. Several authors have extended the range of rare earths substituted for Gd, with Zhou et al. [77] investigating the Pr, Nd, Sm and Gd materials as composites with SDC on a LSGM electrolyte, finding that all materials had ASR values lower than 0.15 Ω cm^2 at 700 °C with a maximum power density of 758 mWcm^{-2} at 800 °C reported by Zhu et al. [78], Fig. 15. Furthermore, Zhu et al. [78] report that the PBCO composition with SDC as a composite cathode achieves 0.035 Ω cm^2 ASR at 750 °C. Peak performance was achieved with a composite with 30 wt % SDC, and improves on the performance of just PBCO. These authors appear to have developed materials that significantly outperform those of Li et al. [70] who reported an ASR of only 0.5 Ω cm^2 at 750 °C for GBCO on a SDC electrolyte. Completing these investigations of double perovskites, Gong et al. [79] extended their studies to include the Eu analogue testing on CGO electrolytes and reporting the ASR as a function of temperature and pO_2, finding that at 700 °C the electrode achieved a low resistance of only 0.095 Ω cm^2.

Due to the structural anisotropy of the double perovskite system it is likely that the conduction is also anisotropic and to investigate this Burriel et al. [80] studied on epitaxial thin films of GBCO. Burriel et al. show that the anisotropic GBCO films on STO substrates achieve an electronic conductivity of 800 Scm^{-1} at the low temperature of 330 °C. Also it is shown that the materials have a remarkable low pO_2 dependence of <1/25, indicating good potential for cathodes. Following these thin film investigations Choi et al. [81] report ionic conductivity in bulk GBCO of 0.01 S cm^{-1} at 1173 K which is pO_2 dependent over the relatively narrow range of $10^{-4} < pO_2$ (atm) < 0.21. From these chemical diffusion studies the authors suggest an activation energy for diffusion of 0.86 eV.

To support these experimental studies and further detail the mechanism of oxygen mobility in the double perovskites molecular dynamics calculations have been reported by Hermet et al. [82] finding the activation energy for diffusion in the *ab* plane is 0.67 eV, agreeing well with reported experimental data.

4 Summary

The findings presented above clearly indicate that the effects of elastic strain on the transport of oxygen in oxide ion conductors are significant. This effect can be either enhancing or retarding depending upon the sign of the imposed strain. The effect of dislocations on the oxygen transport has also been revealed, and the possibility that this effect might be even larger discussed. This may of course be due to an enhancement in the ease with which oxide ions can diffuse in the strain field of the dislocation, which can be very large. This research is still in its early stages and needs much to be understood about the effects of elastic strain on the oxygen migration enthalpy. A possible method for the enhancement of low temperature conductivity in oxygen ion conductors by the introduction of structures, such as heterointerfaces that give rise to either tensile strain or to large number of dislocations holds significant future promise.

Further possible advances, linking both heterostructures and new materials have been identified, particularly with reference to the potential for high performance low temperature cathodes. The series of materials from the Ruddlesden-Popper and layered perovskite families show that there is considerable scope to engineer structures with high levels of ionic and electronic conductivity, exploiting the inherent properties of these materials through thin film technologies and microstructure optimisation. Far from there being no significant materials developments for solid oxide fuel cells, there are now significant opportunities for major advances that will be incorporated in next generation fuel cells.

References

1. J.W. Fergus, Electrolytes for solid oxide fuel cells. J. Power Sources **162**, 30–40 (2006)
2. E.V. Tsipis, V.V. Kharton, Electrode materials and reaction mechanisms in solid oxide fuel cells: a brief review. J. Solid State Electrochem. **12**, 1367–1391 (2008)
3. E.N.S. Muccillo, M. Kleitz, Ionic conductivity of fully stabilised ZrO_2-MGO and blocking effects. J. Eur. Ceram. Soc. **15**, 51–55 (1995)
4. A.M. Stoneham, E. Wade, J.A. Kilner, Model for the fast ionic-diffusion in alumina-doped Lii. Mater. Res. Bull. **14**, 661–666 (1979)
5. H.L. Tuller, Ionic conduction in nanocrystalline materials. Solid State Ionics **131**, 143–157 (2000)
6. N. Sata et al., Mesoscopic fast ion conduction in nanometre-scale planar heterostructures. Nature **408**, 946–949 (2000)
7. I. Kosacki et al., Nanoscale effects on the ionic conductivity in highly textured YSZ thin films. Solid State Ionics **176**, 1319–1326 (2005)
8. A. Karthikeyan, C.L. Chang, S. Ramanathan, High temperature conductivity studies on nanoscale yttria-doped zirconia thin films and size effects. Appl. Phys. Lett. **89**, 183116 (2006)
9. J. Garcia-Barriocanal et al., Tailoring disorder and dimensionality: strategies for improved solid oxide fuel cell electrolytes. Chem. Phys. Chem. **10**, 1003–1011 (2009)
10. X. Guo, Comment on "colossal ionic conductivity at interfaces of epitaxial ZrO_2:Y_2O_3/$SrTiO_3$ heterostructures". Science **324**, 465 (2009)
11. J.A. Kilner, Ionic conductors feel the strain. Nat. Mater. **7**, 838–839 (2008)
12. A. Cavallaro et al. Electronic nature of the enhanced conductivity in YSZ-STO multilayers deposited by PLD. Solid State Ionics **181**, 592–601 (2010)
13. T.J. Pennycook et al., Origin of colossal ionic conductivity in oxide multilayers: interface induced sublattice disorder. Phys. Rev. Lett. **104**, 115901 (2010)
14. A. Kushima, B. Yildiz, Oxygen ion diffusivity in strained yttria stabilized zirconia: where is the fastest strain? J. Mater. Chem. **20**, 4809–4819 (2010)
15. N. Schichtel et al., Elastic strain at interfaces and its influence on ionic conductivity in nanoscaled solid electrolyte thin films-theoretical considerations and experimental studies. Phys. Chem. Chem. Phys. **11**, 3043–3048 (2009)
16. C. Korte et al., Ionic conductivity and activation energy for oxygen ion transport in superlattices—the semicoherent multilayer system YSZ (ZrO_2 + 9.5 mol % Y_2O_3)/Y_2O_3. Phys. Chem. Chem. Phys. **10**, 4623–4635 (2008)
17. A. Peters et al., Ionic conductivity and activation energy for oxygen ion transport in superlattices—The multilayer system CSZ (ZrO_2 + CaO) / Al_2O_3. Solid State Ionics **178**, 67–76 (2007)
18. C. Korte et al., Influence of interface structure on mass transport in phase boundaries between different ionic materials Experimental studies and formal considerations. Monaths. Chem. **140**, 1069–1080 (2009)
19. C.B. Pierce, Effect of hydrostatic pressure on ionic conductivity in doped single crystals of sodium chloride, potassium chloride, and rubidium chloride. Phys. Rev. **123**, 744–754 (1961)
20. W. Araki, T. Adachi, Mechanical effect on oxygen mobility in yttria stabilized zirconia, in *Life-Cycle Analysis for New Energy Conversion and Storage Systems*, ed. by V. Fthenakis, A. Dillon, N. Savage (MRS, Boston, 2008)
21. W. Araki, Y. Imai, T. Adachi, Mechanical stress effect on oxygen ion mobility in 8 mol % yttria-stabilized zirconia electrolyte. J. Eur. Ceram. Soc. **29**, 2275–2279 (2009)
22. W. Araki, Y. Arai, Oxygen diffusion in yttria-stabilized zirconia subjected to uniaxial stress. Solid State Ionics **181**, 441–446 (2010)
23. M. Sillassen et al., Low-Temperature Superionic Conductivity in Strained Yttria-Stabilized Zirconia. Adv. Func. Mater. **20**, 2071–2076 (2010)

24. K. Otsuka et al., Effects of dislocations on the oxygen ionic conduction in yttria stabilized zirconia. Mater. Trans. **45**, 2042–2047 (2004)
25. K. Otsuka et al., Dislocation-enhanced ionic conductivity of yttria-stabilized zirconia. Appl. Phys. Lett. **82**, 877–879 (2003)
26. F. Conchon, A. Boulle, R. Guinebretière, Misfit dislocations in highly mismatched oxide interfaces, an X-ray diffraction study. Phys. Status Solidi A **204**, 2535–2541 (2007)
27. D.C. Parfitt et al., Strain fields and line energies of dislocations in uranium dioxide. J. Phys.-Condens. Matter **22**, 175004 (2010)
28. Y. Nohara et al., Dislocation structures and strain fields in 111 low-angle tilt grain boundaries in zirconia bicrystals. J. Electron Microsc. **59**, S117–S121 (2010)
29. M. Sase et al., Enhancement of oxygen exchange at the hetero interface of (La, Sr)CoO_3/(La, $Sr)_2CoO_4$ in composite ceramics. Solid State Ionics **178**, 1843–1852 (2008)
30. G.J. la'O et al., Catalytic activity enhancement for oxygen reduction on epitaxial perovskite thin films for solid-oxide fuel cells. Angew. Chem.-Intl. Ed. **49**, 5344–5347 (2010)
31. A. Lashtabeg, S.J. Skinner, Solid oxide fuel cells—a challenge for materials chemists? J. Mater. Chem. **16**(31), 3161–3170 (2006)
32. D.J.L. Brett, A. Atkinson, N.P. Brandon, S.J. Skinner, Intermediate temperature solid oxide fuel cells. Chem. Soci. Rev. **37**, 1568–1578 (2008)
33. N.P. Brandon, S. Skinner, B.C.H. Steele, Recent advances in materials for fuel cells. Annu. Rev. Mater. Res. **33**, 183–213 (2003)
34. A.V. Berenov, A. Atkinson, J.A. Kilner, E. Bucher, W. Sitte, Oxygen tracer diffusion and surface exchange kinetics in $La_{0.6}Sr_{0.4}CoO_{3-\delta}$. Solid State Ionics **181**, 819–826 (2010)
35. J. Richter, P. Holtappels, T. Graule, T. Nakamura, L.J. Gauckler, Materials design for perovskite SOFC cathodes. Monaths. Chem. **140**, 985–999 (2009)
36. S. N. Ruddlesden, P. Popper, The compound $Sr_3Ti_2O_7$ and its structure. Acta Cryst. **11**, 54–55 (1958)
37. V.V. Kharton, A.P. Viskup, A.V. Kovalevsky, E.N. Naumovich, F.M.B. Marques, Ionic transport in oxygen-hyperstoichiometric phases with K_2NiF_4-type structure. Solid State Ionics **143**, 337–353 (2001)
38. Q. Li, H. Zhao, L.H. Huo, L.P. Sun, X.L. Cheng, J.C. Grenier, Electrode properties of Sr doped La_2CuO_4 as new cathode material for intermediate-temperature SOFCs. Electrochem. Comm. **9**, 1508–1512 (2007)
39. A.J. Jennings, S.J. Skinner, Thermal stability and conduction properties of the $La_xSr_{2-x}FeO_{4+\delta}$ system. Solid State Ionics **152**, 663–667 (2002)
40. V.V. Kharton, A.P. Viskup, E.N. Naumovich, F.M.B. Marques, Oxygen ion transport in La_2NiO_4-based ceramics. J. Mater. Chem. **9**, 2623–2629 (1999)
41. M. Zinkevich, F. Aldinger, Thermodynamic analysis of the ternary La-Ni-O system. J. Alloys Compd **375**, 147–161 (2004)
42. A.V. Kovalevsky, V.V. Kharton et al., Oxygen permeability, stability and electrochemical behavior of $Pr_2NiO_{4+\delta}$-based materials. J. Electroceram. **18**, 205–218 (2007)
43. R. Sayers, J. Liu, B. Rustumji, S.J. Skinner, Novel K_2NiF_4-Type materials for solid oxide fuel cells: compatibility with electrolytes in the intermediate temperature range. Fuel Cells **8**, 338–343 (2008)
44. G. Amow, S.J. Skinner, Recent developments in Ruddlesden-Popper nickelate systems for soid oxide fuel cell cathodes. J. Solid State Electrochem. **10**, 538–546 (2006)
45. H.S. Kim, H.I. Yoo, Defect chemical analysis of the non-stoichiometry, conductivity and thermopower of $La_2NiO_{4+\delta}$. Phys. Chem. Chem. Phys. **12**, 4704–4713 (2010)
46. S.J. Skinner, J.A. Kilner, Oxygen diffusion and surface exchange in $La_{2-x}Sr_xNiO_{4+\delta}$. Solid State Ionics **135**, 709–712 (2000)
47. M. Yashima, M. Enoki et al., Structural disorder and diffusional pathway of oxide ions in a doped Pr_2NiO_4-based mixed conductor. J. Am. Chem. Soc. **130**, 2762–2763 (2008)
48. J.M. Bassat, P. Odier, A. Villesuzanne, C. Marin, M. Pouchard, Anisotropic ionic transport properties in $La_2NiO_{4+\delta}$ single crystals. Solid State Ionics **167**, 341–347 (2004)

49. M. Burriel, G. Garcia, J. Santiso, J.A. Kilner, R.J. Chater, S.J. Skinner, Anisotropic oxygen diffusion properties in epitaxial thin films of $La_2NiO_{4+\delta}$. J. Mater. Chem. **18**, 416–422 (2008)
50. G.T. Kim, S.Y. Wang, A.J. Jacobson, Z. Yuan, C.L. Chen, Impedance studies of dense polycrystalline thin films of $La_2NiO_{4+\delta}$. J. Mater. Chem. **17**, 1316–1320 (2007)
51. V.V. Kharton, E.V. Tsipis, A.A. Yaremchenko, J.R. Frade, Surface-limited oxygen transport and electrode properties of $La_2Ni_{0.8}Cu_{0.2}O_{4+\delta}$. Solid State Ionics **166**, 327–337 (2004)
52. V.V. Kharton, F.M. Figueiredo, L. Navarro et al., Ceria-based materials for solid oxide fuel cells. J. Mater. Sci. **36**, 1105–1117 (2001)
53. C.N. Munnings, S.J. Skinner, G. Amow, P.S. Whitfield, I.J. Davidson, Stability and reactivity of LSGM electrolytes with nickel-based ceramic cathodes. J. Fuel Cell Sci. Tech. **2**, 34–37 (2005)
54. N. Solak, M. Zinkevich, F. Aldinger, Compatibility of La_2NiO_4 cathodes with $LaGaO_3$ electrolytes: a computational approach. Solid State Ionics **177**, 2139–2142 (2006)
55. A. Aguadero, J.A. Alonso, M.J. Escudero, L. Daza, Evaluation of the $La_2Ni_{1-x}Cu_xO_{4+\delta}$ system as SOFC cathode material with 8YSZ and LSGM as electrolytes. Solid State Ionics **179**, 393–400 (2008)
56. R. Sayers, S.J. Skinner, Evidence for the catalytic oxidation of $La_2NiO_{4+\delta}$. J. Mater. Chem. **21**, 414–419 (2011)
57. R. Sayers, PhD Thesis, Imperial College London, 2010
58. C.N. Munnings, S.J. Skinner, G. Amow, P.S. Whitfield, I.J. Davidson, Structure, stability and electrical properties of the $La_{2-x}Sr_xMnO_{4\pm\delta}$ solid solution series. Solid State Ionics **177**, 1849–1853 (2006)
59. L.P. Sun, L.H. Huo, H. Zhao, Q. Li, C. Pijolat, Substituted Sr_2MnO_4 as a possible cathode material in SOFC. J. Power Sources **179**, 96–100 (2008)
60. M. Rieu , R. Sayers, M.A. Laguna-Bercero, S.J. Skinner, P. Lenormand F. Ansart, Investigation of graded $La_2NiO_{4+\delta}$ cathodes to improve SOFC electrochemical performance. J. Electrochem. Soc. **157**, B477–B480 (2010)
61. C. Laberty, F. Zhao, K.E. Swider-Lyons, A.V. Virkar, High performance solid oxide fuel cell cathodes with lanthanum nickelate based composites. Electrochem. Sol. State Lett. **10**, B170–B174 (2007)
62. C. Lalanne, G. Prosperi, J.M. Bassat et al., Neodymium deficient nickelate oxide $Nd_{1.95}NiO_{4+d}$ as cathode material for anode supported intermediate temperature solid oxide fuel cells. J. Power Sources **185**, 1218–1224 (2008)
63. S. Takahashi, S. Nishimoto, M. Matsuda, M. Miyake, Electrode properties of the Ruddlesden-Popper series $La_{n+1}Ni_nO_{3n+1}$ (n = 1,2 and 3), as intermediate temperature solid oxide fuel cells. J. Am. Ceram. Soc. **93**, 2329–2333 (2010)
64. G. Amow, S.J. Skinner, Recent developments in Ruddlesden-Popper nickelate systems for solid oxide fuel cell cathodes. J. Solid State Electrochem. **10**, 538–546 (2006)
65. M. Respaud, C. Frontera, J.L. Garcia-Munoz et al., Magnetic and magnetotransport properties of $GdBaCo_2O_{5+\delta}$: A high magnetic-field study. Phys. Rev. B **6421**, 214401 (2001)
66. A. Maignan, C. Martin, D. Pelloquin, N. Nguyen, B. Raveau, Structural and magnetic studies of ordered oxygen-deficient perovskites $LnBaCo_2O_{5+\delta}$, closely related to the ″112″ structure. J. Solid State Chem. **142**, 247–260 (1999)
67. A.M. Chang, S.J. Skinner, J.A. Kilner, Electrical properties of $GdBaCo_2O_{5+x}$ for ITSOFC applications. Solid State Ionics **177**, 2009–2011 (2006)
68. A.A. Taskin, A.N. Lavrov, Y. Ando, Achieving fast oxygen diffusion in perovskites by cation ordering. Appl. Phys. Lett. **86**, 091910 (2005)
69. A. Tarancon, S.J. Skinner, R.J. Chater, F. Hernandez-Ramirez, J.A. Kilner, Layered perovskites as promising cathodes for intermediate temperature solid oxide fuel cells. J. Mater. Chem. **17**, 3175–3181 (2007)
70. N. Li, Z. Lu, B.O. Wei., X.Q. Huang, K.F. Chen, Y.Z. Zhang, W.H. Su, Characterization of $GdBaCo_2O_{5+\delta}$ cathode for IT-SOFCs. J. Alloys Compd **454**, 274–279 (2008)

71. J. Pena-Martinez, A. Tarancon, D. Marrero-Lopez, J.C. Ruiz-Morales, P. Nunez, Evaluation of $GdBaCo_2O_{5+\delta}$ as Cathode Material for Doped Lanthanum Gallate Electrolyte IT-SOFCs. Fuel Cells **8**, 351–359 (2008)
72. J.H. Kim, F. Prado, A. Manthiram, Characterization of $GdBa_{1-x}Sr_xCo_2O_{5+\delta}$ ($0 <= x <= 1.0$) double perovskites as cathodes for solid oxide fuel cells. J. Electrochem. Soc. **155**, B1023–B1028 (2008)
73. G. Kim, S. Wang, A.J. Jacobson, L. Reimus, P. Brodersen, C.A. Mims, Rapid oxygen ion diffusion and surface exchange kinetics in $PrBaCo_2O_{5+x}$ with a perovskite related structure and ordered A cations. J. Mater. Chem. **17**, 2500–2505 (2007)
74. G. Kim, S. Wang, A.J. Jacobson, Z. Yuan et al., Oxygen exchange kinetics of epitaxial $PrBaCo_2O_{5+\delta}$ thin films. Appl. Phys. Lett. **88**, 024103 (2006)
75. C.J. Zhu, X.M. Liu, C.S. Yi, D. Yan, W.H. Su, Electrochemical performance of $PrBaCo_2O_{5+\delta}$ layered perovskite as an intermediate-temperature solid oxide fuel cell cathode. J. Power Sources **185**, 193–196 (2008)
76. Q.J. Zhou, T.M. He, Y. Ji, $SmBaCo_2O_{5+x}$ double-perovskite structure cathode material for intermediate-temperature solid-oxide fuel cells. J. Power Sources **185**, 754–758 (2008)
77. Q.J. Zhou, F. Wang, Y. Shen, T.M. He, Performances of $LnBaCo_2O_{5-x}$-$Ce_{0.8}Sm_{0.2}O_{1.9}$ composite cathodes for intermediate-temperature solid oxide fuel cells. J. Power Sources **195**, 2174–2181 (2010)
78. C.J. Zhu, X.M. Liu, C.S. Yi, L. Pei, D.J. Wang, D.T. Yan, K.G. Yao, T.Q. Lu, W.H. Su, High-performance $PrBaCo_2O_{5+\delta}$-$Ce_{0.8}Sm_{0.2}O_{1.9}$ composite cathodes for intermediate temperature solid oxide fuel cell. J. Power Sources **195**, 3504–3507 (2010)
79. W. Gong, M. Yadav, A.J. Jacobson, A comparison of electrochemical performance of double perovskite $REBaCo_2O_{5+x}$ cathodes in symmetrical solid oxide fuel cells. in: *Solid-State Ionics-2008,* eds. by E. Traversa, T.R. Armstrong, K. Eguchi, M.R. Palacin. Materials Research Society, vol 1126 (Warrendale, 2009), pp. 35–40
80. M. Burriel, M. Casas-Cabanas, Zapata et al., Influence of the microstructure on the high-temperature transport properties of $GdBaCo_2O_{5.5+\delta}$ epitaxial films. Chem. Mater. **22**, 5512–5520 (2010)
81. M.B. Choi, S.Y. Jeon, J.S. Lee, H.J. Hwang, S.J. Song, Chemical diffusivity and ionic conductivity of $GdBaCo_2O_{5+\delta}$. J. Power Sources **195**, 1059–1064 (2010)
82. J. Hermet, G. Geneste, G. Dezanneau, Molecular dynamics simulations of oxygen diffusion in $GdBaCo_2O_{5.5}$. Appl. Phys. Lett. **97**, 174102 (2010)
83. S. Sanna, V. Esposito, A. Tebano, S. Licoccia, E. Traversa, G. Balestrino, Enhancement of ionic conductivity in Sm-Doped Ceria/Yttria-stabilized Zirconia Heteroepitaxial Structures. Small **6**, 1863–1867 (2010)
84. I. Kosacki, Nanoscale oxygen conductors for energy conversion: Presentation at Imperial College London. 2006, London
85. S. Azad, O.A. Marina, C.M. Wang, L. Saraf, V. Shutthanandan, D.E. McCready, A. El-Azab, J.E. Jaffe, M.H. Engelhard, C.H.F. Peden, S. Thevuthasan, Nanoscale effects on ion conductance of layer-by-layer structures of gadolinia-doped ceria and zirconia. Appl. Phys. Lett. **86**, 131906 (2005)

Future Fuel Cells

Niels Christiansen

1 Introduction

Fuel cells, aimed for energy production and vehicle propulsion, have attracted much interest in the last decade. The reasons are many: The prospects for high efficiency, low emissions, low noise, scalability, reliability and finally potentially low cost. However, the attractiveness of fuel cells is predominantly their high system efficiency and co-generation networking capabilities even for small capacities and also at part load operation. The solid oxide fuel cell type (SOFC) offers the most interesting combination of these advantages. Intensive R&D activities have been carried out for more than 20 years, especially in the academic world supported by substantial national and international funding. The fuel cell technology is now moving from a research and test phase into industrial demonstration [23]. An enormous fundamental knowledge platform and a large number of promising scientific results have been obtained through the many years of worldwide fuel cell R&D. However, most industrial initiatives have been unsuccessful to effectively and rapidly transfer the concurrent state-of-the-art SOFC development results from science to industrial technology, up-scaling and implementation. Initiatives have often been too fragmented and carried out by groups that are smaller than the critical size to ensure rapid progress, technical commitment and breakthrough. It is notorious that development, optimisation and construction of cells, stacks and systems are more complex than most companies originally envisaged. The wide spectrum of potential market possibilities has caused a lack of focus and weight in the decisive stages of development. Furthermore, sales projections of fuel cells have been very wrong for many years and road maps have been unsuccessful.

N. Christiansen (✉)
Topsoe Fuel Cell A/S, Nymoellevej 66, 2800 Lyngby, Denmark
e-mail: nc@topsoe.dk

J. T. S. Irvine and P. Connor (eds.), *Solid Oxide Fuels Cells: Facts and Figures*,
Green Energy and Technology, DOI: 10.1007/978-1-4471-4456-4_9,

A number of serious critical technical issues still remain to be solved in order to pave the way for fuel cell application such as in the mobile and auxiliary power generation segment as well as for stationary applications. Fuel cells have characteristics which are different from existing technologies. Off-the-shelf components have not been able to fulfil the demanding properties, especially for the high temperature types, and suppliers have not been able to supply components with the right characteristics. Currently, most SOFC demonstration prototypes prove to be quite unreliable or lacking longevity, even though they are over-engineered and therefore quite far away from a mature sustainable marketable product. Lack of a committed proactive integrated R&D and engineering approach is one important factor responsible for this.

The methodology of learning curves, which successfully has been used to predict how the cost of a technology decreases as a function of cumulative production, has also been used in case of fuel cells [26, 27]. The learning rate expresses in percentage with the relative cost reduction followed by each doubling of cumulative produced item. However, in practice, it is difficult to distinguish between different cost reduction sources. The production process is improved partly by acquisition of experience—*learning by doing*, partly by R&D efforts—*learning by researching*. One pitfall is that cost reduction in the pilot manufacturing stage not necessarily represents genuine value added learning such as in product and processing development and optimisation because more trivial learning, such as raw material economics-of-scale and material recycling, may dominate the cost reduction. It is well-known that learning by manufacturing dominates in cases where the technology in question is matured, and has changed from pilot-scale phase to economy-of-scale phase. The later includes automation and standardisation of manufacturing methods, optimised designs and standardised supply chain components. The long and multidisciplinary development chain indicates that a new company entering the SOFC business has a long and steep learning curve to face, where learning curve analyses can provide insight for strategic timing and planning.

The evolution of high-technology industries generated by major technological innovations is well described in the literature [1, 2]. It has been suggested that such industries evolve through three stages: *development stage* (fluid state), *dominant stage* and *mature stage*. In the first stage, the product and the processes may be subject to significant changes due to lack of real product experience and loosely defined processes, at the same time as a large number of different design alternatives compete. A significant effort on R&D is required, and the potential market still remains to be fully identified. In the second stage, a few designs have proven their superiority that provides the standards for the industry to follow. The market uncertainty has been significantly reduced in this stage and emphasis is placed on innovation maximizing performance. In the third stage, manufacturing methods and product designs become standardised based on the dominant design in terms of performance and reliability. In this stage, the unit cost decreases both through increases in manufacturing efficiency, standardisation and supply of components by vertically integrated sub-suppliers. Innovation efforts in the third stage are focusing on further cost reduction.

The right timing and a focused commitment to the primary challenges in each stage during this evolution is a prerequisite for reaching the mature state. Furthermore, a clear understanding of how to treat the transition states in the evolution is decisive. Some of the previous leading SOFC organisations have failed to proceed through these stages which have either caused them to found or compelled them step back to the initial development stage. Based on the models described here the interesting question today is: Which evolution stage the SOFC technology belongs to? The fact is that a wide spectrum of different variants are currently tested and validated worldwide—whether materials (see below), cells (electrolyte supported, anode supported, cathode supported or metal supported, etc.), cell design (tubular, micro-tubular, flat planar, flat tubular, segmented or non-segmented, monolithic, circular, rectangular, thin-film, framed or non-framed, etc.), stack designs (internal manifold, external manifold, size and geometry, compressed or non-compressed, cross-flow, co-flow, counter-flow, tubular modules, sealed or seal-less, etc.) or even a large variety of different system designs. The conclusion is that no dominant superior design exists, at the same time as most subcomponents such at electrode materials, interconnects, coatings and even balance of plant (BOP) components are still subjective to verification, changes and optimisation. This indicates that the SOFC technology still must be in a transition state between the development stage and the dominant stage where progress and evolution is dependent on R&D and innovation capabilities. Learning by researching goes hand-in-hand with learning by doing, and the former may even be the most important learning process.

SOFC supply chains are very complex and far from being mature. The generic and complex characteristics of the SOFC technology indicate that a vertical integrated control of the supply chain is needed and an SOFC company will control the majority of processes ranging from materials to final value innovative products. It is notorious that the low-temperature PEMFC cells are much easier to standardise than SOFC, and an effective supply chain is very different and much faster to establish in case of PEMFC. To draw parallels between the two fuel cell types concerning technology and market evolution is a mistake and potentially damaging, because they are radically different.

Progress in SOFC science and R&D has been significant over the last decade, and recent results offer a range of promising innovative possibilities. An interesting example is novel metal supported cells offering prospects of improved reliability and robustness. Thin film layered components, nano-structured electrodes and radical new improved material and processing solutions make it possible to lower the operation temperature while electrochemical performance and durability is enhanced. New specialty metallic alloys and effective coatings for metallic interconnects have proved that durability of SOFC stacks can be significantly increased. Integrated innovation builds upon disciplines such as tradeoff and risk analysis, rapid validation-iteration loops, synergistic working plans, effective trial and error spirals, concentrated acceleration efforts where they would be most likely to succeed, cooperation and scientific and technical knowledge exchange through effective networks. Furthermore, integrated innovation in fuel cell technology includes innovation in business models for growth and renewal [18]. This approach has an improved potential to increase speed of development, eliminate the potential show stoppers and lead to a targeted breakthrough of the SOFC technology. It helps to identify the important jobs to be done. Commitment to integrated R&D and innovation building up know how, extensive capabilities and partnerships facilitate the establishment of broader market positions, early revenues and a positive margin for high value products, niches and reliable pre-commercial fuel cell products.

2 Fuel Cell Types

Among the most interesting applications are stationary distributed power generation (DG), small residential combined heat and power units (micro CHP) and various transportation applications. Fuel cells can play a large role in future energy supply as being a clean and efficient power and heat generator. With a properly designed fuel processing system (FPS) fuel cells can use all available fuel types. However, the complexity, cost and efficiency of the FPS is strongly dependant on the choice of fuel, fuel processing route and last but not least: fuel cell type and fuel cell design.

Different fuel cells operating on different fuels have very different characteristics and also different stages of development. Nevertheless, there exists a general tendency that the same yardstick is applied to all fuel cells. Interestingly, fuel cell experts are realistic about the technology's limitations, the existing challenges as well as its contributions. Sustainable development and marked introduction of fuel cells can benefit from a continued improvement and effective integration of technical knowledge. Improved accuracy is required in a large number of critical aspects ranging from analysis and definition of technical product characteristics including boundary conditions, and technical tradeoffs to definitions of existing and future costumer/society needs. Because fuel cells are still representing an immature technology, where the marked is still in a development phase surrounded with uncertainties, the concept of marked evaluation and introduction goes far beyond the classical "voice of customer approach". Besides the complexity and versatility of the technology itself part of this complexity is, that regulation, policy and legislation are the key drivers. In some countries, subsidies of up to 80 % for residential fuel cell systems are encouraging the initiatives. Fuel cells are by nature distributed heat and power generators which fit better into a future distributed electricity grid architecture than the existing grid. The different fuel cell types share the same common feature, that they can be applied in a wide range of power generation applications from a fraction of a watt to megawatt level with high electrical efficiency. Although the future fuel cell marked may be considered enormous one could still raise the question: "If fuel cells is the answer, what was

Fig. 1 Fuel flexibility of high temperature fuel cells

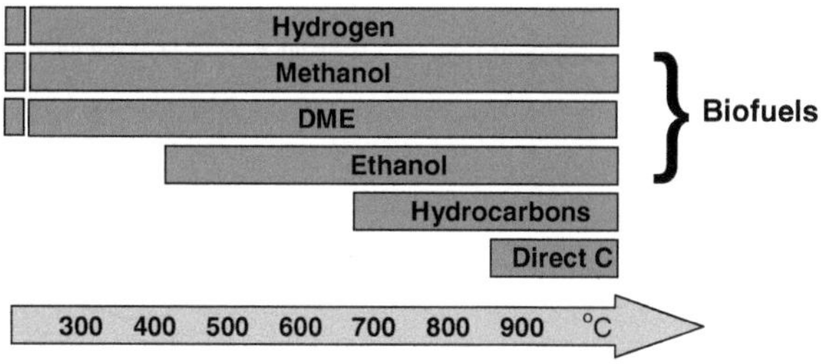

the question?" In other words, there still exists a big challenge to insure a development of reliable competitive fuel cell products toward a sustainable commercial market. Currently, these issues are very well described in the open literature (see [28] and references therein).

Only the high temperature fuel cell types, MCFC and SOFC operating at temperatures above 600 °C, have a significantly fuel flexibility and can reform natural gas directly within their cells or operate directly on CO containing gas or even on ammonia. Hence, the high temperature fuel cells offer a higher fuel to power efficiency than the low temperature fuel cells. In stand-alone systems, MCFC and SOFC have electrical systems efficiencies above 55 % based on natural gas. In a combined cycle system, where the hot exhaust gas from the high temperature fuel cell is used to drive a gas turbine the overall system, electrical efficiency, may reach 70 %. Only SOFC, being a true solid-state fuel cell, offers the possibility of a wide temperature operation interval giving rise to several important advantages. The solid-state nature of SOFC implies enormous future possibilities for further development, advances and breakthroughs in material, design and fabrication processing aiming at even higher cell and system efficiency with a dream of approaching the theoretical limit (Fig. 1).

The wide temperature window from 500 to 900 °C of SOFC makes it possible to design a more simple and optimal BOP for the fuel cell system with more flexible operation possibilities. Comparing operation of different power technologies, the efficiencies of gas turbines and combustion engines decrease at part load, whereas the efficiency of the fuel cell-based systems will be almost independent of part load. On the other hand, such alternatives to fuel cells for DG and CHP are already used widespread and commercially available at a system cost of 700–1,000 US $ per kW, at the same time as power efficiencies of about 45 % can be obtained with current combustion engine power generators of the MW size class. The function of fuel cells on the market is therefore primarily as substitution products by replacing conventional or alternative technologies.

To be competitive, the fuel cells need further technical/economical optimisation. Innovative applications with additional customer benefits have to be identified. On the one hand very few testing standards exist, and on the other hand product characteristics such as power, maximum efficiency or lifetime can be tweaked by varying numerous different product parameters such as fuel composition, fuel utilisation, gas pressure, current/voltage ratio or even materials and

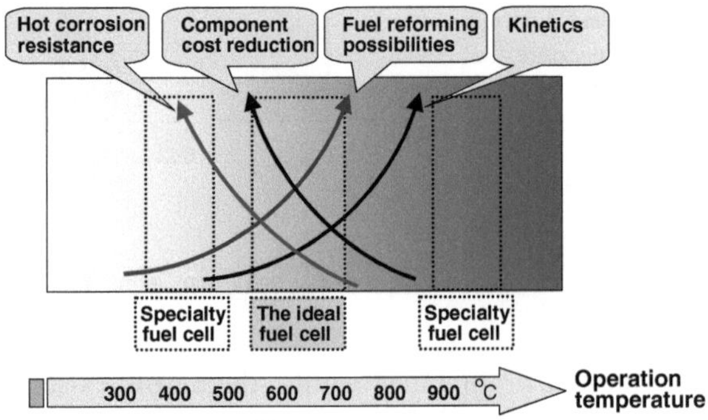

Fig. 2 Fuel cell tradeoffs—medium to high temperature variants

components in question. Although application-oriented criteria have to be fulfilled, the final system should be evaluated based on the services it provides. The assessment of the different possibilities is the task of an expert.

An increased focus and commitment on R&D, including design for cost-effective manufacturing, material development and optimisation, need to be carried out to achieve further and required progress. Further, verification of R&D activities requires adaptation of suitable fabrication techniques on a pilot-scale level approaching industrial mass production. The production capacities have to reach a level that will permit a statistically assessment of functionality and quality of cells and stacks. An integrated innovation approach, building upon combination and synergy between optimal material solutions, design solutions and manufacturing processes, has the potential to lead to innovative cost-effective breakthroughs and pave the way toward commercialisation of SOFC for stationary as well as transportation applications. An overview of the worldwide industrial status of high temperature fuel cells in decentralised generation is given in [28] and references herein.

If the most important fuel cell trade-offs are taken into consideration, the conclusion is that an optimal fuel cell should be operated at 600–800 °C at a relatively high power density without compromising efficiency and durability (see Fig. 2). Generally, the practical fuel cell efficiency is limited by the total internal losses of the overall fuel cell system including cells stacks and system components. The losses of an operating cell are mainly a sum of polarisation and ohmic losses of the anode, cathode and the electrolyte plus gas diffusion losses [20]. The polarisation losses are governed by the kinetics of the electrodes, which are favoured by increased operation temperature. The ohmic losses are governed by material selection, operation temperature and thickness (current path) of the cell components of the fuel cell in question. Figure 3 illustrates the thickness of cell components of different fuel cell types. Due to the current/voltage characteristics of a fuel cell stack a higher power density leads to a lower electrical efficiency, more heat wasted and more fuel and air pumped through the system. This further increases the parasitic losses in fuel and air blowers having negative implications on the overall system economy.

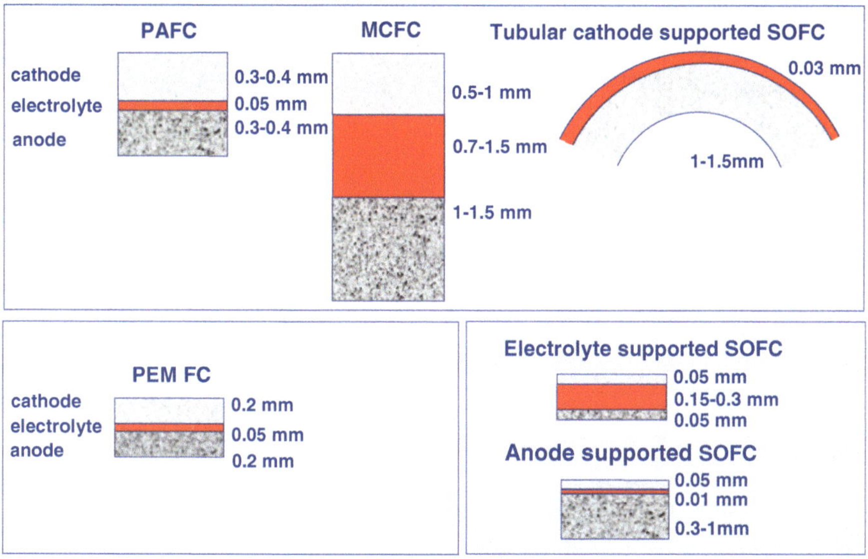

Fig. 3 Fuel cell types, and cell component thickness

One unambiguous way to reduce the operation temperature of high temperature fuel cells, approaching the proposed "ideal" operation regime of 600–700 °C without compromising efficiency and performance, is to minimize electrolyte thickness at the same time as optimal electro-catalytic electrode materials are introduced. Electrolyte thickness of an anode supported cell is only 10–15 microns compared to about 150 microns for the electrolyte-supported cells. This change of cell concept has resulted in a 10-fold reduction of electrolyte loss. Breakthroughs based on structured nano-materials, in combination with advanced ceramic processing and thin film technologies, have paved the way for further electrode and electrolyte improvements for next generation SOFCs. An increasing interest in thin film technologies adopted from the micro-electronic technology has lead to new promising possibilities for small SOFCs in mobile and portable applications [3, 22, 30].

3 SOFC Principles

The principles of a SOFC are described in the literature, as example in [19]. On the fuel side (anode) hydrogen rich gas enters reacting with oxygen ions from the oxygen ion conducting solid electrolyte to form water and at the same time releasing electrons. The electrons pass an external circuit producing power on its way to the air electrode (cathode) where oxygen from the air is reduced to oxygen ions. Individual cells are stacked to yield a sufficient voltage and power output, where a bipolar plate (interconnect) connects electrically one cell with another.

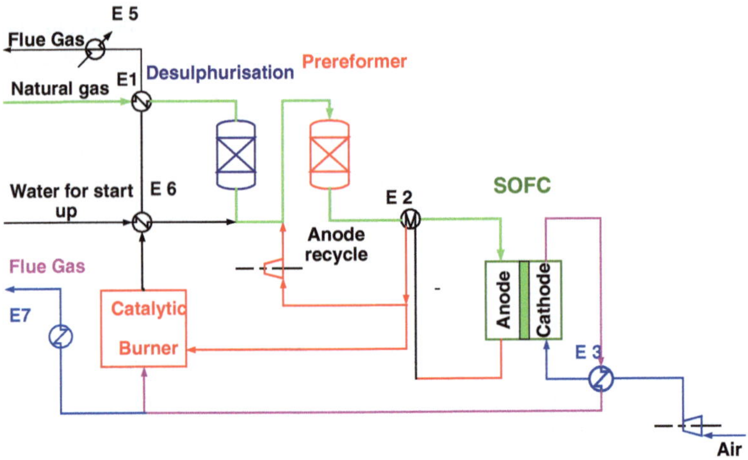

Fig. 4 Basic layout of the 250 kW SOFC based CHP system

An inherent advantage of high temperature fuel cells is that hydrocarbons such as methane may be internally reformed directly in the cell according to the following cell reactions:

$$\text{Reforming: } CH_4 + H_2O \rightarrow CO + 3H_2 \tag{1}$$

$$\text{Shift: } CO + H_2O \rightarrow CO_2 + H_2 \tag{2}$$

$$\text{Oxidation: } O_2(\text{cathode}) + 2H_2(\text{anode}) \rightarrow 2H_2O + \text{heat} \tag{3}$$

Reactions (1) and (2) take place on the anode side. They are catalysed by the anode material and structure.

4 SOFC System Design

The heat consuming internal reforming of methane in combination with the waste heat producing electrochemical cell reaction at high temperatures, lower the demands on the FPS and lead to higher efficiency of SOFC compared with low temperature fuel cells. A very simple base case system design for natural gas fuel consists of the primary unit operations as depicted in Fig. 4 [24].

Natural gas is preheated to 400 °C and de-sulfurised in a zinc oxide bed. A very important criterion for design of FPS systems for fuel cells is the need to avoid carbon formation. The adiabatic pre-reformer quantitatively converts higher hydrocarbons to form a mixture of methane, hydrogen, carbon monoxide and carbon dioxide, and thus eliminates the risk of carbon formation. A typical composition of Danish natural gas is given in Table 1. A final in/out heat

Table 1 Composition of simulated fuel (Danish natural gas)

Component	CH_4	C_2H_6	C_3H_8	C_4H_{10}	C_5H_{12}	C_6H_{14}	N_2	CO_2	H_2S (ppm)
Volume %	88.1	6.4	2.8	1.0	0.18	0.06	0.3	1.3	10

exchanger will increase the inlet fuel gas temperature to around 650 °C. On the cathode side, air is compressed in a blower and preheated in an in/out heat exchanger to about the same temperature. The cathode off-gases, consisting of depleted air, are sent partly to a catalytic burner together with anode-off gases. The rest of the cathode off-gases, together with flue gases, will deliver heat to a heating ventilation and air conditioning (HVAC) system. The burner exhaust gases are also used for steam generation during the start-up phase and for fuel preheating. System simulations of a 250 kW SOFC system based on natural gas fuel, as described above, yield an electric efficiency, including all internal system losses, in the range of 56 % and a total efficiency of 88 %.

In a previous study at Haldor Topsoe A/S [24], it was calculated that approximately half of the waste heat from the oxidation reaction, however, is used to drive the internal reforming reaction, where methane in the anode feed is reformed with water to generate hydrogen. This reaction not only reduces the size of the in/out exchanger E3 and reduces the parasitic loss for air compression, but also "upgrades" waste heat to chemical energy. This is a major reason for the higher electrical efficiency of a stationary SOFC CHP system compared to a stationary low temperature CHP fuel cell such as PEMFC.

The energy flows of the base case are represented in a Sankey diagram; see Fig. 5, showing the energy contents of the streams in kW and kJ/s.

Only part of the enthalpy from reaction (3) is available as external work (exergy or Gibbs free energy). The rest is dissipated in the fuel cell as heat and mainly removed with the cathode air through heat exchanger E3. This heat transfer is the main source of exergy losses in the plant.

5 Fuel Cell Fuels

The complexity and efficiency of the FPS is strongly dependent on the choice of fuel, fuel processing route and fuel cell type. In general, the higher C/H and C/O ratio of the fuel the more difficult the fuel processing, especially taking into account the sulphur most often found in heavier hydrocarbons. Normally, the fuel processing options are limited to those based on steam reforming, cracking, catalytic partial oxidation (CPO) or autothermal reforming.

Table 2 shows a comparison of the system efficiencies calculated for a 250 kW SOFC CHP systems based on different fuels [13, 14]. Methanol and DME can be relatively easily produced from biomass. The numbers for MeOH and DME involve a new process lay-out, developed by Haldor Topsoe A/S.

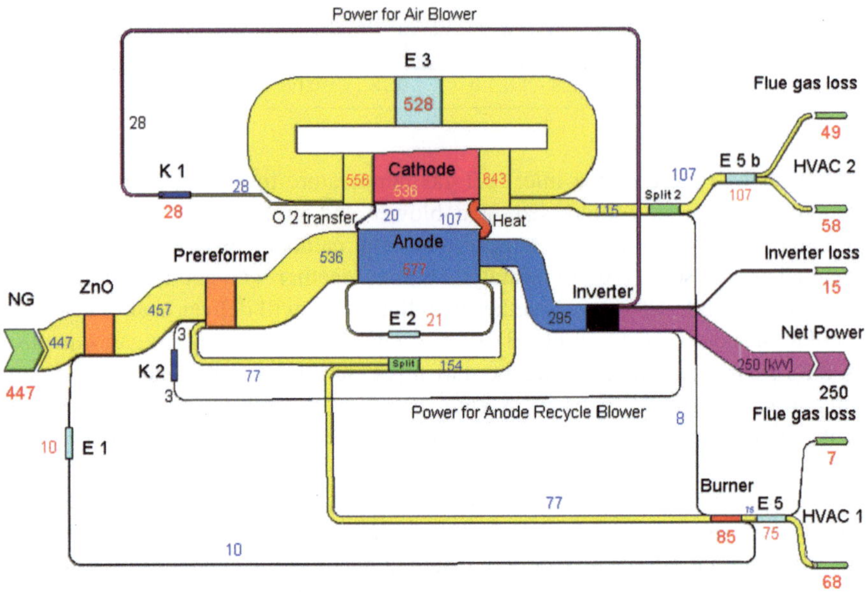

Fig. 5 Energy flows in a 250 kW SOFC CHP NG system in kJ/s

Table 2 SOFC system efficiencies for various fuels

Fuel	Electric efficiency % LHV, net	Total efficiency % (90° C)
Natural gas	55	84
Biogas with 50 % CO_2	54	80
Methanol	53	85
DME	53	83
Ammonia	55	84
Diesel CPO (5 kW)	41	85

6 SOFC Stack Development

State-of-the-art SOFC technology for stationary, as well as for transportation application, is today being demonstrated with either planar or tubular ceramic anode-supported or electrolyte-supported SOFC cells. Already in 2004, Topsoe Fuel Cell A/S (TOFC) demonstrated a record-breaking 13,000 h test of a very compact, material economical, stack design based on planar anode-supported cells. The TOFC multi-layer stack design has proved a record-breaking high specific volume power density of 2.4 kW per litre stack ensuring compactness and economical use of the materials in question. For more information on the SOFC development programme, please refer to [9–11]. Figure 4 shows a picture of a

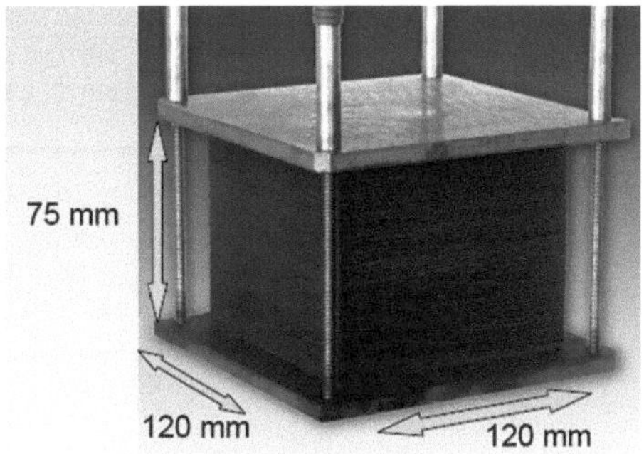

Fig. 6 Compact design of a 75 cell stack with $12 \times 12\ cm^2$ cells

standard TOFC stack containing 75 cells with a footprint of $12 \times 12\ cm^2$. Each repeatable unit of the stack, including a cell and the interconnect plate with its gas distributing channels, has a height of about 1 mm (Fig. 6).

SOFC technology faces many challenges when it comes to up-scaling and commercialisation, since cost reduction, reliability and extended lifetime are required. In order to improve durability and cost efficiency of the cells, stacks and system much of the development has been in the past focused on lower operation temperature, increased power density and material savings based on reduced cell and stack component thickness. Nevertheless, most of the demonstrations with ceramic cells in real system operation have until now revealed problems regarding robustness and reliability, especially in cases where real operation includes dynamic thermo-mechanical loads induced by conditions such as start-up, shut-down, transients and non-steady-state load. Temperature profiles in the cells and stacks during operation are determined by fuel composition, control strategy, flow configuration and cell and stack design. When real fuels under real operation conditions are introduced to the fuel cell, the actual temperature profiles cause mechanical stresses in the cell and stack layer, which represents a critical load condition without tolerance. Due to a general lack of statistical data from real system test under real operation condition, the importance and seriousness of these critical issues have been, in general, underestimated until now.

7 Durability of SOFC Operating at 600–850 °C

The best known type of SOFC anode-supported cells consists of a Ni/Zirconia cermet support layer, a thin functional anode, a thin dense yttria doped zirconia electrolyte layer and a thin Strontium doped lanthanum manganite (LSM) cathode layer (material alternatives are given in Fig. 8). Such cells have proved to be stable and durable for thousands of operation hours when operated at relatively mild conditions [9]. The degradation rate increases with increasing cell polarisation and increasing current density. The polarisation-dependent degradation has been identified to mainly originate from the cathode/electrolyte interface. However,

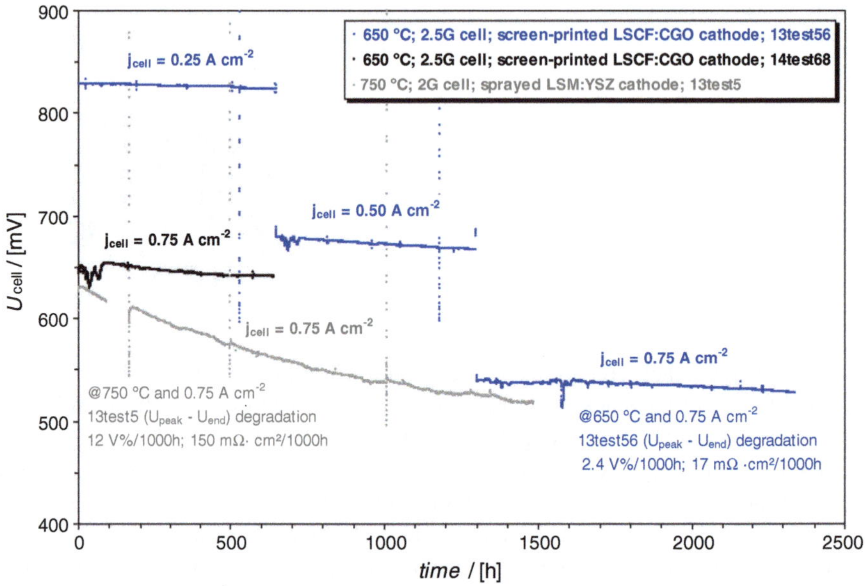

Fig. 7 Cell voltage versus time, with humidified hydrogen as fuel. The *grey curve* represents the cell voltage with time of a standard 2G cell at 750 °C with syngas as fuel

other complex mechanisms contribute to the overall degradation and failure. Efforts are being devoted to develop cells that are also capable of long-term operation under these conditions to enhance the window of operation of the technology. Recently, progress has been achieved in this field with new cathode materials [9, 32].

The results of durability tests of cells with LSCF cathodes (2.5G) are shown in Fig. 7. The objective of the test was to measure the degradation rate of the 2.5G cells at 650 °C at various current densities with a significant electrode polarisation. The results from ageing tests of 2.5G cells are compared to cells with LSM cathodes (2G) at 750 °C, 0.75 A/cm^2, 75–80 % fuel utilisation, 9 % oxygen utilisation in the figure. The current density and degradation rates observed are indicated in the graph.

The test results demonstrate that 2.5G cells show good stability under galvanostatic operation at 650 °C, even at current densities as high as 0.75 A/cm^2. Note that, the observed degradation rates for the 2.5G cells are lower than those observed for the 2G cells at comparable power density.

The interfaces in SOFCs are not in thermodynamically equilibrium during operation, and this may cause formation of electrically insulating phases, e.g. zirconates at YSZ/LSM interfaces, resulting in degradation of the SOFC. It is important to investigate the chemical and microstructural degradation with the aim to identify the detailed mechanisms and to work out counteracting measures. Durability models that relate the degradation rate to particle–particle interaction stresses during thermal cycles and thermal expansion mismatch have been recently investigated [33]. The models include particle sizes, compositions and porosity and predict that small nano-size particles result in improved durability and reliability. To accomplish a practical SOFC solution, it is necessary to study the effect of relevant impurities in the air as well as in the fuel on the initial performance and long-term durability. Recently, a significant improved understanding of these effects has been obtained [6, 7, 12].

Approaching more realistic fuels, minor components and traces have to be included in the performance and durability studies of SOFCs. Sulphur compounds have attracted special attention as they are contained in biogas and gasification gas [15]. Due to strategic targets of developing products for high sulphur logistics fuel, an increased focus is directed on sulphur resistant anodes for anode supported cells. State-of-the-art Ni/YSZ SOFC anodes exhibit a high activity for reforming of methane. However, as in case of conventional reforming catalysts, also these anodes suffer from being sensitive toward sulphur impurities. SOFC with Ni/YSZ and Ni/ScYSZ anodes show similar cell voltage drops when exposed to H_2S concentrations between 2 and 24 ppm at OCV and 850 °C in a fuel consisting of 13 % H_2, 29 % CH_4 and 58 % H_2O. When operating at 1 A/cm^2 current load in the presence of 2 ppm H_2S over 500 h, the cell with the Ni/YSZ anode degraded significantly, mainly due to an irreversible increase of the serial resistance [25]. An improved cell with a Ni/ScYSZ anode did not show any degradation under these conditions [10].

8 Reliability of Cells and Stacks

Until now, all SOFC cell and stack development programmes have relied upon extensive use of ceramic materials whether in the anode-support component or in the supporting electrolyte. Traditional as well as novel alternative materials for various cell components are given in Fig. 8. Numerous studies, on ceramic components for structural as well as functional applications, have proved that components such as anode-supported and electrolyte supported cells inevitably suffer from unpredictable mechanical properties and unreliability due to their brittleness. Lack of thermal shock resistance is further enhanced by their low thermal conductivity, low tensile strength and high Young's modulus. In general, the low fracture toughness of ceramic components makes them extremely sensitive to material defects and unsuspected flaws. The consequence is that the ultimate strength of ceramic cells deviates from their nominal mean value, and over-engineering with large safety factors is needed in order to obtain a reasonable safe design (see Fig. 8). Furthermore, defect sensitivity means that quality control in all manufacturing steps becomes extremely critical and costly.

Recently, a few SOFC organisations have realised the existence of technical and techno-economical barriers in today's conventional SOFC technology. They have, therefore, initiated development programmes on next generation SOFC technology with greater potentials of success and commercial breakthrough.

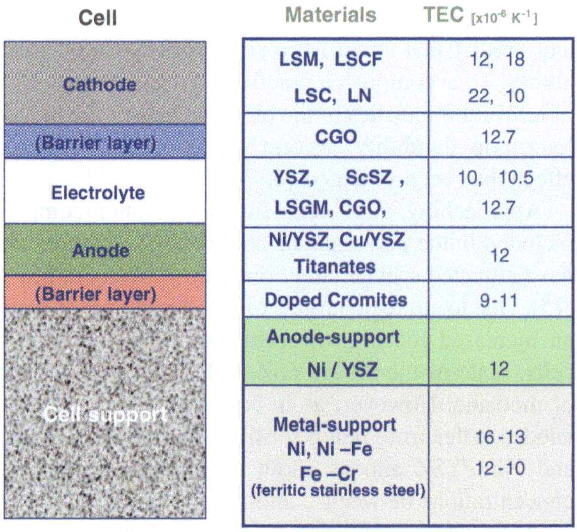

Fig. 8 Selection of cell materials for cells with thin supported electrolytes

Modelling studies [21], as well as recent practical experience, have proved how up-scaling of cells and stacks to larger more industrially relevant sizes generally leads to lower reliability in real system operation and intolerance toward system abuse and operation failures. These observations conform to the statistical distribution of mechanical properties governing the probability of failure of cells based on ceramic materials, whether it is for mobile or for stationary applications.

There are two completely different manifestations of mechanical load, which have important consequences for a cell or a stack in a practical design. One is the extrinsic actual load (F) exerted on the individual components of the stack by the surrounds. The other is the intrinsic maximum load ($F_{\max}$), which is the largest load that the component (such as a cell) can withstand without failure, and this maximum load is a property of the component, a function of its dimensions and material. One interesting development concept aiming at increased robustness and reduced cost is to replace the critical structural components in cells and stacks with metallic materials.

Although the TOFC state-of-the-art anode-supported cells possess a world record mechanical average strength of about 400 MPa [10], (as measured in round robin tests in previous EU projects), statistical analysis of the strength behaviour bears evidence that most cells may behave unreliably under real dynamic operation conditions. Expressed according to Weibull statistical analysis (failure probability as function of applied mechanical stress), the Weibull modulus is in the range of 8–10 (see Fig. 9).

The reliability of a structural component is closely dependent on its fracture toughness K_{1c}. For a ceramic cell the fracture toughness has been measured to be 2–3 MPam$^{1/2}$. For comparison, metal components have fracture toughness more than one order of magnitude higher. The low value of fracture toughness makes ceramic cells extremely sensitive to material defects such as unsuspected flaws

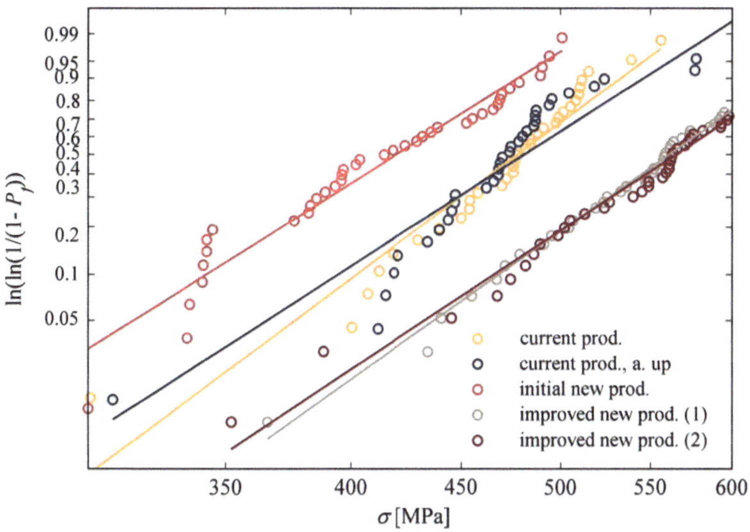

Fig. 9 Weibull plots of the strength of standard cells as well as cells manufactured by a new and more upscalable process. A 17 % increase in average strength, as compared to the standard cells, has been achieved by improved processing

introduced during manufacturing, handling or operation. The consequence is that the ultimate strength of ceramic cells deviates from their nominal intrinsic mean value, and a large safety factor ($n = F_{\max}/F$) typically about 8–10 has to be used in order to obtain a reasonable safe design. Figure 10 shows the statistical distribution of the fracture strength (Weibull distribution) of a typical ceramic anode-supported cell measured at Risø-DTU by 4-point bending. The measured fracture strength has been normalised to an effective cell volume of 300 mm^3, because this volume is typical for a large up-scaled cell to be used in stationary applications. It is seen that the fracture strength of the cell is 146 MPa at a failure probability of 1 %.

In a real stack with planar stack components, the actual loads are more complex and harsher than those applied under a typical bending test as used in this experiment and cannot be predicted with certainty. Furthermore, when ceramic planar cells are assembled in stack structures temperature gradients in three dimensions, thermal expansion mismatch and micro-structural changes of the cell components lead to very complex and critical thermo-mechanical stress distributions. No universal attributes of stress states can be identified which enables failure prediction of stack structures based on a combination of ductile interconnect layers and brittle ceramic cells. A multitude of failure modes are governing the reliability, and no single consistent failure theory is applicable, whereby safety by design becomes a huge challenge. The consequence is, that aiming at a reasonable reliable design, the large variations in ultimate strength of the ceramic cells require the implication of very high safety factors compromising the overall stack economy.

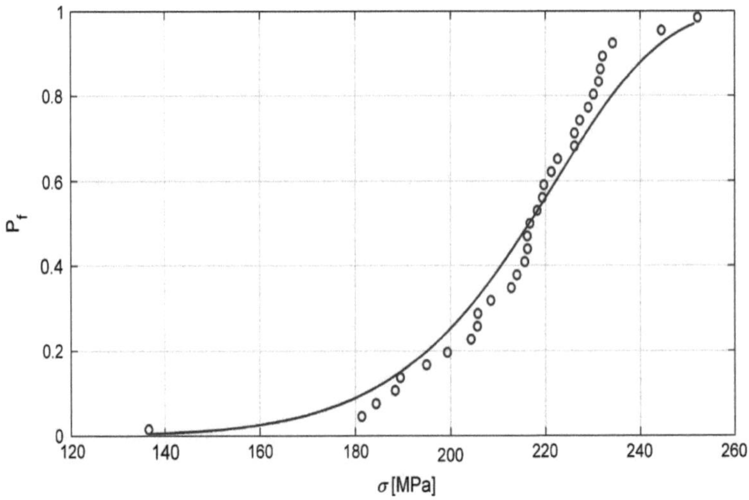

Fig. 10 Failure probability of an anode-supported cell normalised to 300 mm^3 cell volume

Since metal-supported SOFCs have an intrinsically different mechanical behaviour, compared to ceramic components, it makes them potential candidates for withstanding the extrinsic actual loads such as temperature gradients and mechanical stresses whether static or dynamic. The predictable material properties of metals make them more suitable in connection with effective use of FEM modelling in order to obtain optimal cell and stack designs. The numbers of R&D activities related to the so-called metal-supported SOFC (also called third generation SOFC) have increased significantly in recent years [31].

9 Metal-Supported Cells: Next Generation SOFC Technology

A metal-supported cell and stack concept is ideally based upon a low-cost stainless steel cell support and low-cost stainless steel thin metallic sheet interconnect bonded together with a cost-effective metallic brazing or welding technique. The concept substitutes brittle ceramics with ductile metallic components and thereby to achieve a desirable, graceful non-catastrophic failure mode of the cells and stacks. The main issue addressed in the development of this concept is the simultaneous achievement of robustness, reliability and cost-effectiveness offered by the metallic materials as well as the high electrochemical performance offered by next generation electrode development based on nano-structured materials.

TOFC has in collaboration with Risø/DTU a focussed commitment to develop next generation stack technology based on metal-supported cells to improve reliability, cost-effectiveness and functionality including the objective to reach a robustness of SOFC stacks which fulfils the requirements under real system load.

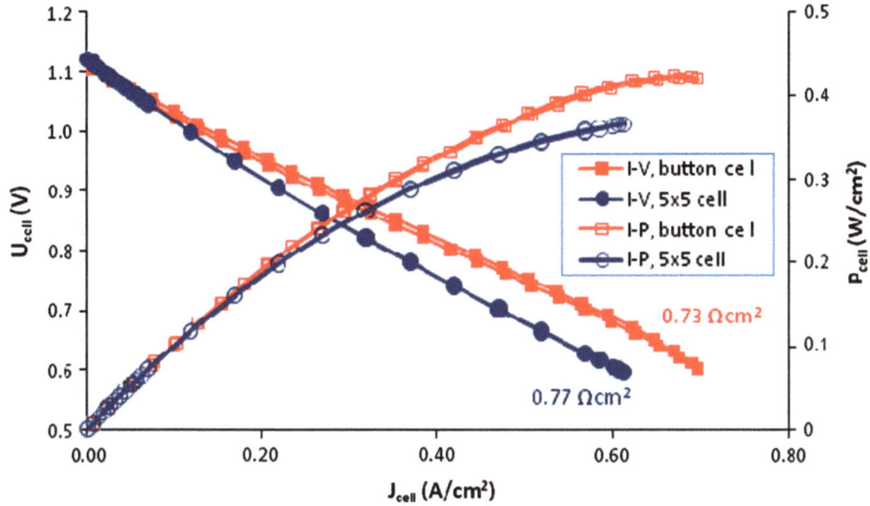

Fig. 11 Electrochemical performance of the metal-supported cells with different footprint sizes at 650 °C (*fuel* 96 % H_2 with 4 % H_2O, *oxidant* air). The inserted numbers are the area specific $ASR_{0.6V,\ secant}$ values calculated for the different cells

The cell design includes an anode structure, in which nano-sized ceria and nickel particles are formed by solution infiltration in the porous microstructure after fabrication of the half cells. The multilayered structure, comprising the metal supported cell, can be obtained by conventional processing techniques such as tape casting, co-sintering and screen printing.

Figure 11 shows polarisation and power curves recorded at 650 °C, comparing a $5 \times 5\ cm^2$ metal-supported cell with a button cell reported in [5]. The area specific resistance (ASR) values were calculated as the secant value at 0.6 V and resulted in 0.73 and 0.77 Ω cm^2 for the button cell and $5 \times 5\ cm^2$ cell, respectively. The polarisation curves were conducted at 650 °C with hydrogen (approximately 4 % H_2O) as fuel and air as oxidant. Galvanostatic durability testing, under a constant current load of 0.25 A/cm^2 at 650 °C for approximately 1,000 h, showed a degradation rate <5 % /1,000 h, for both cell tests, based on the change in cell voltage during the testing period. Further details are given in Ref. [4].

The observed performance and the measured degradation rates over 1,000 h for the button cell test and the $5 \times 5\ cm^2$ test are reasonably close and indicate good correlation between the two different test set-up and contacting protocols. The cell performance results show generally lower or at least comparable ASR values than that of many other metal supported SOFCs presented so far in the literature (see e.g. Ref. [31] and references therein). Recently, the TOFC stack with $12 \times 12\ cm^2$ cell footprint has been modified with metal-supported cells and showed performance results close to those for the single cells.

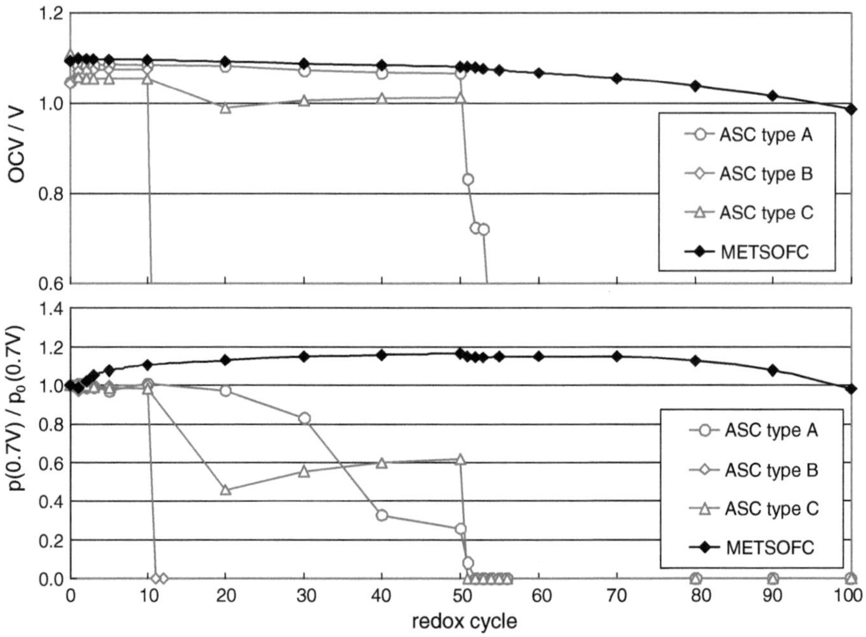

Fig. 12 Redox stability of ASCs (Ni/YSZ anode substrate [5]) and metal supported SOFCs at 800 °C operating temperature: OCV (*top*) and performance degradation (*bottom*)

The developed metal-supported cell design is less sensitive to Ni-agglomeration in the active anode and more robust toward redox cycles. A promising redox tolerance has been revealed as test result in the EU project METSOFC coordinated by TOFC. More than 100 full redox cycles were conducted without detrimental degradation of the cell. Figure 12 displays the performance and OCV degradation of different Ni-YSZ anode-supported cells and a metal-supported cell during 100 Redox cycles [4]. Whereas the anode-supported cells fail after at least the first 10 min redox cycle (cycle no. 51), the metal-supported cell only shows a slight degradation within cycle 51–100 (10 min redox cycle) and even improves during the first 50 cycles (1 min redox cycle). It is obvious that the metal-support is not seriously affected by the re-oxidation. Redox cycling tolerance is important because it allows the cells to withstand cooling without the need for special protection and interruption in fuel supply. The ability to survive these conditions can reduce balance-of-plant complexity and redundancy. The aim of the METSOFC project is to develop a SOFC stack technology which is significantly more robust reliable and cost-effective than current technologies based on anode-supported or electrolyte supported cells.

A crucial issue is the material cost of the current anode-supported cell and stack concepts. For commercial introduction, it is vital that the material cost in case of large scale cell and stack production is reduced. The need of Ni-YSZ materials for the support layer in the current anode-supported cells or YSZ for electrolyte-

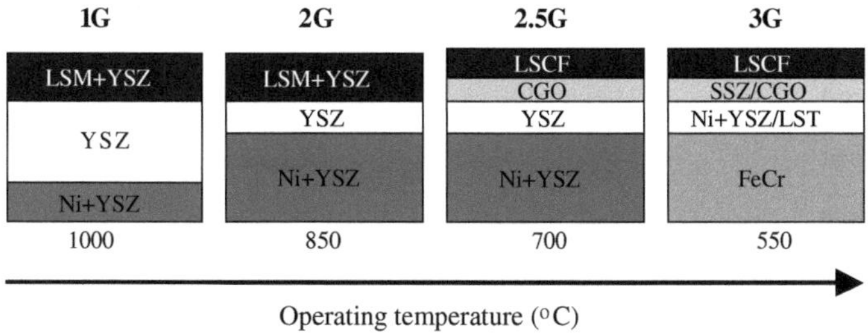

Fig. 13 Various cell generations being developed and produced in the TOFC/Risø consortium and in the METSOFC project

supported cells confines the raw material cost to a level of € 50–80 per kg. The objective of the next generation cell development is to bring the material cost for the major part of the cell down to about € 10–20 per kg. This cost target is brought within reach thanks to the novel metal-supported cell and stack concept.

In case of a metal-supported cell concept, some new important challenges appear that must be overcome primarily concerning materials compatibility, materials processing and interaction, inter-diffusion and vaporisation of metallic alloy elements and finally hot corrosion/oxidation and high temperature creep of the metallic parts. Due to hot corrosion of the metallic layers in the cell the operation temperature has to be lowered to around 600–750 °C, which, on the other hand, leads to new challenges in the development of more active electrode materials. Recently, the METSOFC project has led to a significant progress in performance and durability of the metal-supported cells.

As shown in Fig. 13, the material choice for the new cell type differs from previous cell generations in that porous ferritic steel is used as a ductile, robust cell support and the electrolyte is based on scandia doped zirconia with increased ionic conductivity for higher performance. Novel alternative anodes with improved stability toward sintering, coking, sulphur poisoning and redox cycles can be based on doped titanates (such as LST depicted in Figs. 13 and 8). In addition to cost-effective powder metallurgical processing developed in the METSOFC project some more advanced thin film technologies offer promising possibilities for the next generation SOFC technology such as PVD, CVD and ALD with possibilities to make dense functional coatings with layer thickness in the range of 50–500 nm. Examples are given in the Refs. [8, 22, 30].

10 Novel Electrodes Based on Nano Materials

In order to improve durability and performance of the cells the stacks and the system much of the development have been focused in the past on lower operation temperature and improved electrode materials. EU projects such as REALSOFC and SOFC600 have been targeted at improved SOFC performance with a special

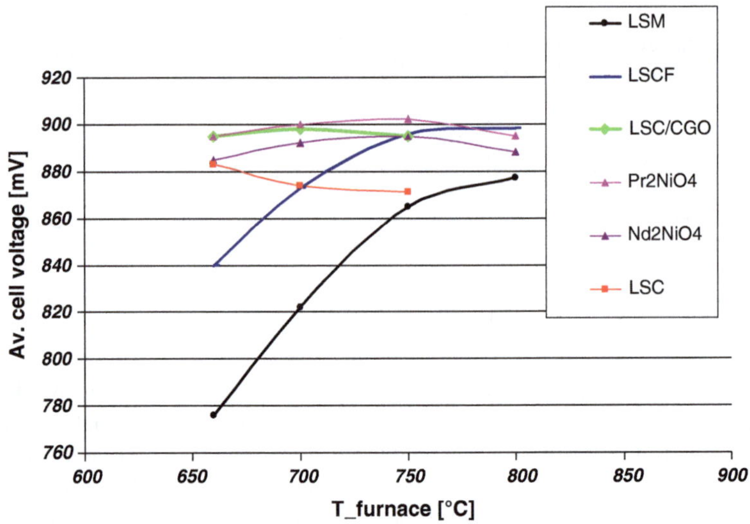

Fig. 14 Stack test with different cathode materials applied on TOFC anode supported 12×12 cm^2 half cells for intermediate temperature operation

focus on durability and reduced operation temperature. In these collaboration projects, TOFC took the initiative to evaluate and verify the most promising novel cell electrode candidates developed by the institute partners. To increase speed of evaluation feedback and overall development of the new electrode materials were applied on TOFC anode-supported half cells assembled and tested in TOFC stacks with 12×12 cm^2 footprint and standard coated metallic interconnects [11]. In the SOFC600 project, the electrodes were developed and manufactured at different European project partners aiming at stack operation around 600–700 °C. For operation below 700 °C the cobaltite and nickelate cathodes were chosen as the best candidates. Pr_2NiO_4 showed the highest performance initially in the stack. However, after 20 thermal cycles, Nd_2NiO_4 based cathodes showed to be more reliable. The LSC cathode-based cells proved to degrade faster (about 2 % per 1,000 h) than the nickelates (about 1 % per 200 h) when operated in the stack at 0.22 A/cm^2 at 650 °C with a cathode air humidity of 2.5 %. A remarkable low dependence on the operation temperature of the stack has been obtained from these cathode materials as shown in Fig. 14, proving that a very wide temperature operation window down to 650 °C with zirconia-based anode supported cells is now feasible.

Another cathode material development work is carried out in the consortium of Risø-DTU and TOFC. $LaCo_{1-x}Ni_xO_{3-d}$ (LCN) has been reported to possess beneficial materials properties for SOFC cathode application [17]. The thermal expansion of the $LaCo_{0.6}Ni_{0.4}O_3$ (LCN60) is better matched to the SOFC anode support as compared to other cathode materials such as LSCF, and especially LSC. Electronic conductivities of about 1,400 S/cm at SOFC operating temperatures have been reported. Symmetrical cell tests at 750 °C and OCV demonstrated no degradation of the polarisation resistance (R_P), in contrast to the increase in R_P of LSM-YSZ and LSC based cells [11, 16]. Figure 15 shows that the cells with porous LCN-CGO cathodes perform substantially better than the LSM-YSZ based cells at 600 °C with an ASR of 0.16–18 Ω cm^2.

Figure 16 illustrates the status of another route to improvement of the cathode performance. Here, a nano-crystalline gadolinia doped ceria catalyst has been added to the electrodes, resulting in improved performance over the comparable reference cells for both LSM/YSZ and LSC/CGO electrodes [29]. Continued development includes investigation of stability of the colloidal impregnated nano-particles as function of time and temperature. Such nano-structured electrodes based on infiltration are expected to play an important role for further improvements of the next generation cell technology.

11 Fuel Processing and System Development

A number of fuel cell organisations including TOFC have studied a range of interesting fuels such as natural gas, LPG, methanol, DME, diesel, ethanol, coal syngas and ammonia as SOFC fuels. As mentioned in Table 2, such studies predict system electrical efficiencies from 40 to 56 % (AC out/LHV fuel in), depending on the fuel used and the size and lay out of the system. Existing catalyst know-how of Haldor Topsoe A/S has proved to lead to new possibilities in fuel processing for SOFC systems. Haldor Topsoe A/S coke resistant methanation catalyst has been introduced in methanol fuelled SOFC APU systems for ships resulting in higher system efficiency and a simplified system lay-out. Further interesting possibilities with similar advantages exist in other cases such as ethanol fuelled or coal syngas fuelled SOFCs [14].

For high efficiency fuel cell systems, it is of paramount importance to have a very tight integration both mechanically and thermally of all the hot components of the system. Several SOFC system developers have, therefore, focused their development on units comprising an integration of fuel processing, stack, feed-effluent heat exchangers and catalytic burner. Today, dynamic system modelling tools have been developed and are available as an effective way to increase speed in the design phase.

12 Market and Technology Demonstration

Today, the fuel cell markets are transitioning from small-scale development to pilot-scale test and demonstration. To meet the growing demand for stacks for testing and demonstration, a number of leading fuel cell organisations are currently commissioning production facilities in the prototype and pre-commercial phase [28].

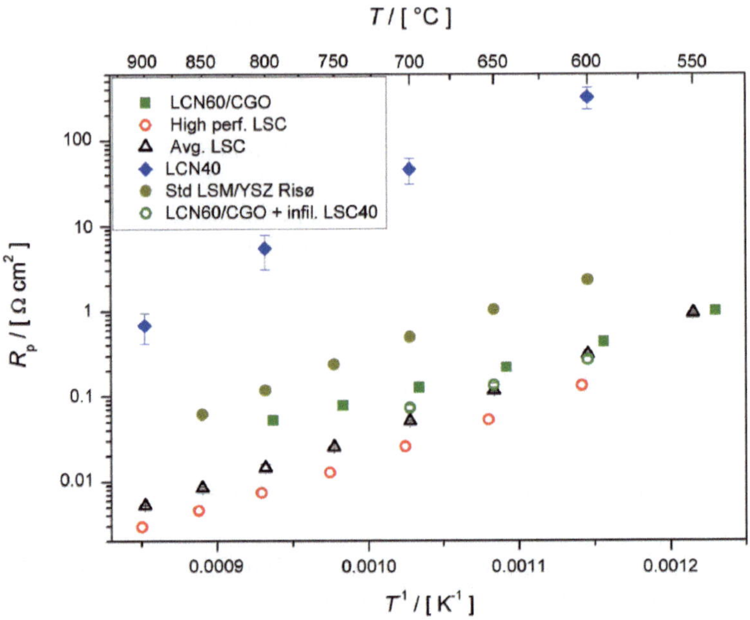

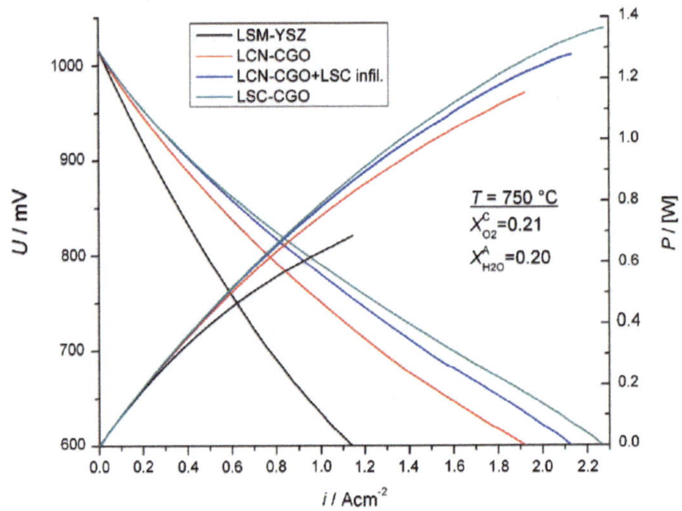

Fig. 15 *Top* Arrhenius graph comparing the R_P of infiltrated and non-infiltrated LCN-CGO cathodes with a selected number of "reference" electrodes. All cathodes have been manufactured at Risø-DTU. *Bottom* Four I/V and power curves for anode supported cells measured at 750 °C

However, demonstration cannot guarantee that the technology and the products are mature.

Future commercial success depends on a continued development where more emphasis is devoted to an integrated engineering approach and innovative

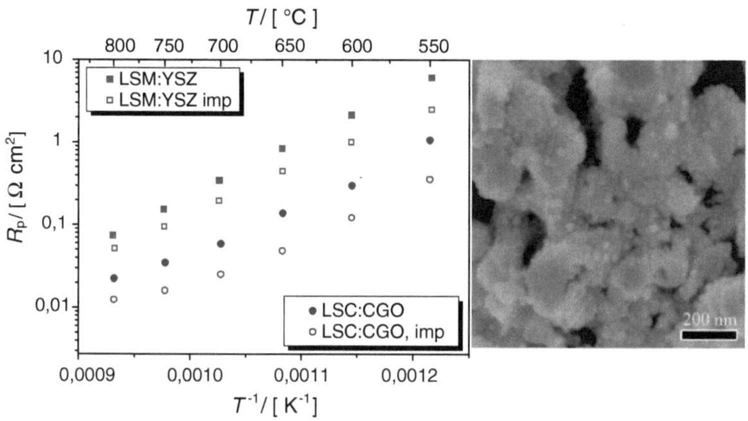

Fig. 16 Effect of adding a nano-crystalline catalyst to screen printed LSM/YSZ and LSC/CGO composite cathodes, polarisation resistance as function of operation temperature

solutions. The cost of stacks and units available today is adequate for prototype technology but not for marketable products. Cost has to be reduced, and this will only be realised by introducing mass manufacturing processes hand-in-hand with development of products designed for manufacturing and performance. It is expected that even radical technology changes will take place to render fuel cells profitable and reliable. The supply chain is still in a development phase, and cost reduction depends on learning by doing as well as improved understanding of real world problems. As understanding increases SOFC systems will become simpler, more optimised and more cost-effective rather than more complex. Further engagement and consolidation will occur in the supply chain as the industry continues to mature. Market possibilities are very wide and diversified which includes the risk of unfocussed development strategies and business models. Fuel cell markets are still emerging, and companies can possibly benefit more from cooperation than competition.

Innovation, that builds upon combination and synergy between optimal material solutions, design solutions and manufacturing processes with a basis in an effective R&D platform and collaboration networks, has the potential to lead to innovative product breakthroughs. The marked opportunities are more clearly analysed by reviewing the different possibilities that SOFC technology can offer in addition to identifying the important problems that it can solve for the end user. An innovative approach integrating innovation in science and technology as well as in business models will more effectively, reliably and affordably pave the way toward sustainable commercialisation of SOFC in real world applications.

13 Future Perspectives for Fuel Cells in Connection with Energy Sources, Energy Carriers and Energy Technology

As described above, the high temperature fuel cells and particularly the SOFC present a better environmental option versus existing fossil fuel technologies. These fuel cells run on natural gas or syngas that is reformed directly within the cell rendering the conversion process highly efficient. When paired with cogeneration applications, the SOFC technology is capable of producing total heat and power efficiencies of more than 80 % which is above traditional coal and natural gas power plants.

The typical source to produce hydrogen, which is also the most economical, has been natural gas according to the steam reforming and shift Eqs. (1) and (2) given above. Coal is also indirectly an alternative in the future. However, both of these processes release CO_2 into the atmosphere.

Hydrogen can be made from electrolysis of water according to the equation:

$$H_2O \rightarrow H_2 + {}^1\!/\!{}_2O_2 \; E_0 = 1.229V \quad (4)$$

A significant amount of heat and electricity is required to drive this water electrolysis reaction which currently makes this technology more than two times as expensive as natural gas steam reforming. Electrolysis of water based on high temperature electrolysers—so-called solid oxide electrolyser cells (SOEC)—possesses a huge future potential, because the high temperature heat loss from the cells helps to balance the energy economy in the heat consuming water splitting and water heating process, and thus complete the electrolysis with lower electricity consumption. Low temperature water electrolysers run at practical efficiency of about 45–60 %, whereas the high temperature electrolysers promise practical efficiencies above 80 %.

In principle, a SOEC is a SOFC operated in the reversed mode. Therefore, a system principally consisting of a heat exchanger and a reversible SOFC/SOEC has a lot of potential advantages compared to other conversion techniques. SOEC has also the potential of splitting carbon dioxide into carbon monoxide and oxygen. This means that electrolysis of a mixture of steam and carbon dioxide results in a mixture of hydrogen and carbon monoxide (syngas). The SOEC materials for the cells and stacks are very similar to those for the SOFC as described above, but further challenges regarding material stability and durability at the required high current densities are obvious. Hence, SOEC may be considered as an interesting future spin-off from the current SOFC development. TOFC has developed and delivered SOEC stacks for hydrogen production in the EU project RelHy. More than 2,000 h at 0.8 A/cm^2 have been demonstrated with stack degradation below 3 % per 1,000 h during hydrogen production from steam electrolysis. The present degradation rate is about a factor three times higher than desired, which shows that further emphasis needs to be directed toward R&D on cells and stacks for SOEC.

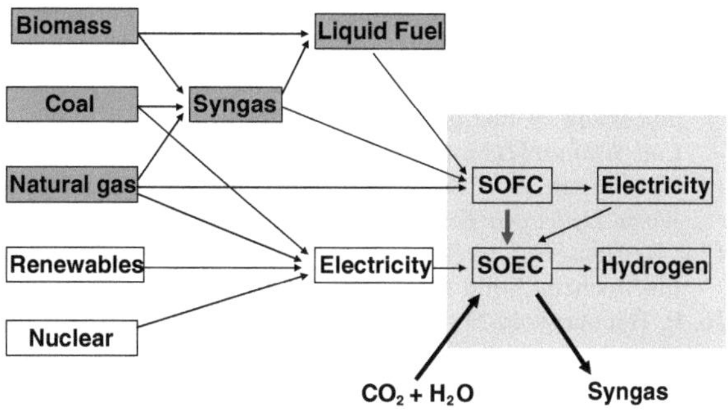

Fig. 17 Diversity of energy sources, carriers and high efficiency power systems

Figure 17 shows the diversity of known energy sources, energy carriers and the potential to use these with high efficiency and CO_2 neutrality based on future SOFC and SOEC technology.

References

1. W.J. Abernathy, J.M. Utterback, Patterns of industrial innovation. Technol. Rev. **80**(7), 40–47 (1978)
2. W.J. Abernathy, K.B. Clark, Innovation: mapping of the winds of creative destruction. Res. Policy **14**, 2–22 (1985)
3. A. Bieberle-Hütter, J.L.M. Rupp, L.J. Gauckler, The effect of etching during microfabrication on the microstructure and electrical conductivity of CGO and YSZ thin films, in *European SOFC Forum, Lucerne*, 2010
4. P. Blennow, J. Hjelm, T. Klemmensø, S. Ramousse, A. Kromp, A. Leonide, A. Weber, Manufacturing and characterization of metal supported SOFCs, in *European SOFC Forum, Lucerne*, 2010
5. P. Blennow, J. Hjelm, T. Klemmensø, Å. Persson, K. Brodersen, A. Srivastava, H.L. Frandsen, M. Lundberg, S. Ramousse, M. Mogensen, Development of planar metal supported SOFC with novel cermet anode. ECS Trans. **25**(2), 701–710 (2009)
6. E. Bucher, A. Egger, M. Yang, W. Sitte, F. Klauser, E. Bertel, In situ studies of long term degradation of the IT-SOFC cathode materials LSCF and NDN, in *European SOFC Symposium, Lucerne*, 2010
7. E. Bucher, W. Sitte, Oxygen exchange Kinetics of $(La,Sr)_{1-z}(Co,Fe)O_{3-\delta}$ in dry and wet atmospheres. Solid State Ionics **191**, 61–67 (2011)
8. Y.-Y. Chen, W.-C.J Wei, Processing and characterization of ultra-thin yttria-stabilized zirconia (YSZ) electrolytic films for SOFC. Solid State Ionics **177**(3–4), 351–357 (2006)
9. N. Christiansen, J.B. Hansen, H. Holm-Larsen, M.J. Jørgensen, L. Theil Kuhn, P.V. Hendriksen, A. Hagen, S. Linderoth, Solid oxide fuel cell research and development at Topsoe fuel cell A/S and Risø/DTU, in *European SOFC Symposium, Lucerne*, 2008
10. N. Christiansen, J.B. Hansen, H.H. Larsen, M.J. Jørgensen, M. Wandel, P.V. Hendriksen, A. Hagen, S. Ramousse, Status of development and manufacture of solid oxide fuel cells at Topsoe Fuel cell A/S and Risø DTU. ECS Trans. **25**(2), 133–142 (2009)
11. N. Christiansen, J.B. Hansen, H.H. Larsen, M.J. Jørgensen, M. Wandel, P.V. Hendriksen, A. Hagen, S. Ramousse, Status of development and manufacture of solid oxide fuel cells at Topsoe Fuel cells A/S and Risø-DTU, in *European SOFC Forum, Lucerne*, 2010

12. A. Hagen, M. Chen, K. Neufeld, Y.L. Liu, Effect of humidity in air on performance and long-term durability of SOFCs. ECS Trans. **25**(2), 439–446 (2009)
13. J.B. Hansen, J. Rostrup-Nielsen, P. Højlund Nielsen, T. Rostrup-Nielsen, Optimum processing of fuel gas from coal gasification for SOFC/gas turbine power plants, in *Fuel Cell Seminar, Honolulu*, 2006
14. J.B. Hansen, N. Christiansen. Solid oxide fuel cell development at Topsoe Fuel Cell A/S, in *World Hydrogen Technology Convention 2007, Montecatini*, 2007
15. J.B. Hansen, Correlating sulfur poisoning of SOFC nickel anodes by temkin isotherm. Electrochem. Solid-State Lett. **11**(10), B178–B180 (2008)
16. P. Hjalmarsson, M. Søgaard, M. Mogensen, Electrochemical behavior of $(La_{1-x}Sr_x)_sCo_{1-y}Ni_yO_{3-\delta}$ as porous SOFC cathodes. Solid State Ionics **180**(26–27), 1395–1405 (2009)
17. P. Hjalmarsson, M. Søgaard, M. Mogensen, Electrochemical performance and degradation of $(La_{0.6}Sr_{0.4})_{0.99}CoO_{3-\delta}$ as porous SOFC-cathode. Solid State Ionics **179**(27–32), 1422–1426 (2008)
18. M.W. Johnson, *Seizing the White Space—Business Model Innovation for Growth and Renewal* (Harvard Business Press, Boston, 2010)
19. J. Larmine, A. Dicks, *Fuel Cell Systems Explained* (Wiley, Chichester, 2001). ISBN 0471490261
20. M. Mogensen, N. Christiansen, Fuel cells—familiar principles for electricity generation. Europhys. News **24**, 7–9 (1993)
21. A. Nakajo, Z. Wuillemin, J.V. Herle, D. Favrat, Simulation of thermal stresses in anode-supported solid oxide fuel cell stacks. Part I: probability of failure of cells. J. Power Sources **193**, 203–215 (2009)
22. H.-S. Noh, H. Lee, B.-K. Kim, J.-H. Lee, H.-W. Lee, J.-W. Son, Parameters affecting the performance of thin film (<1 μm) electrolyte SOFC, in *European SOFC Forum, Lucerne*, 2010
23. M. Noponen, T. Hottinen, WFC20 biogas unit operation, in *European SOFC Forum, Lucerne*, 2010
24. J. Pålsson, J.B. Hansen, N. Christiansen, J.U. Nielsen, S. Kristensen, Solid oxide fuel cells-assessment of the technology from an industrial perspective, in *Proceedings 2003 Risø Energy Conference*, 2003
25. J.F.B. Rasmussen, A. Hagen, The effect of H_2S on the performance of Ni–YSZ anodes in solid oxide fuel cells. J. Power Sources **191**(2), 534–541 (2009)
26. K. Scroots, F. Ferioli, G.J. Kramer, B.C.C. van der Zwaan, Learning curves for Hydrogen Production Technology—an assessment of observed cost reductions. Int. J. Hydrogen Energy **33**, 2630–2645 (2008)
27. K. Schoots, G.J. Kramer, B.C.C. vander Zwaan, Technology learning for fuel cells: an assessment of past and potential cost reduction. Energy Policy (2010). doi:10.1016/j.enpol.2010.01.022
28. R. Steinberger-Wilckens, N. Christiansen, in *High-Temperature Fuel Cells in Decentralized Generation*, ed. by D. Stolten. Hydrogen and Fuel Cells—Fundamentals Technologies and Applications (Wiley, Weinheim, 2010)
29. M. Søgaard, T.Z. Sholklapper, M. Wandel, L.C. De Jonghe, M. Mogensen, in *European SOFC Symposium, Lucerne*, 2008
30. P.C. Su, C.C. Chao, J.H. Shim, R. Fasching, F.B. Prinz, Solid oxide fuel cell with corrugated thin film electrolyte. Nano Lett. **8**, 2289–2292 (2008)
31. M.C. Tucker, Progress in metal-supported solid oxide fuel cells: a review. J. Power Sources **195**, 4570–4582 (2010)
32. M. Wandel, J.R. Bowen, M. Mogensen, Performance, degradation and microstructure of LSCF/CGO composite cathodes, in *European SOFC Symposium, Lucerne*, 2008
33. Y. Zhang, C. Xia, A durability model for solid oxide fuel cell electrodes in thermal cycle processes. J. Power Sources **195**, 6611–6618 (2010)

Index

J. T. S. Irvine and P. Connor (eds.), *Solid Oxide Fuels Cells: Facts and Figures*,
Green Energy and Technology, DOI: 10.1007/978-1-4471-4456-4,

MIX
Papier aus verantwortungsvollen Quellen
Paper from responsible sources
FSC® C105338

If you have any concerns about our products,
you can contact us on
ProductSafety@springernature.com

In case Publisher is established outside the EU,
the EU authorized representative is:
Springer Nature Customer Service Center GmbH
Europaplatz 3, 69115 Heidelberg, Germany

Printed by Libri Plureos GmbH
in Hamburg, Germany